Lecture Notes in Computer Scien

Commenced Publication in 1973
Founding and Former Series Editors:
Gerhard Goos, Juris Hartmanis, and Jan van Leeuwen

Umeshwar Dayal   Johann Eder
Jana Koehler   Hajo A. Reijers (Eds.)

Business Process Management

7th International Conference, BPM 2009
Ulm, Germany, September 8-10, 2009
Proceedings

 Springer

Volume Editors

Umeshwar Dayal
Hewlett-Packard Laboratories
1501 Page Mill Rd., Palo Alto, CA 94304, USA
E-mail: umeshwar.dayal@hp.com

Johann Eder
Alps Adria University Klagenfurt
Universitätsstr. 65, 9020 Klagenfurt, Austria
E-mail: johann.eder@uni-klu.ac.at

Jana Koehler
IBM Zurich Research Laboratory
Saeumerstr. 4, 8803 Rüschlikon, Switzerland
E-mail: koe@zurich.ibm.com

Hajo A. Reijers
Eindhoven University of Technology
PO Box 513, 5600 MB Eindhoven, The Netherlands
E-mail: h.a.reijers@tue.nl

Library of Congress Control Number: 2009932360

CR Subject Classification (1998): J.1, H.5.3, D.2.9, F.3.2, K.6

LNCS Sublibrary: SL 3 – Information Systems and Application, incl. Internet/Web and HCI

ISSN	0302-9743
ISBN-10	3-642-03847-6 Springer Berlin Heidelberg New York
ISBN-13	978-3-642-03847-1 Springer Berlin Heidelberg New York

springer.com

© Springer-Verlag Berlin Heidelberg 2009
Printed in Germany

Typesetting: Camera-ready by author, data conversion by Scientific Publishing Services, Chennai, India
Printed on acid-free paper   SPIN: 12739387   06/3180   5 4 3 2 1 0

Preface

The BPM (Business Process Management) Conference series has the ambition to be the premier forum for researchers in the area of process-aware information systems. It has a record for attracting contributions in innovative research of the highest quality related to all aspects of business process management including theory, frameworks, methods, techniques, architectures, and empirical findings.

BPM 2009 was the 7th instantiation of this series. It took place in Ulm, Germany, September 8–10, 2009, organized by the Institute of Databases and Information Systems of the University of Ulm. This volume contains 17 contributed research papers and two contributed industrial papers selected from 116 submissions from 31 countries. The thorough reviewing process—each paper was reviewed by three to five Program Committee members—was extremely competitive as the acceptance rate of 16% indicates. In addition to the contributed papers, these proceedings contain two papers and an outline documenting the invited keynote talks. Furthermore, a report is included on the collaboration structure in BPM research derived from an analysis of papers accepted for all past BPM conferences.

In conjunction with the main conference, nine international workshops took place the day before the conference. These workshops fostered the exchange of fresh ideas and experiences between active BPM researchers, and stimulated discussions on new and emerging issues in line with the conference topics. The proceedings with the papers of all workshops will be published in a separate volume of Springer's *Lecture Notes in Business Information Processing* series.

This is the place to express our gratitude to all those who made BPM 2009 possible by generously and voluntarily sharing their knowledge, skills and time: the General Chairs Peter Dadam and Manfred Reichert and the Organization Chairs Jens Kolb and Rüdiger Pryss for providing an excellent environment for the conference, and all other colleagues holding offices. In particular we thank the senior and regular Program Committee members as well as the additional reviewers for devoting their expertise and time to ensure the high quality of the conference in an extensive review and discussion process. And last but not least, we are grateful to all the authors who showed their appreciation of the conference by submitting their valuable work to it.

September 2009

Umeshwar Dayal
Johann Eder
Jana Koehler
Hajo Reijers

Conference Organization

General Chairs

Peter Dadam, Germany
Manfred Reichert, Germany

Program Chairs

Umeshwar Dayal, USA
Johann Eder, Austria
Hajo Reijers, The Netherlands

Industry Chair

Jana Koehler, Switzerland

Local Organization

Jens Kolb, Germany
Rüdiger Pryss, Germany

Workshop Chairs

Frank Leymann, Germany
Stefanie Rinderle-Ma, Germany
Shazia Sadiq, Australia

Workshop Chairs

Ana Karla Alves de Medeiros, The Netherlands
Barbara Weber, Austria

Tutorial/Panel Chairs

Joachim Herbst, Germany
Gerti Kappel, Austria

Publicity Chair

Heiko Ludwig, USA

Senior Program Committee

Wil van der Aalst, The Netherlands
Gustavo Alonso, Switzerland
Boualem Benatallah, Australia
Fabio Casati, Italy
Peter Dadam, Germany
Joerg Desel, Germany
Marlon Dumas, Estonia
Schahram Dustdar, Austria
Gregor Engels, Germany
Claude Godart, France

Rick Hull, USA
Stefan Jablonski, Germany
Frank Leymann, Germany
Manfred Reichert, Germany
Michael Rosemann, Australia
Amit Sheth, USA
Jianwen Su, USA
Arthur ter Hofstede, Australia
Kees van Hee, The Netherlands
Mathias Weske, Germany

Program Committee

Ana Karla Alves De Medeiros,
    The Netherlands
Pedro Antunes, Portugal
Joonsoo Bae, South Korea
Hyerim Bae, South Korea
Alistair Barros, Australia
Catriel Beeri, Israel
Djamal Benslimane, France
M. Brian Blake, USA
Christoph Bussler, USA
Jorge Cardoso, Germany
Malu Castellanos, USA
Valeria De Antonellis, Italy
Jan Dietz, The Netherlands
Maria Grazia Fugini, Italy
Avigdor Gal, Israel
Dimitrios Georgakopoulos, USA
Peter Green, Australia
Paul Grefen, The Netherlands
Daniela Grigori, France
Thomas Gschwind, Switzerland
Manfred Hauswirth, Ireland
Marta Indulska, Australia
Leonid Kalinichenko, Russia
Gerti Kappel, Austria
Ekkart Kindler, Denmark
Jana Koehler, Switzerland
Agnes Koschmider, Germany
John Krogstie, Norway
Jochen Kuester, Switzerland

Akhil Kumar, USA
Lea Kutvonen, Finland
Selma Limam Mansar, Qatar
Chengfei Liu, Australia
Ling Liu, USA
Bertram Ludscher, USA
Heiko Ludwig, USA
Zongwei Luo, Hong Kong
Axel Martens, USA
Jan Mendling, Australia
Bela Mutschler, Germany
John Mylopoulos, Canada
Andreas Oberweis, Germany
Aris Ouksel, USA
Cesare Pautasso, Switzerland
Barbara Pernici, Italy
Olivier Perrin, France
Calton Pu, USA
Frank Puhlmann, Germany
Krithivasan Ramamritham, India
Jan Recker, Australia
Berthold Reinwald, USA
Wolfgang Reisig, Germany
Stefanie Rinderle, Germany
Shazia Sadiq, Australia
Mohand Said-Hacid, France
Heiko Schuldt, Switzerland
Karsten Schultz, Australia
Timos Sellis, Greece
Juliane Siegeris, Germany

Stefan Tai, USA
Farouk Toumani, France
Aphrodite Tsalgatidou, Greece
Jan Vanthienen, Belgium
Hagen Voelzer, Switzerland

Barbara Weber, Austria
Petia Wohed, Sweden
Andreas Wombacher, The Netherlands
Xiaohui Zhao, Australia

External Reviewers

Antonia Albani
Syaiful Ali
Samuil Angelov
George Athanasopoulos
Micheal Axelsen
Joseph Barjis
Sami Bhiri
Devis Bianchini
Wassim Derguech
Joao Ferreira
Nadine Froehlich
Christian Gerth
Christian Gierds
Mati Golani
Pieter Van Gorp
Peter Green
Armin Haller
Susan Hickl
Bjrn Keuter
Paul El Khoury
Kostas Kontogiannis
Rob Kusters
Christoph Langguth
Philipp Liegl
Maya Lincoln

Niels Lohmann
Linh Thao Ly
Peter Massuthe
Michele Melchiori
Thorsten Moeller
Hamid Motahari
Kreshnik Musaraj
Oanea Olivia
Michael Pantazoglou
Daniel Ried
Al Robb
Martina Seidl
Dimitrios Skoutas
Jos Trienekens
Jochem Vonk
Gabriela Vulcu
Qingyang Wang
Daniela Weinberg
Manuel Wimmer
Jiajie Xu
Ustun Yildiz
Jianwei Yin
Sira Yongchareon
Marco Zapletal
Maciej Zaremba

Table of Contents

Editorial

Invited Talks

Modeling I

Managing Processes

Process Mining I

Processes and Services

Modeling II

Verification and Compliance

Process Mining II

A Collaboration and Productiveness Analysis of the BPM Community

Hajo A. Reijers[1], Minseok Song[1], Heidi Romero[1], Umeshwar Dayal[2], Johann Eder[3], and Jana Koehler[4]

[1] Eindhoven University of Technology, P.O. Box 513, 5600 MB Eindhoven, The Netherlands
{h.a.reijers,m.s.song,h.l.romero}@tue.nl
[2] HP Laboratories, 1501 Page Mill Road, Palo Alto, 94304, USA
umeshwar.dayal@hp.com
[3] University of Klagenfurt, Universittsstrae 65-67, 9020 Klagenfurt, Austria
johann.eder@uni-klu.ac.at
[4] IBM Zurich Research Laboratory CH-8803 Rueschlikon, Switzerland
koe@zurich.ibm.com

Abstract. The main scientific event for academics working in the field of Business Process Management is the International BPM Conference. In this paper, social network analysis techniques are used to unveil the *co-authorship networks* that can be derived from the papers presented at this conference. Links between two researchers are established by their co-authorship of a paper at one of the conference editions throughout the years 2003-2008. Beyond the relations between individual authors, aggregated analyses are presented of the interactions between the institutes that the authors are affiliated with as well as their country of residence. Additionally, the output of individual authors is measured. All analyses are carried out for the individual conference years and at cumulative levels. In this way, this paper identifies the hotbeds of BPM research and maps the progressive collaboration patterns within the BPM community.

1 Introduction

In the introduction of the first proceedings of the BPM conference series[1], Business Process Management (BPM) has been characterized as the study of those methods, techniques, and software that can be used to design, enact, control, and analyze operational processes involving humans, organizations, applications, documents and other sources of information [1]. In line with this view, the technological perspective and the attention for formal methods have consistently been important ingredients of the papers presented in the series. Recently, an inflow of papers and keynotes have appeared that also deal with BPM as a management philosophy instead of a pure technological approach [2,3]. Additionally, papers have been included that concentrate not so much on the design but rather on the empirical evaluation of methods and techniques [4,5].

---

[1] For an overview of all its editions, see http://www.bpm-conference.org/

U. Dayal et al. (Eds.): BPM 2009, LNCS 5701, pp. 1–14, 2009.
© Springer-Verlag Berlin Heidelberg 2009

Consequently, the character of the BPM conference extends over a diverse array of subjects and draws from disciplines such as computer science, information systems, management science, artificial intelligence, industrial engineering, software engineering, and economics. In this multidisciplinary context, it is evidently important for BPM researchers to foster and extend collaborations to stay up to date of state-of-the-art developments in their own field and acquire access to complementary fields of expertise.

This paper aims at analyzing the collaborations between individual researchers, their institutes, and the hosting countries as tied to contributions to the BPM conference series. For this purpose, all papers that have been included in the proceedings of the six conference editions between 2003 and 2008 have been considered. The biographical attributes of these papers have been subjected to various Social Network Analysis techniques [6,7,8]. The social network miner and analyzer [9], as part of the ProM framework [10], have been applied to carry out this analysis. The collaboration networks that are the main results identify the hotbeds of BPM research activity and disclose the interconnections between them. The presented analysis of these networks over time also gives an indication of how the BPM community extends and becomes more interwoven.

The structure of this paper is now as follows. In Section 2, we will describe the methodology to collect, conceptualize, analyze and verify the used data. Section 3 will present the results from this analysis. The paper ends with a discussion and a conclusion.

2   Methodology

The methodology employed to identify the underlying relationship between researchers followed several steps that will be detailed in this section. Those steps are described in four subsections: (1) Data collection, (2) Conceptualization of the event logs, (3) Log verification, and (4) Network generation using ProM.

2.1   Data Collection

Our data collection has consisted of the compilation of all the references from the conferences publications. The proceedings of these conferences were published by Springer as volumes in the Lecture Notes in Computer Science series (See Table 1).

The references were obtained from SpringerLink[2], the online access point to Springer's Lecture Notes in Computer Science series, and stored in Reference Manager[3], a bibliographical database system. Entries were extracted from SpringerLink in the RIS format using the "RIS FORMAT (Include ID)" filter available in the ReferenceManager software system. All references were classified as Book Chapters and included the chapter title, authors, start and end page, abstract and a unique ID (DOI). After all entries had been entered, the entire

---

[2] http://www.springerlink.com/
[3] http://www.refman.com/

Table 1. BPM Conference Proceedings

Conference	LNCS Volume	Venue	Dates	ISBN
BPM 2003	2678	Eindhoven	June 26-27, 2003	3-540-40318-3
BPM 2004	3080	Potsdam	June 17-18, 2004	3-540-22235-9
BPM 2005	3649	Nancy	September 5-8, 2005	3-540-28238-6
BPM 2006	4102	Vienna	September 5-7, 2006	3-540-38901-6
BPM 2007	4714	Brisbane	September 24-28, 2007	978-3-540-75182-3
BPM 2008	5240	Milan	September 2-4, 2008	978-3-540-85757-0

database has been exported to an Excel Sheet, which was a convenient format to translate the data into the form of a MXML file, which is readable by ProM.

2.2 Conceptualization

To arrive at an MXML file that could be analyzed with the social network analysis toolset [9], a decision needed to be made on how to translate the previously described information into a form that is both convenient and meaningful for analysis. It should be noted that the main purpose of the ProM framework, which was selected for the derivation and analysis of the collaboration networks because of its ease of use and analytical power, is to support the analysis of an *event log*. This is a collection of events that is stored by any kind of transactional information system, e.g., ERP, CRM, or workflow management system [11]. Clearly, the occurrences of successive editions of a conference series is not a business process in the traditional sense, so a mapping of the domain concepts on the business process concepts that are common to ProM had to be established.

In essence, each publication has been considered as a separate case instance that could be processed by one or more authors. Each author performs a unique activity in dealing with a case, as to avoid conflicts in situations where one author participates in more than one publication in the same year. Because we only have the year information and no more fine-grained time notion, it would otherwise be difficult to distinguish these. This simple conceptualization leads to a total of 383 authors and the same amount of activities.

As mentioned, ProM can read files in the MXML format [12], a generic XML format appropriate to representing event log data (see Figure 1). It includes the definition of the process instance with the audit trails including attributes like Workflow Model Element, Event type, Timestamp and Originators. The exact mapping of bibliographical elements to the elements of the MXML format that is used by ProM can now be specified as follows.

The process instance is defined as a publication (book chapter), and each of these has an identical Id which is the DOI, related to the book that this book chapter belongs to. The description linked to every Process Instance is the name of the publication, for instance "Modeling Medical e-Services". For each Process Instance an attribute was defined as a Data item called "ConferenceYear", to facilitate filtering the publications related only to a specific year. For each process

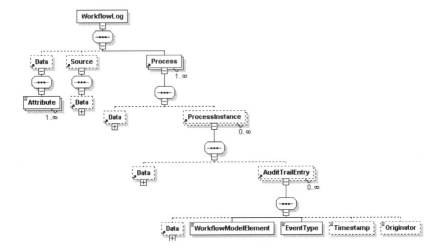

Fig. 1. The MXML format

instance there is an Audit Trail Entry related to each activity followed in the process. We define an activity as the contribution of every author to the publication. This means that a publication performed by two authors will contain two audit trails, each one with a Workflow Model Element that contains the author's name, and only containing "Complete" event types. The timestamp was not defined in the log because the information needed is the year related to the conference in which the paper was published. The latter information is already provided by the Data attribute mentioned earlier. Finally, the Originator is the name of the author.

In a similar way an organizational model file [9] (without log) was created, to relate the authors to their university and countries. The university was identified as the Role and the country as the Organizational Unit. The university or institution associated to each author is where the author works or is affiliated with. The country being considered is the country where the university resides. We found authors that changed their working place during the time of the study. In those few cases, the author was assigned to the university where they published the most and, in case of a tie, was assigned arbitrarily to one.

2.3   Verification

To verify whether no errors have been introduced in the creation of the MXML file, the basic ProM feedback was used to assess high level errors. ProM reports that the log contains one process with 190 cases (publications) and 383 originators (authors). It further indicates a minimum number of events that equals 1, a mean number of 2 and a maximum of 7, representing the number of authors per publication. This information is consistent with the figures that could be obtained in the Excel sheet. Next, we used the "ConferenceYear" attribute to check for potential mistakes in the assignment of the year to cases. We manually inspected each entry, for example, by applying the filter using the ConferenceYear 2003, it is expected to see only the instances with the Id 3-540-44895-0. In case that a different code appears it means that there is a publication from another year misclassified. For this study every filter was manually verified to avoid this error, before further analysis.

Another important point of verification is whether the organizational model links every author to exactly one country and one university. To inspect this aspect we used the Filter / Advanced and selected Replacement Filters to generate separate lists of originators, representing the countries and universities. In this way it was possible to detect for each author a missing assignment or an undesirable swap, e.g. a country that was used as a university.

2.4   Network Generation

The social networks that were generated are all based on the working together metric [13, Definition 4.8]. Informally stated, the metric expresses for each pair of performers whether or not they have performed activities for the same case. In our domain, this means that two authors have worked on the same paper. Clearly, the relation is symmetrical. While our analysis considers only undirected graphs, the figures in this paper show graphs where the relation is show as two arrows with opposed directions. It should be noted that such a visual pair of arrows will account for only a single link. We have established the network on the basis of the working together metric for each individual conference year (leading to six networks) and for each of the cumulative increments for the conferences (leading to another six networks, e.g. for 2003, 2003+2004, 2003+2004+2005, etc.). Additionally, the results on the working together metric between individuals were exported in a matrix representation to NetMiner [14], which is another tool to analyze the social networks. With NetMiner the following analysis measurements were derived for each network:

1. *Density* measures the level of connectedness among the nodes in a network. This measurement is important to compare networks of different sizes. It defined as follows.
   $D = \frac{2|A|}{|N|(|N|-1)}$, where $|A|$ denotes number of arcs and $|N|$ denotes number of nodes.

2. *Inclusiveness* represents a measurement of the connectivity between nodes in the network. It defined as follows.
   $I = \frac{|N|-|N_i|}{|N|} * 100$, where $|N|$ denotes number of nodes and $|N_i|$ denotes number of isolated nodes.
   The isolated nodes are those with a connection degree equals to zero.

Finally, these networks and analysis measures were also established by aggregating the authors to the level of *research institutes* and *countries*. For example, a co-publication at the BPM 2005 conference between an author from Macquarie University in Australia and another author from Tilburg University in the

Netherlands leads to a relation between the respective universities in the institute network for 2005, as well as a relation between Australia and the Netherlands in the countries network of 2005.

It should be noted that many more relations exist to create social networks, as well as many more analysis measurements to evaluate these. For an overview, see [13]. However, the working together metric is the most useful for our purpose. The networks resulting from our analysis will be presented in the next section.

3   Results

The results in this section will be successively presented on the individual (author) level, the institute level, and the country level.

3.1   Individual Level

To provide an initial idea of the networks that were created, Figure 2 represents the social networks on the level of co-authors for publications in the year 2003, while Figure 3 represents the same type of social network considering all publications of the years 2003 to 2008 combined. While it is clear that the network has expanded considerably over the years, the data in Table 2 gives a better insight into this development.

Over the years, the social network has grown as a whole, both with respect to the publications (going up from 23 in 2003 to 190 accumulated over all the years)

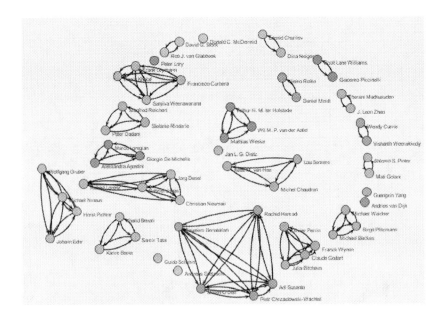

Fig. 2. Social network on individual level for 2003 (working together metric)

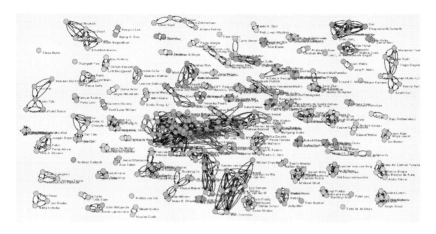

Fig. 3. Social network on individual level for 2003-2008 (working together metric)

Table 2. Cumulative statistics individual level (working together metric)

	2003	2003-2004	2003-2005	2003-2006	2003-2007	2003-2008
Number of publications	23	45	86	128	158	190
Number of authors	60	99	190	271	335	383
Number of links	59	103	217	372	463	555
Inclusiveness (%)	88.3	91.9	95.8	96.3	96.7	96.3
Network Density	0.032	0.021	0.012	0.010	0.008	0.008

and individual authors (going up from 60 in 2003 to 383 authors accumulated over all the years). The factors of growth are over 8 and 6 respectively. The inclusiveness of the cumulative networks also grows, from 88.3% to around 96%. In other words, the relative number of authors purely publishing by themselves drops as a relative measure, but it should be noted that this was already a small minority from the start. At the same time, the density of the network can be seen to drop, from 0.032 to 0.008. This means that the inflow of papers by new authors is not matched by a corresponding increase of new collaborations that become possible.

From all the 190 authors that have published one or more papers in the conference series, it can be determined how many papers they contributed overall. The authors that published 4 papers or more are shown in Table 5. It can be seen that the top contributors generate a considerable but not excessive share of the overall number of papers, which can be seen as a sign of academic health for the conference series.

3.2   Institute Level

To provide an understanding of the evolution in collaboration patterns between the institutes, several cumulative networks are now presented in succession.

Table 3. Contributions individual authors

Originator	Publications (absolute)	Publications (relative to total)
Wil M.P. van der Aalst	12	6.32%
Manfred Reichert	9	4.73%
Hajo A. Reijers	8	4.21%
Mathias Weske	8	4.21%
Gero Decker	6	3.16%
Marlon Dumas	6	3.16%
Claude Godart	6	3.16%
Arthur H.M. ter Hofstede	6	3.16%
Jan Mendling	6	3.16%
Stefanie Rinderle	6	3.16%
Peter Dadam	5	2.63%
Kees M. van Hee	5	2.63%
Monique H. Jansen-Vullers	4	2.11%
Chengfei Liu	4	2.11%
Olivier Perrin	4	2.11%
Natalia Sidorova	4	2.11%
Marc Voorhoeve	4	2.11%
Xiaohui Zhao	4	2.11%

Table 4. Cumulative statistics institute level (working together metric)

	2003	2003-2004	2003-2005	2003-2006	2003-2007	2003-2008
Number of institutes	33	54	91	114	140	154
Number of links	13	29	49	79	104	128
Inclusiveness (%)	51.5	66.7	71.4	72.8	75.7	76.6
Network Density	0.024	0.020	0.011	0.012	0.010	0.010

Figures 4, 5, 6, and 7 respectively show the network in 2003, the cumulative network over the years 2003 to 2006, the cumulative network over the years 2003 to 2007, and the cumulative network over the years 2003 to 2008.

In 2003, after the first conference, 7 collaboration groups could be identified (see Figure 4). By 2006, the number of groups had increased to 27 groups (see Figure 5). Among them, five big groups can be identified: two Dutch-German groups, one French group, one German-USA group, and one multi-continental group. By 2007, one of the Dutch-German group was notably extended (see Figure 6). The situation by 2008 can be seen in Figure 7. By that time, the two Dutch-German groups were merged and Eindhoven University of Technology became the bridge between these two formerly separate groups. The German-USA group had also grown by this time and the IBM JT Watson research center had become the center of it.

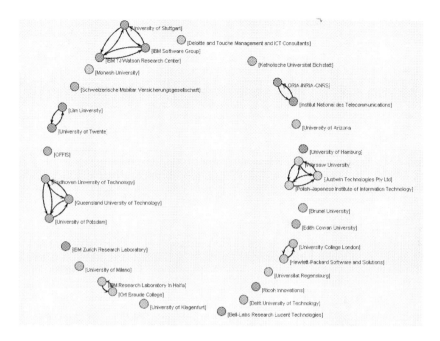

Fig. 4. Social network on institute level for 2003 (working together metric)

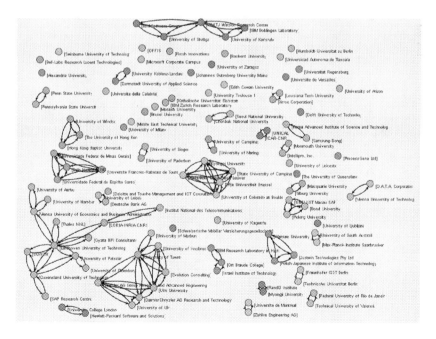

Fig. 5. Social network on institute level for 2003 - 2006 (working together metric)

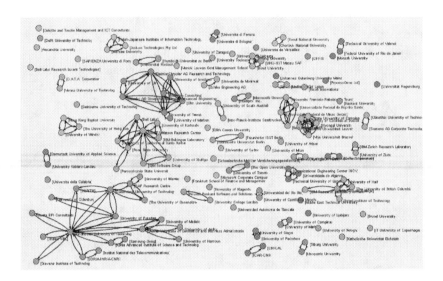

Fig. 6. Social network on institute level for 2003 - 2007 (working together metric)

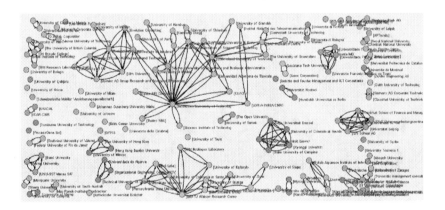

Fig. 7. Social network on institute level for 2003-2008 (working together metric)

Table 4 gives the data for the cumulative years. What can be noted is that the number of institutes grows considerably over the years (from 33 to 154), a factor greater than 4, but this is not as strong an increase as the number of papers or new authors over this period (see the previous subsection). Yet, the growth of the inclusiveness of the network, from 51.5% to 76.6%, is more impressive than the network on the individual level discussed previously. This means that the cooperation between people from different institutes grows stronger than the cooperation between individuals. Similar to the analysis level of individual authors, these increasing figures are accompanied by a drop of the network density throughout this development (from 0.024 to 0.010). This also suggests that the

Table 5. Contributions institutes

Originator	Publications (absolute)	Publications (relative to total)
Eindhoven University of Technology	72	37.89%
Queensland University of Technology	27	14.21%
LORIA-INRIA-CNRS	23	12.10%
University of Potsdam	21	11.05%
University of Ulm	20	10.53%
Swinburne University of Technology	15	7.89%
IBM Zurich Research Laboratory	13	6.84%
Humboldt-Universitaet zu Berlin	11	5.79%
University of Stuttgart	11	5.79%
SAP Research Centre	10	5.25%
The University of Queensland	10	5.26%

entry of new institutes in the BPM field is not matched with a corresponding growth of new collaborations on this level.

Finally, Table 5 shows the institutes that have generated 10 or more contributions to the various editions of the conference series. Here, it can be seen that the institute leading the table, Eindhoven University of Technology, has a very large part in the overall production of papers with a production that is over twice as big as that of the number two, Queensland University of Technology, and over three times as big of the number three, LORIA-INRIA-CNRS.

3.3   Country Level

Collaborations between authors from institutes in different countries are visualized with the network in Figure 8, while the quantitative data of the various cumulative networks are given in Table 6.

Over the years, the number of participating countries has more than doubled, growing from 11 countries in 2003 to 25 accumulated over the whole conference series. The inclusiveness of the network has also grown, from 54.5% to 88.0%, which is comparable to the growth of the network at the institute level over the same period (see the previous subsection). Interestingly, and in contrast to the networks that have been discussed previously for the other analysis levels,

Table 6. Cumulative statistics country level (working together metric)

	2003	2003-2004	2003-2005	2003-2006	2003-2007	2003-2008
Number of countries	11	15	17	23	25	25
Number of links	6	8	11	24	32	39
Inclusiveness (%)	54.5	66.7	70.6	73.9	80.8	88.0
Network Density	0.109	0.076	0.081	0.094	0.106	0.130

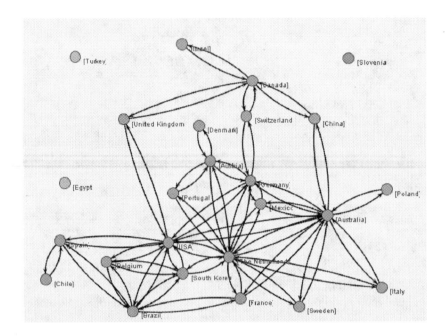

Fig. 8. Social network on country level for 2003-2008 (working together metric)

Table 7. Contributions countries

Originator	Publications (absolute)	Publications (relative to total)
Germany	127	66.84%
The Netherlands	87	45.78%
Australia	75	39.47%
USA	46	24.21%
France	33	17.37%
Italy	22	11.58%
Switzerland	18	9.47%
Austria	16	8.42%
Israel	15	7.89%
South Korea	14	7.37%
Belgium	12	6.32%
Spain	11	5.79%
Brazil	10	5.26%

the density of the network is rather stable over the years. In other words, the inflow of contributions from authors of 'new' countries goes along with an actual exploitation of such inter-country collaborations opportunities.

As a final analysis result, the countries are shown in Table 7 of which the hosted BPM researchers have contributed the most papers to the conference series. Note that only countries with 10 or more generated publications are shown. From this table, Germany, The Netherlands, and Australia emerge as dominant providers of content for the BPM conference series. This could well be expected from the results shown previously in Table 5, which mainly shows institutes located in these countries. What is somewhat surprising is the position of the USA in fourth place, while no American institute appears in Table 5. This seems to suggest that the BPM research in the USA is much more scattered over various institutes than is the case in Europe or Australia.

4   Discussion and Conclusion

The analysis of the collaboration patterns behind the papers on the BPM conference series shows a growth of collaborations at all the levels of analysis: author, institute, and country level. On the country level the increase of entrants result in a utilization of the new research ties that potentially become available with the new entrants. This can be taken as a sign that the BPM conference series appears as a genuinely international forum. At the individual and institute level this type of expansion is not so apparent. This can perhaps be more or less expected on an individual level, where young researchers will be entering this field with a small collaboration network. However, the lack of growth in cooperation patterns between the institutes over the years can be seen as somewhat worrisome if this is taken as an indication of researchers at the various institutes of favoring "in-house" research.

From the analysis of the numbers of contributions, one can establish that Germany, The Netherlands and Australia are the leading countries in their participation in this research area. Several universities are actively participating in Germany including the University of Potsdam, Ulm University, Humboldt-University of Berlin, and the University of Stuttgart, with shares of 5 up to 11% of the total number of papers. In the Netherlands, the leading university is Eindhoven University of Technology, with strong contributions of several authors amounting to a share of publications in the range of 38%. In Australia, a broader array of universities is contributing to the field, notably Queensland University of Technology, Swinburne University of Technology, and The University of Queensland, with respective shares of 14.21%, 7.89% and 5.26%.

To conclude, the co-authorship networks derived from the publications in the BPM international conferences 2003-2008 help to recognize how the community of researchers is built around this research area, and what influence is exerted by their universities and the countries that they are affiliated with. The paper is also a nice illustration of the versatility of the BPM tools that are developed by the same community. After all, it is the ProM tool that was developed for supporting various process mining techniques that was used for this bibliographic study. It is our hope that the presented results provide the members of the BPM community with some new insights and will encourage them to link up with others, in pursuit of breakthrough research.

References

1. van der Aalst, W.M.P., ter Hofstede, A.H.M., Weske, M.: Business process management: A survey. In: van der Aalst, W.M.P., ter Hofstede, A.H.M., Weske, M. (eds.) BPM 2003. LNCS, vol. 2678, pp. 1–12. Springer, Heidelberg (2003)
2. Willaert, P., Van den Bergh, J., Willems, J., Deschoolmeester, D.: The process-oriented organisation: A holistic view developing a framework for business process orientation maturity. In: Alonso, G., Dadam, P., Rosemann, M. (eds.) BPM 2007. LNCS, vol. 4714, pp. 1–15. Springer, Heidelberg (2007)
3. Harmon, P.: Business process management: Today and tomorrow. In: Dumas, M., Reichert, M., Shan, M.-C. (eds.) BPM 2008. LNCS, vol. 5240, pp. 1–1. Springer, Heidelberg (2008)
4. Reijers, H.A., Song, M., Jeong, B.: On the performance of workflow processes with distributed actors: Does place matter? In: Alonso, G., Dadam, P., Rosemann, M. (eds.) BPM 2007. LNCS, vol. 4714, pp. 32–47. Springer, Heidelberg (2007)
5. Siegeris, J., Grasl, O.: Model driven business transformation – an experience report. In: Dumas, M., Reichert, M., Shan, M.-C. (eds.) BPM 2008. LNCS, vol. 5240, pp. 36–50. Springer, Heidelberg (2008)
6. Burt, R., Minor, M.: Applied Network Analysis: A Methodological Introduction. Sage, Newbury Park (1983)
7. Scott, J.: Social Network Analysis. Sage, Newbury Park (1992)
8. Wasserman, S., Faust, K.: Social Network Analysis: Methods and Applications. Cambridge University Press, Cambridge (1994)
9. Song, M., van der Aalst, W.M.P.: Towards comprehensive support for organizational mining. Decision Support Systems 46(1), 300–317 (2008)
10. van Dongen, B.F., de Medeiros, A.K.A., Verbeek, H.M.W(E.), Weijters, A.J.M.M.T., van der Aalst, W.M.P.: The proM framework: A new era in process mining tool support. In: Ciardo, G., Darondeau, P. (eds.) ICATPN 2005. LNCS, vol. 3536, pp. 444–454. Springer, Heidelberg (2005)
11. van der Aalst, W.M.P., Reijers, H.A., Weijters, A.J.M.M., van Dongen, B.F., Medeiros, A., Song, M., Verbeek, H.M.W.: Business Process Mining: An Industrial Application. Information Systems 32(5), 713–732 (2007)
12. Guenther, C., van der Aalst, W.M.P.: A Generic Import Framework For Process Event Logs. BPM report (2006),
    http://is.tm.tue.nl/staff/wvdaalst/BPMcenter/reports.htm
13. van der Aalst, W.M.P., Reijers, H.A., Song, M.: Discovering Social Networks from Event Logs. Computer Supported Cooperative work 14(6), 549–593 (2005)
14. NetMiner: NetMiner Tool User's Guide (2009), http://www.netminer.com

BPM 3.0

August-Wilhelm Scheer and Joerg Klueckmann

IDS Scheer AG, Altenkesseler Str. 17,
66115 Saarbrücken, Germany
joerg.klueckmann@ids-scheer.com

Abstract. Business Process Management (BPM) is an established management discipline. Since today's organizations expect every employee to think and act like an entrepreneur, i.e., like a manager, BPM is also increasingly becoming part of everyday operations. But merely adopting a process-based approach across the enterprise is not enough to enable BPM at every level. What is needed is a combination of organizational forms and technologies that support distributed BPM initiatives while simultaneously consolidating them company-wide. Every employee must be empowered to model and optimize their own processes. At the same time, the entire BPM community needs a platform that brings together all the individual initiatives. This is the only way to leverage the full potential of process-oriented management. In the following article, the authors describe the trends in BPM development that are turning users into process managers and supporting the creation of a BPM community.

Keywords: Business Process Management, Governance, Community, Crowdsourcing.

1 BPM Guerrilla vs. BPM Governance

BPM initiatives often start with a guerrilla approach—individual departments launch improvement initiatives to remedy inefficient workflows, with the choice of software tools being based on departmental preferences. To avoid time-consuming procurement and approval processes, the choice falls on low-cost tools bought online. Team members gradually evolve their own methodology for describing processes, which is then developed in ad hoc fashion. When there are more than two or three process modelers involved, the manual effort needed to maintain model consistency rises sharply. The lack of standards means that content is frequently interpreted differently, leading to decisions being made on incorrect data. If the BPM project grows virally across departmental boundaries, further problems arise: other teams have already recorded their processes using different technology, enterprise-wide process chains cannot be created, and teams find it difficult to share their insights. Accordingly, the guerrilla BPM initiative fails before it can deliver real value.

Many organizations recognize that BPM initiatives tend to come into being virally, as described above, rather than due to strategic management decisions from the center. Such projects should be encouraged rather than prevented, since process orientation is beneficial to businesses. Having said that, there needs to be sufficient upfront intervention to ensure that the results of individual BPM teams can be

U. Dayal et al. (Eds.): BPM 2009, LNCS 5701, pp. 15–27, 2009.

combined. At the same time, it is important to maintain momentum: after all, no one has the time or inclination to wait months for the internal procurement department to finally come up with BPM software. The guerrilla approach thus calls for products that can be bought over the Internet and are easy to install. ARIS Express is just such a tool. It can be downloaded online free of charge and installed on every workstation. The integrated event-driven process chain (EPC) method ensures a consistent basis for interpreting and evaluating process models. Models can be combined to form cross-departmental process chains, despite the ARIS Express users being scattered across the organization (Fig. 1).

ARIS Express also supports other types of models in addition to EPCs, such as organizational charts, value chains, BPMN, and IT environments. Unlike the majority of modeling tools, the software is thus not limited to a process view. This flexibility is a key requirement for a holistic approach to Business Process Management that takes full account of business reality.

If more specialized functionality is needed, such as model versioning, complex evaluations across multiple process chains, or process simulation, all content can be transferred from ARIS Express into IDS Scheer's professional ARIS tools. The unstructured BPM guerrilla approach can therefore be transitioned into structured, sustained, holistic Business Process Management [1].

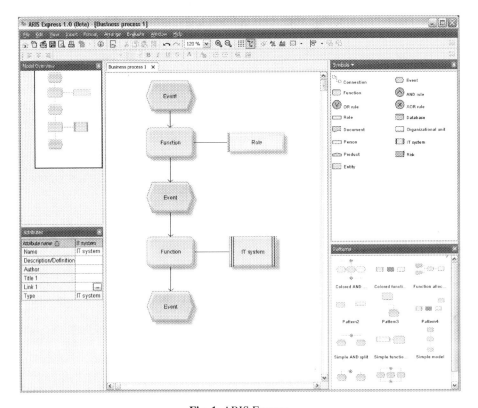

Fig. 1. ARIS Express

BPM governance plays an important role in facilitating the transition from unstructured to structured BPM. Organizations with a large number of BPM users and high workflow flexibility requirements need to establish structures and policies that take the process management process itself to a professional level. When this degree of BPM maturity has been achieved, technologies, such as ARIS Governance Engine, must be deployed to define and implement BPM structures and policies. Thanks to the model-driven approach behind ARIS Governance Engine, individual departments can create and implement their own governance processes by way of the ARIS method. A dialog designer provides a simple drag-and-drop interface to enable creation of input templates for subsequent automation. Thanks to predefined form fields, users can quickly create the screens in a structured manner. The data flow needed for process control can be created via a graphical data flow designer (Fig. 2). This greatly simplifies linking of dialog box elements with the underlying data for non-expert users.

The process desktop provides a custom, role-based task view with a list of all open, ongoing, delegated, and completed tasks, thus delivering a quick overview of the user's area of responsibility within the process. Users are guided step-by-step through each task from this desktop via the dialog screens created earlier. Replacement cover and escalation mechanisms deliver extra flexibility and help ensure that day-to-day workflows run smoothly.

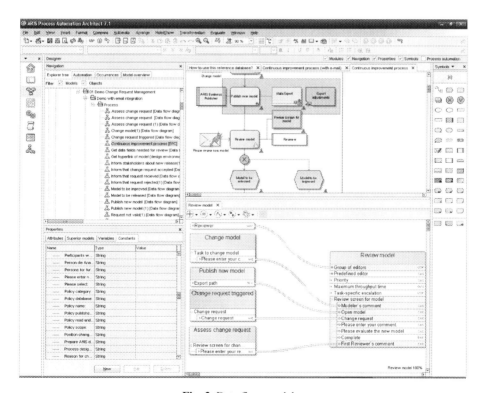

Fig. 2. Data flow modeler

With ARIS Governance Engine, business process modelers equipped with the necessary rights can become BPM governors; no IT support is needed. In change management projects, BPM managers gain the ability to optimize and accelerate their change processes and make them more transparent, while the role of IT architecture managers is to implement professional approval processes and optimize IT architecture and portfolio management activities. Governance managers can tailor their policies to meet compliance needs and roll them out across the organization. The overall result is more efficient and effective BPM organization.

Companies wanting to establish their level of BPM maturity can access Professor Scheer's Advanced BPM Assessment (http://www.professor-scheer-bpm.com/). A BPM application wheel and BPM technology wheel display the various phases and technologies on the path to a holistic BPM approach.

2 The Power of the Community

In recent years, there has been an explosion of online communities covering a wide range of subjects, with Facebook and LinkedIn being particularly prominent examples. This process has been driven by a desire to communicate with like-minded people and learn from them. The success of business communities, such as LinkedIn, shows that business users and BPM teams also want to interact with other users—both within the organization and beyond. This is not exactly a new trend because user groups have existed for a number of years, usually relating to a specific technology or product family, such as SAP, ARIS, or UNIX. Such groups also tend to be locally based. Today's online communities, however, are global and characterized by a virtualization of relationships. Regular face-to-face meetings are less important, with their place being taken by thriving online forums. Participation of employees in a BPM community yields a number of benefits. These include acquiring a process mindset in an informal peer group setting, thus helping to establish a process culture within the enterprise. External advice and comparisons with other companies become available without the need to invest in costly consulting projects. And by actively participating in a BPM community, employees acquire the knowledge they need to create a successful process culture.

The new ARIS Community (Fig. 3) allows BPM evangelists and project teams to acquire process knowledge and stay updated on the latest developments. Personal connections can be established with BPM experts on the provider side and in other organizations. Web 2.0 technologies make it possible for BPM teams to join forces, exchange process knowledge, and engage in collaborative development work, regardless of location. Since collective results and decisions are better than individual ones, communities improve the effectiveness and efficiency of BPM initiatives [2]. The ARIS Community also offers a growing number of BPM courses and training materials that enable new BPM users to get up to speed quickly on the subject.

Fig. 3. ARIS Community

The ARIS Community (www.ariscommunity.com) also has plenty to offer ARIS users, who can submit questions to IDS Scheer experts about specific ARIS products in a special forum. Frequently asked questions are documented and accessible via a user-friendly search function. In addition, ARIS Express can be downloaded from the ARIS Community site. This software features the standards required by the community to ensure that results are compatible and can be interpreted in a consistent fashion. ARIS Community members based in different locations worldwide can work together on best practices. The established ARIS blog (www.arisblog.com) is integrated into the Community, as is ARIS TV (www.youtube.com/aristv), the first BPM channel on YouTube. A Twitter feed (http://twitter.com/ariscommunity) updates members on new contributions and keeps them informed about the ongoing development of Community functions.

ARIS Campus (www.ariscampus.com) is a special community with a BPM focus established by IDS Scheer for universities. Students and lecturers benefit from online campus access to BPM knowledge, software, best practice examples, and the latest research results. The PROWIT research project (Process-Oriented Web 2.0-based Integrated Telecommunications service) uses ARIS Campus to examine the potential impact of Web 2.0 on BPM [3]. The ADiWa project is conducting research into the potential components of semantic BPM (see Sneak Preview: Semantic BPM). This project also uses ARIS Campus to bring together the various teams online and encourage interaction.

3 Crowdsourcing – Making Products and Processes Democratic

Consumers are increasingly demanding more say in how products are designed. In the auto industry, mass customization has long been established practice, with customers able to modify the specification of their chosen car right up to shortly before production commences. Other industries have picked up on this trend: Customers of Lego Factory (http://factory.lego.com/), for example, can create new models from standard Lego bricks using a form of CAD software, publish them in a community, and order the relevant bricks. Other members are then able to add to the models. The creativity of the community is greater than that of any design department, and it helps Lego to produce many new models based on standard elements. Tasks previously performed by company employees are now outsourced to a large, undefined group of people organized in the form of a community. This approach is referred to as "crowdsourcing" [4].

The software industry has also experienced increased demand for customer input into products. Customers know best what range of functionality a software product should have and what the underlying processes they need to support look like. The obvious answer is to provide customers with standardized components from which they can create their own applications. The IDS Scheer response to such demands includes ARIS MashZone, which enables departmental users to create mashups (composite applications) for visualizing and evaluating a range of data. Every user department holds huge amounts of data that can only be accessed and visualized with a great deal of effort. This data is often derived from different sources, and users have a huge need to create connections and establish patterns. Because such data is often very volatile, static reports are no longer sufficient. In fact, most reports are obsolete by the time they reach the recipient—a situation that many companies cannot tolerate in times of crisis.

Like in the Lego Factory scenario, ARIS MashZone users can perform their own situation-based evaluations without needing any programming knowledge. This in turn reduces the workload of IT staff. With every consumer of information now a mashup producer, the long wait for analysis results becomes a thing of the past. Employees gain better insights faster and have more freedom when it comes to preparing/presenting key performance indicators relevant to the decision-making process. A mashup combines internal data (ARIS Process Performance Manager, ARIS Process Event Monitor, ARIS Business Optimizer, Excel, ERP systems, CRM systems, data warehouse systems, etc.) with Web data (Google Maps, statistical databases, financial tickers, etc.). To do this, it taps data sources and turns them into feeds (Fig. 4). Consistency is ensured at all times because data remains in its original source location. The feeds are then assembled into mashups in a visual composer. ARIS MashZone offers a host of visualization components for this purpose

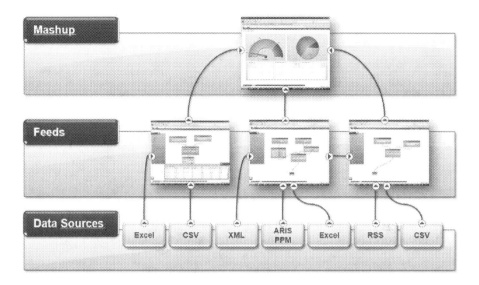

Fig. 4. ARIS MashZone

(bar graphs, pie charts, pyramid charts, funnel charts, maps, etc.). Interactive display enables intuitive analysis and filtering of data in the mashup.

Typical ARIS MashZone scenarios include evaluation of marketing campaigns, competitor analysis, customer satisfaction analysis, and financial analysis. Fig. 5 shows a mashup of United Motors Group (UMG), IDS Scheer AG's demo company. The mashup visualizes the results of a roadshow marketing campaign to launch a new vehicle. A CRM system (sales for last 12 months) is linked to e-mail marketing software (invitations), plus two Excel lists (registrations, participants), bringing together all the data. Users can display considerable amounts of detail, such as showing the sales figures for an area 200 km around the roadshow location in an integrated Google Maps view. Integrating sales process indicators from ARIS Process Performance Manager (ARIS PPM) makes it possible to use the mashup to check whether objectives are being achieved. If sales are below target, the user can go directly to ARIS PPM to analyze the live processes.

The Web 2.0 approach behind ARIS MashZone means that employees can share their mashups with colleagues and/or redeploy them (user-generated content). Thus process information from ARIS products can also be distributed to employees who do not use ARIS. The attractive way in which process KPIs (e.g., from ARIS PPM) are presented in mashups increases user acceptance, making Business Process Management fun and exciting. Thanks to a multidimensional combination of data, new information can be acquired, generating new knowledge in the process. As a result, the intellectual and creative potential of every member of the organization is realized.

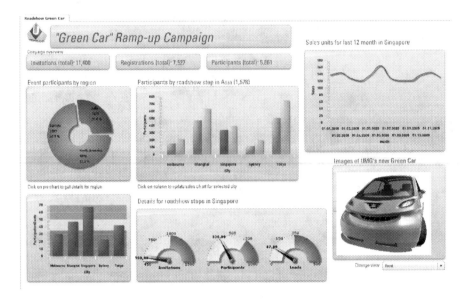

Fig. 5. Mashup of United Motors Group

4   The Search Function – A Gateway to Everything

Few technologies have attracted as much attention in recent years as search functionality. Much of this can be attributed to Google, which was founded by students at Stanford University. The Google search engine revolutionized not only the way Web information is accessed, but also user behavior. Instead of trawling through sitemaps or complex menu structures in software products, more and more users are using search functionality to access the information or functions they need. The major benefit of this kind of search is finding things you never knew existed. Conversely, if information or functions are "invisible," they are worthless.

The same is true of BPM projects: best practices, policies, and process standards are worthless if employees cannot find them. New ARIS Rocket Search therefore gives all company staff—regardless of their process and methodology knowledge—easy access to all forms of process content, ranging from workflow descriptions to templates for documents associated with the workflow and decision-making aids. Rather than navigating through a series of hierarchies to identify a process, all users have to do is enter the description of the process in the search box. Every letter entered reduces the number of hits returned. Fig. 6 shows a search for a sales process in the UMG process repository. Entering "SAP" as an additional search term would return all sales processes supported by an SAP system. Users can go directly from the search screen to the relevant process or object. ARIS Rocket Search also makes work easier for modelers working on process models, giving them the ability to select objects in the process chain and search for them in the ARIS repository. A filter is available to restrict the search to an exact combination of objects or to each individual object (Fig. 7).

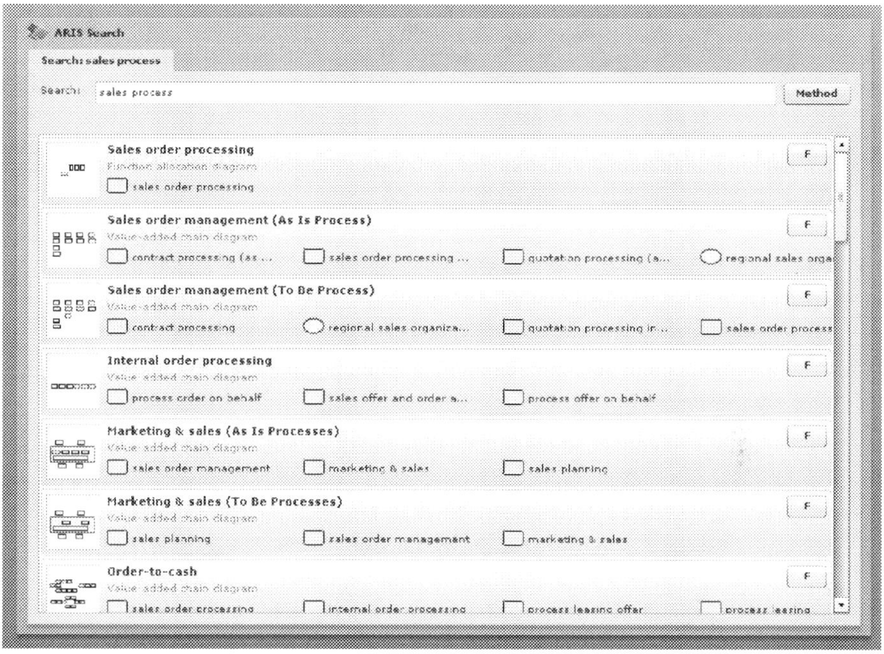

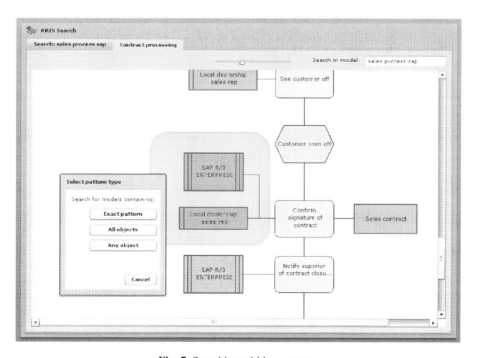

Fig. 6. Searching for a sales process

Fig. 7. Searching within a process

To make working in ARIS easier, it is also possible to search menu structures, interfaces, and the help system. For example, to find out how to run a report, users can simply enter "Run report," and the function is executed immediately. Since BPM content is often stored in different databases, ARIS Rocket Search can search any number of process, architecture, or service repositories and associated document management systems. Hit accuracy improves with every search because ARIS Rocket Search tracks the results most frequently used. Key information is identified, categorized, and prioritized during the next search, aided by the use of semantic information from the process repository. Searches for a job description also return processes, manuals, IT systems, and employees (social network) associated with the job. Benefits include the ability to create training programs quickly and easily. The search results can be displayed in list form as well as graphically, offering statistics and intuitive visualization in greater detail. ARIS Rocket Search also benefits from sophisticated input recognition functionality, enabling number ranges and time spans to be recognized and searched for. Whether the approach is BPM guerrilla or BPM governance, intuitive search functions that deliver fast, accurate hit rates are set to become the standard access mechanism for BPM information.

5 BPM and the Cloud

One of the hottest topics in the BPM community is currently BPM in the cloud. Using the software-as-a-service (SaaS) model, this involves offering and consuming BPM services via the Internet (cloud). Users can access these services and execute the application on the Web via a thin client. Because processing takes place online, an almost complete BPM environment is available. But before an organization decides to take to the cloud with its BPM initiative, it needs to distinguish between core processes and context processes and identify the relevant risks and opportunities.

Core processes are a company's essential value-adding processes. As such, they are the operational expression of corporate strategy. If the strategy is one of price leadership, core processes will be geared toward cost minimization. If high service quality is the strategy, the focus will be on smooth operation and ongoing optimization of all processes with customer touchpoints. In a word: core processes carry an organization's DNA and are crucial to business success. Clearly, such processes should not be placed in the cloud without careful consideration—competitors could access them and gain an advantage. Core processes are also not candidates for public exchange of information on freely accessible platforms, so there is no real case for making them available across corporate boundaries.

The situation is different when it comes to context processes, which tend to play a supporting role (HR processes, support processes, etc.). Best practices are frequently exchanged publicly and documented as part of various standardization initiatives across a broad range of industries. For example, the business processes of telecommunications companies are documented in eTOM [5].

IDS Scheer supports both scenarios, enabling organizations to use Java-based ARIS versions to launch BPM initiatives in a "private cloud" for users in different locations around the world. All data is stored and consolidated in the central ARIS repository. ARIS Governance Engine enables transparent design and efficient

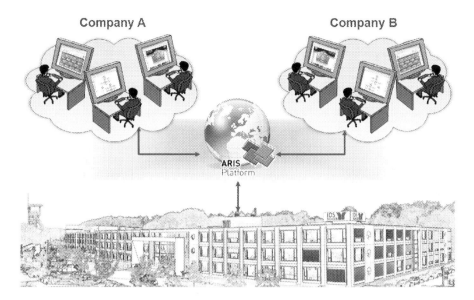

Fig. 8. Modeling as a service

implementation of change processes. IDS Scheer also offers modeling as a service, which enables customers to use an installation operated within IDS Scheer's data center, instead of having to buy hardware or install software. Rights- and role-based access to the system is provided via the Web (Fig. 8). Context processes, on the other hand, can be published and discussed on ARIS Community, where the collective intelligence of members enables continuous process improvement. Of course, not-for-profit organizations can also have their core processes analyzed and improved by the community, in the same way that open source software, such as Linux, is developed collectively.

But many enterprises have a much more basic problem: they have no real knowledge of their core and context processes. This can lead to huge problems in times of economic turmoil. If core processes are compromised by cost cutting and restructuring activities, customers will quickly switch to competitors. At the same time, care must be taken that any outsourcing decision only applies to context processes, while continued investment in core processes is essential to retain existing customers and acquire new ones.

6   Sneak Preview: Semantic BPM

To implement Business Process Management at the IT level, business processes need to be transformed into technical processes. For some years now, the technology of choice here has been service-oriented architectures (SOAs). All too often, however, SOA projects fail to deliver the desired results, often because of a lack of adequate cooperation between user department (SOA customer) and IT department (SOA provider). The barrier between business and IT is due in large part to the different language used by the two groups. The SUPER research project (Semantics Utilized for Process Management within and between Enterprises) seeks to bridge this gap through a combination of semantic Web services and BPM [6]. The result is a methodology for semantic BPM covering the entire BPM lifecycle. Process models are semantically annotated, which increases reuse of process fragments. The models are then translated into the BPEL4SMS standard and executed. Users can use semantic queries at runtime to search for and execute new software services [7].

As an example of a semantic BPM scenario, take a food manufacturer, who has to comply with new EU regulations stipulating that certain foods must now be sold within a shorter sell-by date than before. If the organization already has a BPM initiative, it needs to identify the processes relating to shipping of the relevant products—a task that can only be performed by a human being. With semantic BPM, the business processes and EU regulations would be defined in a way that allows automatic execution of the required process changes. Technologies used here include reasoners, ontologies, and mediators, which have all been developed to make the semantic Web a reality. In the semantic Web, machines can understand and categorize Web content [8].

Although semantic BPM currently only exists in the research labs of universities and a limited number of BPM technology providers, such as IDS Scheer, this technology is key to the future of BPM. For it to succeed, organizations must have a high level of BPM maturity, but given the rapid progress made recently by BPM at the technology and organizational levels, it can only be a matter of a few years before semantic BPM becomes reality.

7 Summary

The BPM guerrilla approach and BPM governance are not conflicting choices. Rather, they are two different organizational forms that represent a natural progression as an organization's BPM maturity increases. Both depend on technology to deliver the benefits of a process-driven organization. It needs to be possible to combine the output of independent BPM teams and interpret it consistently. Teams also need the ability to form communities and share insights. This is the only way to propagate the process mindset across the enterprise. Whether search functionality or mashup technology are used, access to process information must be quick and easy. Increased viral distribution of process information enables generation of greater knowledge. Although prospects for the emergence of a self-learning BPM system are encouraging, it should be remembered that processes will always be modeled, optimized, implemented, and monitored by humans. Technology is simply a means to an end—it is always people who drive growth.

References

1. Davis, R., Brabänder, E.: ARIS Design Platform – Getting Started with BPM. Springer, London (2008)
2. Tapscott, D., Williams, A.D.: How Mass Collaboration Changes Everything, Portfolio, USA (2008)

3. Wagner J.: BPM in the Web 2.0 era,
   http://www.arisblog.com/2009/03/30/bpm-in-the-web-20-era/
4. Howe, J.: Crowdsourcing: How the Power of the Crowd is Driving the Future of Business. Crown Business, USA (2008)
5. TM Forum, http://www.tmforum.org/browse.aspx
6. IDS Scheer, http://www.ariscampus.com/community/research
7. Van Lessen, T., Wetzstein, B., Nitzsche, J., Ma, Z., Karastoyanova, D., Leymann, F.: Geschäftsprozessmanagement Meets Semantic Web, Stuttgart,
   http://www.taval.de/INPROC-2007-60%20BPMmeetsSW.pdf
8. Stein, S., Stamber, C.: Semantic Business Process Management. In: Kuropka, D., Tröger, P., Staab, S., Weske, M. (eds.) Semantic Service Provisioning, pp. 127–143. Springer, Berlin (2008)

Change in Control

John Hoogland

Pallas Athena
Piet Joubertstraat 4, 7315 AV Apeldoorn, The Netherlands
johnhoogland@pallas-athena.com

Outline

Change is a popular word since Barack Obama so successfully used it in his campaign to become president of the US. Change almost became a synonym for "yes, we can". Change, and things will improve. Suddenly change is sexy. We don't seem to suffer from resistance to change anymore. People now LIKE to change?

Control usually is considered to almost be opposite to change. Control is not sexy. It is associated with accountants, budget restrictions, penalties, rules. It is considered to be counterproductive. You need it for compliance, but how easy would it be to change without control?

BPM is stuck in the middle. It used to be positioned as a technology that improves efficiency, effectiveness and quality. Cheaper, faster and better. But suddenly BPM is all about Agility. BPM for change. BPM is sexy too! At the same time, BPM is considered to be a tool for compliance, risk management and a means to guarantuee SLA's, KPI's, etc. Control!

So how can this be? BPM does not bring agility by itself. Nothing is more agile than processes that are not under control of any system. BPM is only called agile, because it gives more flexibility than built-in process control in legacy applications. At least, that is what we all believe (or say we believe). We call this the flexibility paradox. Process automation adds control, takes away flexibility and at the same time brings agility.

What do analysts, vendors and consultants say about BPM and agility?

Gartner says: "Business Process Management is a discipline that enables business agility in three important ways. It allows faster and better-informed decisions, reduces the process revision cycle time, and promotes consensus for rapid adoption of change"[1].

It is very important to realize that two different types of agility are mixed here. First of all, there is the ability to make better change decisions based on better data and communicated in a better way. Second, there is the ability to execute changes easier and faster. In this position paper, the emphasis is on the second form of agility.

Most vendors of BPM suites connect the concept of agility to two BPM functionalities:

1. Round trip BPM, meaning that the iterative process cycle from design to execution, reporting and analysis is done in one integrated environment, allowing business users to be heavily involved.

U. Dayal et al. (Eds.): BPM 2009, LNCS 5701, pp. 28–30, 2009.

2. Rules engines, allowing to change business rules without the need to change the process definition.

It is really surprising to see that hardly any vendor or analyst explains HOW a BPM system delivers better agility. It is almost considered to be common sense, like you don't have to explain why you get warm in the sun. But it is not that obvious at all. So how can agility be achieved through executing and controlling business processes by automating them?

Flexibility in business processes can be achieved in two ways:

1. Intrinsic flexibility thanks to a rich BPM engine able to support many work-flow patterns, resource patterns or even exception handling patterns [2,3,4]
2. Flexibility through adaptability of process behavior.

The main difference between the two is that in the first case any change in the behavior of the system needs a change in the process definition. That is why better engines offer more flexibility. The goal is to prevent changes in the definition of processes, because however easy it is to make these changes, every change will require releasing a new version of the process model. And such a version upgrade will need to be tested, accepted and therefore planned. This is especially true when the process is highly integrated with other applications or services in the infrastructure. Based on experience, one might state that these changes cannot be effectuated more than three times per year. This process cycle is often called the *outer circle* of process management.

Therefore, support for many patterns, operational flexibility that allows end users to defer from the process definition in a controlled way (based on senior-ity for instance), exception handling capabilities like skipping steps, roll back processes with compensation actions, ad-hoc activities, etc. etc., all bring about functionality that prevent releasing new versions. This concept is implemented in a number of BPM tools, but hardly communicated as such. Many vendors even with these capabilities in their product, still market agility in relation to easier modeling, round trip modeling (from process design to execution without intermediate translation). And this is strange, since round trip modeling is not a differentiator, while the mentioned capabilities are!

In the second case, which can be referred to as the *inner circle*, changes in the behavior of the BPM system can be achieved through parameterization in rules, distribution, process variables, resource allocation, web services, etc. The goal is that in a separated configuration environment, business users are able to adapt the process, without the need to release a new version. The cycle time of these kind of changes is short (often real-time), and preferably the configura-tion capabilities are decentralized in business domains, and hierarchical layers, ensuring in this way that business managers can only make changes in business areas for which they are responsible or authorized. Inner circle support brings true agility. Work distribution can be changed ad hoc and real time, without any IT support. Business rules can be adapted without the need to release new versions. Process thresholds for control steps, escalation steps, can be lowered or increased in order to cope with unexpected workload, deadlines and SLA's can temporarily be adapted for the same reason.

And this is how we find the equilibrium between change and control. The outer circle brings control where needed, the inner circle offers the flexibility to manoeuver between clear boundaries. Change in control.

References

1. Hill, J., Melenovsky, M.: Achieving Agility: BPM Delivers Business Agility through New Management Practices; Gartner. Gartner Report G00137553 (2006)
2. Russell, N., Ter Hofstede, A., van der Aalst, W., Mulyar, N.: Workflow control-flow patterns: A revised view. BPM Center Report BPM-06-22, BPMcenter. org, 06–22 (2006)
3. Russell, N., Hofstede, A., Edmond, D., Aalst, W.: Workflow Resource Patterns. BETA Working Paper Series, WP 127, Eindhoven University of Technology, Eindhoven (2005)
4. Russell, N., van der Aalst, W., ter Hofstede, A.: Exception handling patterns in process-aware information systems. BPM Center Report BPM-06-04, BPMcenter. org, 06–04 (2006)

Scientific Workflows: Business as Usual?*

Bertram Ludäscher[1,2], Mathias Weske[3], Timothy McPhillips[1], and Shawn Bowers[1]

[1] Genome Center, University of California Davis, USA
{ludaesch,tmcphillips,bowers}@ucdavis.edu
[2] Department of Computer Science, University of California Davis, USA
[3] Hasso-Plattner-Institute University of Potsdam, Germany
weske@hpi.uni-potsdam.de

Abstract. Business workflow management and business process modeling are mature research areas, whose roots go far back to the early days of office automation systems. Scientific workflow management, on the other hand, is a much more recent phenomenon, triggered by (i) a shift towards data-intensive and computational methods in the natural sciences, and (ii) the resulting need for tools that can simplify and automate recurring computational tasks. In this paper, we provide an introduction and overview of scientific workflows, highlighting features and important concepts commonly found in scientific workflow applications. We illustrate these using simple workflow examples from a bioinformatics domain. We then discuss similarities and, more importantly, differences between scientific workflows and business workflows. While some concepts and solutions developed in one domain may be readily applicable to the other, there remain sufficiently many differences that warrant a new research effort at the intersection of scientific and business workflows. We close by proposing a number of research opportunities for cross-fertilization between the scientific workflow and business workflow communities.

1 Introduction

Whether scientists explore the limits and origins of the observable universe with ever more powerful telescopes, probe the invisibly small through particle accelerators, or investigate processes at any number of intermediate scales, scientific knowledge discovery increasingly involves large-scale data management, data analysis, and computation. With researchers now studying complex ecological systems, modeling global climate change, and even reconstructing the evolutionary history of life on Earth via genome sequencing and bioinformatics analyses, science is no longer "either physics or stamp collecting"[1]. Instead, science is increasingly driven by new and co-evolving observational and experimental methods, computer simulations, and data analysis methods. Today's scientific experiments happen in large parts *in silico*, i.e., in the computer [8]. In the UK, the term *e-Science* [1] was coined to describe computationally and data intensive science, and a large e-Science research program was started there in 2000. Similarly, in the US, the National Science Foundation created a new Office for Cyberinfrastructure (OCI) to advance computer science and informatics technologies in support

---

* This research was conducted while the second author was on sabbatical leave at UC Davis.
[1] *"All science is either physics or stamp collecting."* – Ernest Rutherford [11].

U. Dayal et al. (Eds.): BPM 2009, LNCS 5701, pp. 31–47, 2009.

of e-Science. As a result, the new opportunities of data-driven and compute-intensive science have introduced the new challenges of managing the enormous amounts of data generated and the complex computing environments provided by cluster computers and distributed Grid environments.

Given these developments, domain scientists face a dilemma. Scientific progress in their fields relies ever more on complex software systems, high-performance computing environments, and large-scale data management. For example, advanced computational science simulations involve all of the above [46]. But employing all of these resources is a time-consuming and labor-intensive task, made all the more challenging by the high rate at which new technologies, services, and applications appear. Understandably, many scientists would prefer to focus on their scientific research and not on issues related to the software and platforms required to perform it. As a result, interest in the area of *scientific workflow management* has increased significantly in recent years [25, 28, 30, 38–40, 49, 51], and many projects are now employing or developing scientific workflow technology [5, 19, 26, 27, 37, 45].

One goal of scientific workflows is to support and whenever possible automate what would be otherwise error-prone, repetitive tasks, e.g., data access, integration, transformation, analysis, and visualization steps [39]. Thus, scientific workflows are often used to chain together specialized applications and new data analysis methods. However, as is the case in business workflow management, scientific workflows are not only about workflow enactment and execution; modeling, design, analysis, and reuse of workflows are also becoming increasingly important in this area. The main goals of scientific workflows, then, are (i) to save "human cycles" by enabling scientists to focus on domain-specific (science) aspects of their work, rather than dealing with complex data management and software issues; and (ii) to save machine cycles by optimizing workflow execution on available resources.

In this paper, we provide an introduction and overview of scientific workflows, and compare and contrast with the well-established, mature area of business workflows. The outline and contributions of this paper are as follows. In Section 2 we provide an overview of the scientific workflow life cycle and common use cases. Section 3 describes some key concepts and emerging approaches for addressing the technical challenges encountered in developing and deploying scientific workflows. A family of bioinformatics workflows is then used in Section 4 to further illustrate some of the use cases and technical issues. In Section 5, we compare and contrast scientific workflow concepts and issues with those in found in the business workflow arena. Finally, in Section 6 we propose areas of future research and opportunities for cross-fertilization between the scientific workflow and business workflow communities.

2   The Scientific Workflow Life Cycle

Figure 1 depicts a high-level view of the scientific workflow life cycle. Starting from a scientific hypothesis to be tested, or some specific experimental goals, a **workflow design** phase is initiated. During this phase, scientists often want to reuse pre-existing workflows and templates or to refine them. Conversely, they can decide to share a (possibly revised and improved) workflow design, or make workflow products (derived data,

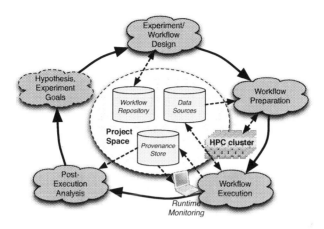

Fig. 1. Scientific Workflow Life Cycle

new components, subworkflows, *etc.*) available via a public repository or shared **project space**. Scientific workflow design differs significantly from general programming, with analysis libraries, available web services, and other pre-existing components often being "stitched together" (similar to scripting approaches [48]) to form new data analysis pipelines.

During **workflow preparation**, data sources are selected and parameters set by the user. Workflows may require the scheduling of high-performance computing (HPC) resources such as local cluster computers, or remote (Grid or cloud computing) resources; also data may have to be staged, i.e., moved to certain locations where the compute jobs running on the HPC cluster(s) expect them.

During **workflow execution**, input data is consumed and new data products created. For large-scale computational science simulations (running on hundreds or thousands of nodes; for hours, days, or weeks at a time), *runtime monitoring* is critically important: intermediate data products and special provenance information are often displayed on a web-based monitoring "dashboard" to inform the scientist about progress and possible problems during execution. Depending on this information, the scientist may decide to abort a simulation or workflow run.

Scientists often need to inspect and interpret workflow results in a **post-execution analysis** phase to evaluate data products (*does this result make sense?*), examine execution traces and data dependencies (*which results were "tainted" by this input dataset?*), debug runs (*why did this step fail?*), or simply analyze performance (*which steps took the longest time?*). Depending on the workflow outcomes and analysis results, the original hypotheses or experimental goals may be revised or refined, giving rise to further workflow (re-)designs, and a new iteration of the cycle can begin.

The workflow life cycle typically involves users in different **roles**: Domain scientists often act as the (high-level) *workflow designers* and as the *workflow operators*, i.e., they execute and possibly monitor the workflow after having prepared the run by selecting datasets and parameters. Depending on the complexity of the target workflows and the skills required to compose these in a particular system, *workflow engineers*[2] commonly are also involved in implementing the workflow design.

Types of Scientific Workflows. There is no established scientific workflow classification yet. Indeed, there seems to be no single set of characteristic features that would uniquely define what a scientific workflow is and isn't. To get a better grasp of the meaning and breadth of the term 'scientific workflow', we have identified a number of dimensions along which scientific workflows can be organized.

In many disciplines, scientists are designers and developers of new experimental protocols and data analysis methods. For example in bioinformatics, the advent of the next generation of ChIP-Seq[3] protocols and the resulting new raw data products are leading to a surge in method development to gain new knowledge from the data these experiments can produce. Scientific workflows in such realms are often *exploratory* in nature, with new analysis methods being rapidly evolved from some some initial ideas and preliminary workflow designs. In this context, it is crucial that scientific workflows be easy to reuse and modify, e.g., to replace or rearrange analysis steps without "breaking" the analysis pipeline. Once established, *production workflows*, on the other hand, undergo far fewer changes. Instead, they are executed frequently with newly acquired datasets or varying parameter settings, and are expected to run reliably and efficiently.

Scientific workflow designs can also differ dramatically in the types of steps being modeled. For example, we may distinguish *science-oriented workflows* [42], in which the named steps of the workflow spell out the core ideas of an experimental protocol or data analysis method, from lower-level *engineering* (or "plumbing") *workflows*, which deal with data movement and job management [46]. Another category along this dimension are *job-oriented workflows*, typically expressed as individual compute jobs for a cluster computer, whose job (i.e., task) dependencies are modeled as a DAG [22].

3   Scientific Workflow Concepts and System Features

In order to address the various challenges encountered throughout the scientific workflow life cycle, and in light of the vastly different types and resulting requirements of scientific workflows, a number of concepts have been and are being developed. In the following, we use terminology and examples from the Kepler scientific workflow system [2, 37] (similar concepts exist for other systems, e.g., Taverna [3], Triana [4], *etc.*)

Integrated Workflow Environment. Many (but not all) scientific workflow systems aim at providing an integrated 'problem-solving environment'[4] to support the workflow life cycle illustrated in Figure 1. For workflow design, a *visual programming interface* is often used for wiring up reusable workflow components (or *actors*). To facilitate rapid workflow development and reuse, *actor libraries* (containing executable code) and *workflow repositories* (e.g., myExperiment [31]) can be used. Similarly, a *metadata catalog* may be used to locate relevant datasets from distributed data networks. The

---

[2] i.e., software engineers with workflow system expertise.

[3] Chromatin ImmunoPrecipitation with massively parallel DNA Sequencing.

[4] A term that has been used earlier in the context of computational science simulations [47].

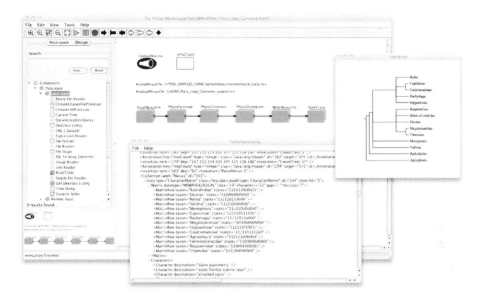

Fig. 2. A Kepler scientific workflow for inferring evolutionary relationships using morphological data. The windows labeled *Drawgram* and *CollectionDisplay* show the resulting phylogenetic tree and reveal the nested data streaming through the actors, respectively.

screenshot in Figure 2 depicts the Kepler user interface, including the workflow canvas and actor library.

Workflow Preparation and Execution Support. Many scientific experiments require multiple workflow runs using different parameter settings, data bindings, or analysis methods. In such cases, *parameter sweeps* [5] can be used to simplify these experiments. When running workflows repeatedly with varying parameter settings, data bindings, or alternative analysis methods, a "smart rerun" capability is often desirable [37] to avoid costly recomputation. This can be achieved, e.g., by using a data cache and analysis of dataflow dependencies (when parameters change, only downstream computations need to be re-executed [6]). For long-running workflows, some systems offer capabilities for runtime monitoring, e.g., using web-based *dashboards*, which display and visualize key variables (e.g., of large-scale fusion simulation experiments [34]). Similarly, long-running workflows require support for *fault-tolerance*, e.g., in the case of actor-, service-, or other workflow-failures, a "smart resume" capability avoids re-execution of previously successful steps, either by a form of (application-dependent) checkpointing [38, 46], or by employing suitable logging and provenance information recorded by the workflow system [21]. In so-called "grid workflows", workflow jobs need to be scheduled and mapped onto the distributed computing resources [26].

Data-Driven Models of Computation. While scientific workflow designs visually emphasize processing steps, the actual computation is often *data-driven*. Indeed, without workflow system support, scientists often spend much of their time reading,

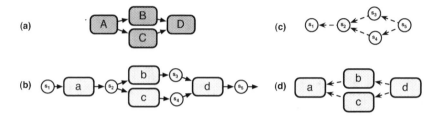

Fig. 3. Basic workflow and provenance models: (a) workflow definition (here, a DAG) with four actors A, ..., D; (b) example flow graph with process invocations a, ..., d and atomic data $s_1, ..., s_5$; (c) data dependencies and (d) invocation dependencies inferred from (b).

reformatting, routing, and saving datasets. Instead, *dataflow-oriented* and *actor-oriented* models of computation [36] for scientific workflows [12] emphasize the central role of data. What passes between workflow steps is not just control (the triggering of a subsequent step in response to all prior steps having completed), but *data streams*[5] that flow between and through actors (either physically or virtually [52]) and that drive the computation.

Consider the simple workflow DAG in Figure 3(a). In a business workflow model (and in some job-oriented scientific workflow models [22]), we would view A as an AND-split, followed by two task-parallel steps B and C and an AND-join D. Dataflow is often implicit or specified separately. In contrast, in a data-driven model of computation, the tokens emitted by a workflow step drive the (often repeated) invocations of downstream steps. Figure 4 illustrates that there are (often implicit) *data queues* between workflow steps to trigger multiple process invocations of an actor. Such dataflow-oriented models of computation are also beneficial for (i) *streaming workflows* (e.g., for continuous queries and window-based aggregates over sensor data streams [9]), and (ii) *pipeline-parallel* execution of scientific workflows [46]; see Figure 4 and below.

Data Provenance. In recent years, research and development activities relating to data *provenance* (or *data lineage*) and other forms of provenance information have increased significantly, in particular within the scientific workflow community [13, 18, 23, 24, 43]. Information about the processing history of a data product, especially the dependencies on other, intermediate products, workflow inputs, or parameter settings, can be valuable for the scientist during virtually all phases of the workflow life cycle, including workflow execution (e.g., for fault-tolerant [21] or optimized execution) and post-execution analysis (i.e., to validate, interpret, or debug results as described above).

Consider the *flow graph* in Figure 3(b). It captures relevant provenance information, e.g., that in a particular workflow run, the actor A consumed an input data (structure s_1) and produced output data (s_2); the linkage between inputs and outputs is given via an *invocation* a of A. The *data lineage graph* in Figure 3(c) is a view of the graph in (b), and shows how the final workflow output s_5 depends on the input s_1 via intermediate data products s_2, s_3, s_4. The *invocation dependency graph* in Figure 3(d) highlights how actor invocations depended on each other during a run.

---

[5] The so-called "information packets" in Flow-based Programming [44].

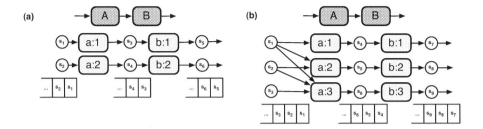

Fig. 4. The process network (PN) model supports *streaming* and *pipelined execution*: (a) step A of the workflow (top) yields two *independent* invocations (a:1, a:2) within the flow graph (bottom), possibly executed concurrently over the input stream $s_1, s_2, \dots$; (b) a variant of (a) where A is *stateful*, preserving information between invocations a:*i*, resulting in additional dependencies.

Different Models of Computation and Provenance. The model of provenance (MoP) to be used for a scientific workflow may depend on the chosen model of computation (MoC) that is used to describe the workflow execution semantics [15, 41]. For example, for a simple MoC that views a workflow specification as a DAG [22], the associated MoP need not distinguish between multiple invocations a:1, a:2, ... of an actor A, simply because each actor is invoked no more than once. For the same reason, it is not meaningful to distinguish *stateless* from *stateful* actors. In contrast, in MoCs that (i) allow loops in the workflow definition, and/or (b) support pipeline parallel execution over data streams, multiple invocations need to be taken into account, and one can distinguish stateful from stateless actors. Consider Figure 4(a), which depicts a simple workflow pipeline consisting of two steps A and B. In a dataflow MoC with *firing semantics* [35], each data token s_i on a channel[6] may trigger a separate invocation; here: a:1, a:2, ..., and b:1, b:2, ... With the appropriate MoP, the provenance graph indicates that two (or more) *independent instances* of the workflow were executing. This is because the actors A and B are stateless, i.e., each invocation is independent of a prior invocation (e.g., A might convert Fahrenheit data tokens to Celsius). On the other hand, Figure 4(b) shows a provenance graph that reveals that the multiple invocations of A are *dependent* on each other, i.e., A is stateful. Such a stateful actor might, e.g., compute a running average, where a newly output token depends on more than one previously read token.

Workflow Modeling and Design. Experiment and workflow designs often start out as napkin drawings or as variants or refinements of existing workflows. Since workflows have to be executable to yield actual results, various *abstraction* mechanisms are used to deal with the complex design tasks. For example, Kepler inherits from Ptolemy II [17] the capability to nest *subworkflows* as composite actors inside of workflows, possibly adopting a different model of computation (implemented via a separate *director*) for the nested subworkflow. Top-level workflows are often coarse-grained process pipelines, where each step may be running as an independent process (e.g., executing a web service or R script), while lower-level "workflows" might deal with simple, fine-grained

---

[6] In the process network model [33] actors (processes) communicate via unbounded queues.

steps such as the evaluation of arithmetic expressions. Thus, it can be beneficial to employ different MoCs at different levels [46], e.g., Kahn process networks at the top-level (implemented via a so-called PN director) and synchronous dataflow[7] for lower levels.

Actor-oriented modeling and design of scientific workflows can also benefit from the use of *semantic types* [10, 12], where data objects, actors, and workflows can be annotated with terms from a controlled vocabularly or ontology to facilitate the design process. Finally, *collection-oriented modeling and design* [14] can be seen as an extension of actor-oriented modeling which takes into account the *nested collection structure* frequently found in scientific data organization to obtain workflow designs that are easier to understand, develop, and maintain [42].

4   Case Study: Phylogenetics Workflows in Kepler

Here we illustrate the challenges, use cases, concepts, and approaches described above using concrete examples of automated scientific workflows implemented in the Kepler scientific workflow system. The example computational protocols come from the field of phylogenetics, which is the study of the tree-like, evolutionary relationships between natural groups of organisms. While phylogenetics methods and data comprise a narrow sub-domain of bioinformatics, they are broadly relevant to the understanding of biological systems in general.

The pPOD Extension to Kepler. The pPOD project[8] is addressing tool and data integration challenges within phylogenetics through a workflow automation platform that includes built-in mechanisms to record and maintain a continuous processing history for all data and computed results across multiple analysis steps. Like many other science domains, these steps currently are carried out using a wide variety of scripts, standalone applications, and remote services. Our solution, based on the Kepler system, automates common phylogenetic studies, routing data between invocations of local applications and remote services, and tracking the dependencies between input, intermediate, and final data objects associated with workflow runs [16]. The immediate goal of the current version of the system is to provide researchers an easy-to-use desktop application that enables them to create, run, and share phylogenetic workflows as well as manage and explore the provenance of workflow results. The main features of the system include: (i) a library of reusable workflow components (i.e., actors) for aligning biological sequences and inferring phylogenetic trees; (ii) a graphical workflow editor (via Kepler) for viewing, configuring, editing, and executing scientific workflows; (iii) a data model for representing phylogenetic artifacts (e.g., DNA and protein sequences, character matrices, and phylogenetic trees) that can facilitate the conversion among different data and file formats; (iv) an integrated provenance recording system for tracking data and process dependencies created during workflow execution; and (v) an interactive provenance browser for viewing and navigating workflow provenance traces (including data and process dependencies).

---

[7] Such subworkflows execute in a single thread, statically scheduled by an SDF director.
[8] http://www.phylodata.org

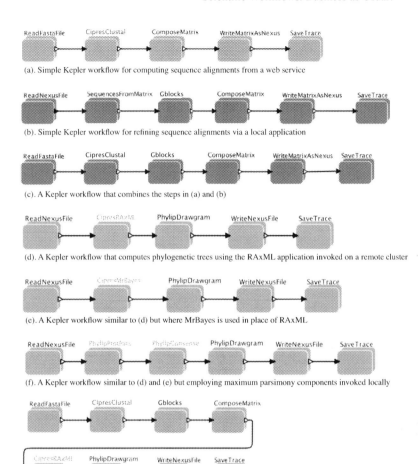

Fig. 5. Common data analyses used in phylogenetics implemented in the *Collection-Oriented Modeling and Design* (COMAD) framework of Kepler, highlighting benefits of actor reuse and workflow composition enabled through scientific workflow systems.

Figure 5 illustrates a number of workflows that can be constructed readily using the actors included with the pPOD extension to Kepler. The workflow shown in Figure 5(a) reads one or more files containing DNA or protein sequences in the FASTA file format. Each sequence represents a different group of organisms (e.g., species). The workflow then employs the Clustal application to align these sequences to each other (thereby inferring which positions in each sequence are related by evolution to positions in the other sequences). From this multiple sequence alignment, the workflow composes a phylogenetic character matrix and saves it to the researcher's disk in the Nexus format. The final actor saves a record of the workflow run containing the provenance information required to later reconstruct the derivation history of data products.

The Importance of Data Management in Tool Integration. The workflow in Figure 5(a) defines a relatively simple computation protocol: only the CipresClustal step performs a "scientifically meaningful" task, whereas the rest of the workflow simply automates the reading, reformatting, routing, and saving of data sets, and records the provenance of new data. However, even in this case, scientific workflow systems provide a number of critical benefits. In the absence of a framework such as this, researchers must run the Clustal program by hand, supplying it input data and instructions in an appropriate (but highly idiosyncratic) manner. They must either install the program on their own computers, have someone else install it for them, or else run Clustal via one of several web-based deployments of the application. In any event, they should (ideally) record precisely how they use the application each time they use it: what parameters they used, what input data files served as input, and what the immediate outputs of the program were. Each of these steps is labor-intensive and error-prone. Moreover, researchers typically carry out operations such as these *many times* on the same input data sets, varying the values of parameters given to the applications, alternately including or excluding subsets of input data sets, and repeatedly comparing and evaluating the results of all these variations. Further, because a protocol such as this occurs not in isolation, but as part of a larger set of workflows that comprise a scientific study, these kinds of variations in the upstream computational protocols cascade to the later protocols in a study, further multiplying the number of times a particular protocol must be carried out on what is conceptually the same data set. Consequently, managing data files, converting formats, and otherwise massaging scientific data in preparation for use with particular tools takes considerable time and effort, and must generally be done—again and again—by hand.

Immediate Advantages over Standard Scripting Approaches. Scientists typically define programs for automating analyses using scripting languages, e.g., when manual operation of tools such as these becomes too onerous. However, employing scripting languages in this way has serious limitations that scientific workflow systems directly aim to address. For example, a significant weakness of scripting languages is their lack of built-in provenance recording facilities. Further, the use of scripting languages for automating scientific protocols often involves *ad hoc* approaches for wrapping and executing external applications, whereas scientific workflow systems can provide users with uniform access to computational components (e.g., in Kepler through the actor model). The result is that external applications are typically only incorporated, or wrapped, into a workflow system once, making analyses easier to construct and components easier to reuse and adopt in new protocols. The limitation that scientists run into the most, however, is the difficulty of using a single script to automate a process spanning multiple compute nodes, heterogeneous communication protocols, and disparate job scheduling systems. A scientist wanting to run a scripted protocol on a local cluster rather than on her laptop must be ready to rewrite the script to take into account the associated job scheduling software, and be prepared to manually move data to and from the cluster by hand. To employ a web-based application to carry out one or more steps, she may also need to develop additional, often custom programs to send data to the service, supply it with the desired parameter values, invoke it, and wait for it to complete (which, e.g., often involves either polling or waiting for an e-mail message). Scripting languages are cumbersome platforms for this kind of distributed, heterogeneous process automation; and the situation becomes much harder when researchers wish to mix and match different kinds of applications and service protocols in a single script.

For instance, when the CipresClustal actor in Figure 5(a) is invoked, it implicitly calls a web service that runs the application remotely. By using systems such as Kepler to invoke these services, a researcher can easily repeat a protocol on the same or different data set, using all of the same parameter values used previously, or make variants of the workflows with different parameterizations. Furthermore, the researcher not only may create workflows employing multiple such services in the same workflow, but combine local applications and remote services in a single workflow. In this case, e.g., Kepler automatically routes data to and from the underlying compute resources as needed, waiting for services to complete, retrying failed service invocations, and dealing transparently with the different ways applications must be invoked on a local machine, on a Linux cluster, or at a supercomputer center.

As a further example, the workflow in Figure 5(b) can be used to refine a sequence alignment produced by the workflow in Figure 5(a) using the (locally installed) Gblocks application included with the pPod extension to Kepler. Both workflows can easily be concatenated to yield a protocol that uses heterogeneous computing resources without any effort on the part of the researcher (Figure 5c). Additional variants of workflows can easily be created without regard to how and where the particular steps of a protocol are carried out. For instance, the workflow in Figure 5(d) invokes the Cipres RAxML service at the San Diego Supercomputer Center [20] to infer phylogenetic trees from a provided character matrix, and a researcher can easily swap out this maximum likelihood method for tree inference with one based on Bayesian methods simply by replacing the CipresRAxML actor with the CipresMrBayes actor (as shown in Figure 5e). As shown, no other actors need be reconfigured. Similarly, the researcher may modify the workflow to employ a maximum parsimony method for tree inference by inserting two actors into the workflow (Figure 5f). Again, the workflow of Figure 5(d) can easily be concatenated with the workflow of Figure 5(c) to yield the workflow of Figure 5(g), which invokes two remote services and runs two local applications in the course of its execution. The ease with which new workflows can be composed, reused, repurposed, and deployed on heterogeneous resources–and later redeployed on *different* resources– is one of the major benefits of scientific workflow modeling.

5   Scientific Workflows *vs.* Business Workflows

In the following, we compare features of scientific workflows and business workflows. Even within each family, there seem to be few (if any) characteristic features that would yield a universally accepted, unambiguous classification without exceptions. Rather, it seems that workflows are related via a series of overlapping features, i.e., they exhibit a form of *family resemblance* [50]. Despite the fact that there are few sharp, categorical bounderies, the comparison below should help in assessing commonalities and typical differences between scientific workflows and business workflows.

Implementation *vs.* Modeling. The primary goal of business process modeling is to develop a common understanding of the process that involves different persons and various information systems. Once a business process model is developed and agreed upon (and in many cases improved or optimized), it can serve as a blueprint for implementing the process, all or in part, in software. Business workflows are the automated parts of these business processes. Scientific workflows, on the other hand, are developed with executability in mind, i.e., workflow designs can be viewed as executable specifications. In recent years, the modeling aspect in scientific workflows is receiving some more attention, e.g., to facilitate workflow evolution and reuse [42].

Experimental *vs*. Business-Driven Goals. A typical scientific workflow can be seen as a computational experiment, whose outcomes may confirm or invalidate a scientific hypothesis, or serve some similar experimental goals. In contrast, the outcome of a business workflow is known before the workflow starts. The goal of business workflows is to efficiently execute the workflow in a heterogeneous technical and organizational environment and, thereby, to contribute to the business goals of the company.

Multiple Workflow Instances. It is common that business workflows handle large numbers of cases and independent workflow instances at any given time. For example, each instance of an order workflow makes sure that the particular customer receives the ordered goods, and that billing is taken care of. In scientific workflows, truly independent instances are not as common. Instead, large numbers of *related* and interdependent instances may be invoked, e.g., in the context of parameter studies.

Users and Roles. Business workflows (in particular human interaction workflows) usually involve numerous people in different roles. A business workflow system is responsible for distributing *work* to the human actors in the workflow. In contrast, scientific workflows are largely automated, with intermediate steps rarely requiring human intervention. Moreover, the nature of these interactions is usually different, i.e., no work is assigned, but runtime decisions occasionally require user input (e.g., to provide an authentication information for a remote resource, an unknown parameter value, or to select from multiple execution alternatives).

Dataflow *vs*. Control-Flow Focus. An edge A → B in a business workflow typically means B can only start after A has finished, i.e., the edge represents *control-flow*. Dataflow is often implicit or modeled separately in business workflows. In contrast, A → B in a scientific workflow typically represents *dataflow*, i.e., actor A produces data that B consumes. In dataflow-oriented models of computation, execution control flows implicitly with the data, i.e., the computation is data-driven. The advantage of "marrying" control-flow with dataflow is that the resulting model is often simpler and allows stream-based, pipeline-parallel execution. The disadvantage is that certain workflow patterns (e.g., for conditional execution or exception handling) can be awkward to model via dataflow.

Dataflow Computations *vs*. Service Invocations. In scientific workflows data is often streamed through independent processes. These processes run continuously, getting input and producing output while they run. The input-output relationships of the activities are the dataflow. As a result, a sequence of actors A → B → C can provide pipelined concurrency, since they work on different data items at the same time. In business workflows, there are usually no data streams. An activity gets its input, performs some action, and produces output. An order arrives, it is checked, and given to the next activity in the process. In typical enterprise scenarios, each activity invokes a service that in turn uses functionality provided by some underlying enterprise information system.

Different Models of Computation. Different scientific workflow systems support different models of computation. For example, Pegasus/DAGMan [22, 26] workflows are job-oriented "grid workflows" and employ a DAG-based execution model without loops, in which each workflow step is executed only once. Branching and merging in these workflow DAGs corresponds to AND-splits and AND-joins in business workflows, respectively. Other workflow systems such as Taverna [3] and Triana [4] have different computation models that are dataflow-oriented and support loops; Kepler [2] supports multiple models of computation, including PN (Kahn's dataflow process network), SDF (Synchronous Dataflow, for fine-grained, single-threaded computations) and COMAD (for collection-oriented modeling and design). Given the vast range of scientific workflow types (job-oriented grid workflows, streaming workflows, collection-oriented workflows, *etc.*) there is no single best or universal model of computation that fits all needs equally. Even so, dataflow-based models are widespread among scientific workflows. In business workflows, on the other hand, Petri nets are used as the underlying foundation; BPMN is the de facto standard of an expressive process modeling language; WS-BPEL is used to specify workflows whose steps are realized by web services.

6 The Road Ahead

In this paper we have given an introduction and overview to scientific workflows, presented a bioinformatics case study, and compared features in scientific workflows with those in business workflows. Compared to the well-established area of business workflows, scientific workflow management is a fairly recent and active area of research and development.

For example, *workflow modeling and design* has not yet received the attention it deserves in scientific workflows. Workflow designs should be easy to reuse and evolve. They should be resilient to change, i.e., not break if some components are removed, added, or modified [42]. Techniques and research results from the business workflow community but also from the databases, programming languages, and software engineering communities will likely provide opportunities for future research in this area.

The business workflow community has embraced Petri nets as the unifying foundation for describing and analyzing workflows. The situation in scientific workflows is less uniform. In addition to Petri nets (e.g., combined with a complex object model [32]), there are other underlying models, e.g., well-established formalisms such as dataflow process networks [33, 35, 36], and new, specialized dataflow extensions, e.g., for nested data [42]. For optimizing streaming workflows, techniques from the database community for efficiently querying data streams look promising as well.

A very active area of research in scientific workflows is *provenance*, in particular techniques for capturing, storing, and querying not only data provenance [7] but also workflow evolution provenance (a form of versioning for configured workflows) [29]. In this context, statically analyzable dependencies between steps in a workflow can be used, for instance, to optimize data routing [52], or to check whether each step will eventually receive the required data. This is interesting, since the business workflow community has developed a set of soundness criteria for a given process model based on control-flow, disregarding data dependencies to a large extent. The integration of workflow analysis methods based on dataflow and on control-flow is a promising new area of research and cross-fertilization between the communities that can yield new results and insights for both scientific workflows and business workflows.

Acknowledgements. Work supported through NSF grants IIS-0630033, OCI-0722079, IIS-0612326, DBI-0533368, ATM-0619139, and DOE grant DE-FC02-01ER25486.

References

1. Defining e-Science (2008), www.nesc.ac.uk/nesc/define.html
2. The Kepler Project (2008), www.kepler-project.org
3. The Taverna Project (2008), www.mygrid.org.uk/tools/taverna
4. The Triana Project (2008), www.trianacode.org
5. Abramson, D., Enticott, C., Altinas, I.: Nimrod/K: Towards Massively Parallel Dynamic Grid Workflows. In: ACM/IEEE Conference on Supercomputing (SC 2008). IEEE Press, Los Alamitos (2008)
6. Altintas, I., Barney, O., Jaeger-Frank, E.: Provenance collection support in the Kepler scientific workflow system. In: Moreau, L., Foster, I. (eds.) IPAW 2006. LNCS, vol. 4145, pp. 118–132. Springer, Heidelberg (2006)
7. Anand, M., Bowers, S., McPhillips, T., Ludäscher, B.: Exploring Scientific Workflow Provenance Using Hybrid Queries over Nested Data and Lineage Graphs. In: Intl. Conf. on Scientific and Statistical Database Management (SSDBM), pp. 237–254 (2009)
8. Anderson, C.: The End of Theory: The Data Deluge Makes the Scientific Method Obsolete. WIRED Magazine (June 2008)
9. Babcock, B., Babu, S., Datar, M., Motwani, R., Widom, J.: Models and issues in data stream systems. In: PODS, pp. 1–16 (2002)
10. Berkley, C., Bowers, S., Jones, M., Ludäscher, B., Schildhauer, M., Tao, J.: Incorporating Semantics in Scientific Workflow Authoring. In: 17th Intl. Conference on Scientific and Statistical Database Management (SSDBM), Santa Barbara, California (June 2005)
11. Birks, J.B.: Rutherford at Manchester. Heywood (1962)
12. Bowers, S., Ludäscher, B.: Actor-oriented design of scientific workflows. In: Delcambre, L.M.L., Kop, C., Mayr, H.C., Mylopoulos, J., Pastor, Ó. (eds.) ER 2005. LNCS, vol. 3716, pp. 369–384. Springer, Heidelberg (2005)
13. Bowers, S., McPhillips, T., Ludäscher, B., Cohen, S., Davidson, S.B.: A model for user-oriented data provenance in pipelined scientific workflows. In: Moreau, L., Foster, I. (eds.) IPAW 2006. LNCS, vol. 4145, pp. 133–147. Springer, Heidelberg (2006)

14. Bowers, S., McPhillips, T., Wu, M., Ludäscher, B.: Project histories: Managing data provenance across collection-oriented scientific workflow runs. In: Cohen-Boulakia, S., Tannen, V. (eds.) DILS 2007. LNCS (LNBI), vol. 4544, pp. 122–138. Springer, Heidelberg (2007)

15. Bowers, S., McPhillips, T.M., Ludäscher, B.: Provenance in Collection-Oriented Scientific Workflows. In: Moreau, Ludäscher [43]

16. Bowers, S., McPhillips, T., Riddle, S., Anand, M.K., Ludäscher, B.: Kepler/pPOD: Scientific workflow and provenance support for assembling the tree of life. In: Freire, J., Koop, D., Moreau, L. (eds.) IPAW 2008. LNCS, vol. 5272, pp. 70–77. Springer, Heidelberg (2008)

17. Brooks, C., Lee, E.A., Liu, X., Neuendorffer, S., Zhao, Y., Zheng, H.: Heterogeneous Concurrent Modeling and Design in Java (Volume 3: Ptolemy II Domains). Technical Report No. UCB/EECS-2008-37 (April 2008)

18. Cheney, J., Buneman, P., Ludäscher, B.: Report on the Principles of Provenance Workshop. SIGMOD Record 37(1), 62–65 (2008)

19. Churches, D., Gombas, G., Harrison, A., Maassen, J., Robinson, C., Shields, M., Taylor, I., Wang, I.: Programming Scientific and Distributed Workflow with Triana Services. In: Fox, Gannon [28]

20. Cyberinfrastructure for Phylogenetic Research, CIPRES (2009), www.phlyo.org

21. Crawl, D., Altintas, I.: A provenance-based fault tolerance mechanism for scientific workflows. In: Freire, J., Koop, D., Moreau, L. (eds.) IPAW 2008. LNCS, vol. 5272, pp. 152–159. Springer, Heidelberg (2008)

22. Directed Acyclic Graph Manager, DAGMan (2009), www.cs.wisc.edu/condor/dagman

23. Davidson, S.B., Boulakia, S.C., Eyal, A., Ludäscher, B., McPhillips, T.M., Bowers, S., Anand, M.K., Freire, J.: Provenance in Scientific Workflow Systems. IEEE Data Eng. Bull. 30(4), 44–50 (2007)

24. Davidson, S.B., Freire, J.: Provenance and Scientific Workflows: Challenges and Opportunities (Tutorial Notes). In: SIGMOD (2008)

25. Deelman, E., Gannon, D., Shields, M., Taylor, I.: Workflows and e-Science: An overview of workflow system features and capabilities. Future Generation Computer Systems 25(5), 528–540 (2009)

26. Deelman, E., Singh, G., Su, M.-H., Blythe, J., Gil, Y., Kesselman, C., Mehta, G., Vahi, K., Berriman, G.B., Good, J., Laity, A., Jacob, J., Katz, D.: Pegasus: A framework for mapping complex scientific workflows onto distributed systems. Scientific Programming 13(3), 219–237 (2005)

27. Fahringer, T., Prodan, R., Duan, R., Nerieri, F., Podlipnig, S., Qin, J., Siddiqui, M., Truong, H., Villazon, A., Wieczorek, M.: ASKALON: A grid application development and computing environment. In: IEEE Grid Computing Workshop (2005)

28. Fox, G.C., Gannon, D. (eds.): Concurrency and Computation: Practice and Experience. Special Issue: Workflow in Grid Systems, vol. 18(10). John Wiley & Sons, Chichester (2006)

29. Freire, J.-L., Silva, C.T., Callahan, S.P., Santos, E., Scheidegger, C.E., Vo, H.T.: Managing rapidly-evolving scientific workflows. In: Moreau, L., Foster, I. (eds.) IPAW 2006. LNCS, vol. 4145, pp. 10–18. Springer, Heidelberg (2006)

30. Gil, Y., Deelman, E., Ellisman, M., Fahringer, T., Fox, G., Gannon, D., Goble, C., Livny, M., Moreau, L., Myers, J.: Examining the Challenges of Scientific Workflows. Computer 40(12), 24–32 (2007)

31. Goble, C., Roure, D.D.: myExperiment: Social Networking for Workflow-Using e-Scientists. In: Workshop on Workflows in Support of Large-Scale Science, WORKS (2007)

32. Hidders, J., Kwasnikowska, N., Sroka, J., Tyszkiewicz, J., den Bussche, J.V.: DFL: A dataflow language based on Petri nets and nested relational calculus. Information Systems 33(3), 261–284 (2008)

33. Kahn, G.: The Semantics of a Simple Language for Parallel Programming. In: Rosenfeld, J.L. (ed.) Proc. of the IFIP Congress 74, pp. 471–475. North-Holland, Amsterdam (1974)

34. Klasky, S., Barreto, R., Kahn, A., Parashar, M., Podhorszki, N., Parker, S., Silver, D., Vouk, M.: Collaborative Visualization Spaces for Petascale Simulations. In: Intl. Symposium on Collaborative Technologies and Systems (CTS), May 2008, pp. 203–211 (2008)

35. Lee, E.A., Matsikoudis, E.: The Semantics of Dataflow with Firing. In: Huet, G., Plotkin, G., Lévy, J.-J., Bertot, Y. (eds.) From Semantics to Computer Science: Essays in memory of Gilles Kahn. Cambridge University Press, Cambridge (2008)

36. Lee, E.A., Parks, T.M.: Dataflow Process Networks. Proceedings of the IEEE, 773–799 (1995)

37. Ludäscher, B., Altintas, I., Berkley, C., Higgins, D., Jaeger, E., Jones, M., Lee, E.A., Tao, J., Zhao, Y.: Scientific Workflow Management and the Kepler System. Concurrency and Computation: Practice & Experience 18(10), 1039–1065 (2006)

38. Ludäscher, B., Altintas, I., Bowers, S., Cummings, J., Critchlow, T., Deelman, E., Freire, J., Roure, D.D., Goble, C., Jones, M., Klasky, S., Podhorszki, N., Silva, C., Taylor, I., Vouk, M.: Scientific Process Automation and Workflow Management. In: Shoshani, A., Rotem, D. (eds.) Scientific Data Management: Challenges, Existing Technology, and Deployment. Chapman and Hall/CRC (to appear, 2009)

39. Ludäscher, B., Bowers, S., McPhillips, T.: Scientific Workflows. In: Özsu, M.T., Liu, L. (eds.) Encyclopedia of Database Systems. Springer, Heidelberg (to appear, 2009)

40. Ludäscher, B., Goble, C. (eds.): ACM SIGMOD Record: Special Issue on Scientific Workflows, vol. 34(3) (September 2005)

41. Ludäscher, B., Podhorszki, N., Altintas, I., Bowers, S., McPhillips, T.M.: From computation models to models of provence: The RWS approach, vol. 20(5), pp. 507–518

42. McPhillips, T., Bowers, S., Zinn, D., Ludäscher, B.: Scientific Workflow Design for Mere Mortals. Future Generation Computer Systems 25, 541–551 (2009)

43. Moreau, L., Ludäscher, B. (eds.): Concurrency and Computation: Practice & Experience – Special Issue on the First Provenance Challenge. Wiley, Chichester (2007)

44. Morrison, J.P.: Flow-Based Programming – A New Approach to Application Development. Van Nostrand Reinhold (1994), www.jpaulmorrison.com/fbp

45. Oinn, T., Greenwood, M., Addis, M., Alpdemir, M.N., Ferris, J., Glover, K., Goble, C., Goderis, A., Hull, D., Marvin, D., Li, P., Lord, P., Pocock, M.R., Senger, M., Stevens, R., Wipat, A., Wroe, C.: Taverna: Lessons in Creating a Workflow Environment for the Life Sciences. In: Fox, Gannon [28]

46. Podhorszki, N., Ludäscher, B., Klasky, S.A.: Workflow automation for processing plasma fusion simulation data. In: Workshop on Workflows in Support of Large-Scale Science (WORKS), pp. 35–44. ACM Press, New York (2007)

47. Rice, J.R., Boisvert, R.F.: From Scientific Software Libraries to Problem-Solving Environments. IEEE Computational Science & Engineering 3(3), 44–53 (1996)

48. Stajich, J.E., Block, D., Boulez, K., Brenner, S.E., Chervitz, S.A., Dagdigian, C., Fuellen, G., Gilbert, J.G., Korf, I., Lapp, H., Lehvaslaiho, H., Matsalla, C., Mungall, C.J., Osborne, B.I., Pocock, M.R., Schattner, P., Senger, M., Stein, L.D., Stupka, E., Wilkinson, M.D., Birney, E.: The BIOPERL Toolkit: Perl Modules for the Life Sciences. Genome Res. 12(10), 1611–1618 (2002)

49. Taylor, I., Deelman, E., Gannon, D., Shields, M. (eds.): Workflows for e-Science: Scientific Workflows for Grids. Springer, Heidelberg (2007)
50. Wittgenstein, L.: Philosophical Investigations. Blackwell Publishing, Malden (1953)
51. Yu, J., Buyya, R.: A Taxonomy of Scientific Workflow Systems for Grid Computing. In: Ludäscher, Goble [40]
52. Zinn, D., Bowers, S., McPhillips, T., Ludäscher, B.: X-CSR: Dataflow Optimization for Distributed XML Process Pipelines. In: 25th Intl. Conf. on Data Engineering (ICDE), Shanghai, China (2008)

Graph Matching Algorithms for Business Process Model Similarity Search

Remco Dijkman[1], Marlon Dumas[2], and Luciano García-Bañuelos[2,3]

[1] Eindhoven University of Technology, The Netherlands
r.m.dijkman@tue.nl
[2] University of Tartu, Estonia
marlon.dumas@ut.ee
[3] Universidad Autonoma de Tlaxcala, Mexico
lgbanuelos@gmail.com

Abstract. We investigate the problem of ranking all process models in a repository according to their similarity with respect to a given process model. We focus specifically on the application of graph matching algorithms to this similarity search problem. Since the corresponding graph matching problem is NP-complete, we seek to find a compromise between computational complexity and quality of the computed ranking. Using a repository of 100 process models, we evaluate four graph matching algorithms, ranging from a greedy one to a relatively exhaustive one. The results show that the mean average precision obtained by a fast greedy algorithm is close to that obtained with the most exhaustive algorithm.

1 Introduction

As organizations reach higher levels of Business Process Management (BPM) maturity, repositories with hundreds of business process models become increasingly common [18]. For example, the SAP reference model contains over 600 business process models. A similar number of process models can be found in the reference model for Dutch Local Governments [6]. On a larger scale, tool vendors distribute reference model repositories (e.g. the IT Infrastructure Library – ITIL) with over a thousand process models each[1]. These models are used, for example, to document and to communicate internal procedures or to enable the re-design and automation of business processes. In order to effectively fulfil these tasks, tool support is needed to retrieve relevant models from such repositories.

In this paper, we focus on the problem of similarity search in process model repositories: Given a process model or fragment thereof (the *search model*), find those process models in the repository that most closely resemble the search model. The need for similarity search arises in multiple scenarios. For example, when adding a new process model into a repository, similarity search allows one to detect duplication or overlap between the new and the existing process models. Meanwhile, in the context of reference process model repositories, such

---

[1] See for example CaseWise's ITIL repository (http://www.casewise.com/Gateway/)

U. Dayal et al. (Eds.): BPM 2009, LNCS 5701, pp. 48–63, 2009.

as ITIL, similarity search allows one to retrieve reference models that overlap with an existing "as is" process model.

Answering a similarity search query involves determining the degree of similarity between the search model and each model in the repository. In this context, similarity can be defined from several perspectives, including the following.

- Text similarity: based on a comparison of the labels that appear in the process models (task labels, event labels, etc.), using either syntactic or semantic similarity metrics, or a combination of both.
- Structural similarity: based on the topology of the process models seen as graphs, possibly taking into account text similarity as well.
- Behavioural similarity: based on the execution semantics of process models.

In previous work, we evaluated several similarity metrics across all three perspectives [5,19]. We found that a structural similarity metric based on graph matching achieved the highest retrieval quality (precision and recall). However, the operationalization of this metric is hindered by the fact that the underlying graph matching problem, namely the graph-edit distance problem, is NP-complete [14]. This is not only a theoretical limitation, but a practical one: our experiments show that for real-life process models with more than 20 nodes, exhaustive graph matching algorithms lead to combinatorial explosion. Therefore, heuristics are needed that strike a tradeoff between computational complexity and precision. This paper presents and compares four heuristic algorithms for calculating the similarity of business process models based on graph matching.

The rest of the paper is structured as follows. Section 2 formulates the problem and introduces the structural similarity metric studied in the paper. Section 3 presents four algorithms that provide alternative operationalizations of the structural similarity metric. Section 4 presents an experimental evaluation of these algorithms. Section 5 discusses related work and Section 6 concludes.

2   Preliminaries

This section defines the notion of business process used in this paper and formulates the structural similarity metric used for comparing pairs of process models.

2.1   Business Process

A business process is a collection of related tasks that lead to a specified goal. Many modeling notations are available to capture business processes, including Event-driven Process Chains (EPC), UML Activity Diagrams and the Business Process Modeling Notation (BPMN) [20]. In this paper, we seek to abstract as much as possible from the specific notation used to represent process models, to allow for measuring similarity of business processes modeled using different notations. Accordingly, we adopt an abstract view in which a process model is a directed attributed graph, as captured in the following definition.

Definition 1 (Business process graph, Pre-set, Post-set, Source, Sink).
*Let $\mathcal{L}$ be a set of labels and $\mathcal{T}$ be a set of types of nodes. A business process graph
is a tuple (N, E, τ, λ), in which:*

- *N is the set of nodes;*
- *$E \subseteq N \times N$ is the set of edges; and*
- *$\tau : N \to \mathcal{T}$ is a function that maps nodes to types.*
- *$\lambda : N \to \mathcal{L}$ is a function that maps nodes to labels.*

*Let $G = (N, E, \tau, \lambda)$ be a graph and $n \in N$ be a node: $\bullet n = \{m | (m, n) \in E\}$ is
the pre-set of n, while $n\bullet = \{m | (n, m) \in E\}$ is the post-set of n. Source nodes
are nodes with an empty pre-set and sink nodes are nodes with an empty post-set.*

Function τ serves to distinguish between types of nodes. The available types
of nodes depend on the notation. In EPCs we can distinguish between at least
three types of nodes: functions ('f'), events ('e') and connectors ('c'). Similarly,
in BPMN we can distinguish between activities ('a'), events ('e') and gateways
('g'). We could also distinguish between different types of BPMN gateways
and events, but it is not the intention of this paper to be exhaustive in this
respect.

When abstracting a process model as a process graph, we may drop certain
types of nodes. Figure 1 shows two process models (one EPC and one BPMN
diagram) and two ways of abstracting them as process graphs. The left column
shows the original process models. The middle column shows the corresponding
process graphs after the events are abstracted away. Each node is annotated
with a pair indicating the node type and the node label. The right column
shows the process graphs after events and connectors/gateways are abstracted
away. As discussed later, this connector-less abstraction lifts one of the sources of
combinatorial explosion when comparing process models using graph matching.

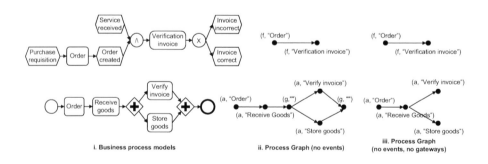

Fig. 1. Two processes and their graphs

2.2   Business Process Similarity Metric

To compare pairs of process graphs, we define a metric based on on the notion of graph edit distance [4]. The graph edit distance between two graphs is

the minimal cost of transforming one graph into the other. Transformations are captured as sequences of elementary transformation operations. Each elementary operation has a cost, which is given by a cost function. Conceptually, a graph-edit distance algorithm must try possible combinations of transformation operations and return the one with the minimal total cost. We consider the following elementary transformation operations.

- Node substitution: a node from one graph is substituted for a node from the other graph.
- Node insertion/deletion: a node is inserted into or deleted from a graph.
- Edge insertion/deletion: an edge is inserted into or deleted from a graph.

We consider cost functions that return a constant value for insertion and deletion of nodes and edges (e.g. a cost of 0.5 for edges and 0.2 for nodes). Meanwhile, we assume that the cost of a node substitution is one minus the similarity of the nodes. The similarity of nodes is determined by the similarity the node labels and types. We introduce a predicate cs ('can substitute') that holds iff one type of node can substitute another type of node (e.g. an EPC function can substitute a BPMN activity). For a given pair of nodes, if cs does not hold, the similarity of these nodes is undefined ($\bot$). If cs holds, their similarity is determined using the string-edit distance of the node labels as defined below.

Definition 2 (String edit distance, Node similarity). *Let s and t be two strings and let $|x|$ be the length of a string x. The string edit distance of s and t, denoted $ed(s,t)$ is the minimal number of atomic string operations needed to transform s into t or vice versa. The atomic string operations are: inserting a character, deleting a character or substituting a character for another.*
Let $G_1 = (N_1, E_1, \tau_1, \lambda_1)$ and $G_2 = (N_2, E_2, \tau_1, \lambda_2)$ be two graphs and $n_1 \in N_1$ and $n_2 \in N_2$ two nodes. The similarity of n_1 and n_2 is:

$$Sim(n_1, n_2) = \begin{cases} 1.0 - \dfrac{ed(\lambda_1(n_1), \lambda_2(n_2))}{\max(|\lambda_1(n_1)|, |\lambda_2(n_2)|)} & \text{if } cs(\tau_1(n_1), \tau_2(n_2)) \\ \bot & \text{otherwise} \end{cases}$$

For example, if 'f' and 'a' can substitute each other, then the string edit distance between 'Verify invoice' and 'Verification invoice' from figure 1 is seven; substitute 'y' for 'i' and insert 'cation'. Consequently, the string edit similarity is $1.0 - \frac{7}{20}$. Algorithms for computing the string edit distance are well known [9].

String-edit distance is only one possible similarity metric between labels. In separate work, we studied other label similarity metrics based on word stemming and synonym relations [5]. However, the purpose of the present paper is not to evaluate label similarity metrics, but rather to evaluate algorithms that, given a label similarity metric, compute a similarity measure between pairs of process models. Therefore, the choice of label similarity metric is secondary.

Given the above, we define the graph edit distance as follows.

Definition 3 (Graph edit distance). *Let $G_1 = (N_1, E_1, \tau_1, \lambda_1)$ and $G_2 = (N_2, E_2, \tau_1, \lambda_2)$ be two graphs. Let $M : N_1 \nrightarrow N_2$ be a partial injective mapping* *that maps nodes in G_1 to nodes in G_2. Let $\mathrm{dom}(M) = \{n_1|(n_1, n_2) \in M\}$ be the domain of M and $\mathrm{cod}(M) = \{n_2|(n_1, n_2) \in M\}$ be the codomain of M.*

Given an $n \in N_1 \cup N_2$, n is substituted iff $n \in \mathrm{dom}(M)$ or $n \in \mathrm{cod}(M)$. subn is the set of all substituted nodes. A node $n_1 \in N_1$ is deleted from G_1 (or inserted in G_2) iff it is not substituted. A node that is deleted from G_2 (or inserted in G_1) is defined similarly. skipn is the set of all inserted and deleted nodes.

Let $(n_1, m_1) \in E_1$ be an edge in E_1. (n_1, m_1) is deleted from G_1 (or inserted in G_2) if and only if there do not exist mappings $(n_1, n_2) \in M$ and $(m_1, m_2) \in M$ and edge $(n_2, m_2) \in E_2$. Edges that are deleted from G_2 (or inserted in G_1) are defined similarly. skipe is the set of all inserted and deleted edges. An edge is substituted if it is not inserted or deleted.

The graph edit distance that is induced by the mapping M is:

$$|\mathrm{skipn}| + |\mathrm{skipe}| + 2 \cdot \Sigma_{(n_1, n_2) \in M}(1 - \mathrm{Sim}(n_1, n_2))$$

The graph edit distance of the two graphs is the minimal possible distance induced by some mapping.

For example, given the two process graphs in figure 1.iii, we can create a mapping from 'Order' to 'Order', and from 'Verification invoice' to 'Verify invoice'. The graph edit distance induced by this mapping is: $2.0 + 4.0 + 2.0 \cdot (0.0 + 0.35) = 6.7$ (2 inserted nodes, 4 deleted/inserted edges and 2 substituted nodes).

Finally, we define the graph edit similarity metric as follows.

Definition 4 (Graph edit similarity). *Let $G_1 = (N_1, E_1, \lambda_1)$ and $G_2 = (N_2, E_2, \lambda_2)$ be two graphs. Let $M : N_1 \nrightarrow N_2$ be a partial injective mapping that maps nodes in G_1 to nodes in G_2 and let subn, skipn and skipe be the sets of substituted nodes, inserted or deleted nodes and inserted or deleted edges as defined in definition 3. Furthermore, let $0 \leq \mathrm{wsubn} \leq 1$, $0 \leq \mathrm{wskipn} \leq 1$ and $0 \leq \mathrm{wskipe} \leq 1$ be the weights that we assign to substituted nodes, inserted or deleted nodes and inserted or deleted edges, respectively.*

The fraction of inserted or deleted nodes, denoted fskipn, the fraction of inserted or deleted edges, denoted fskipe and the average distance of substituted nodes, denoted fsubn, are defined as follows.

$$\mathrm{fskipn} = \frac{|\mathrm{skipn}|}{|N_1| + |N_2|} \quad \mathrm{fskipe} = \frac{|\mathrm{skipe}|}{|E_1| + |E_2|} \quad \mathrm{fsubn} = \frac{2.0 \cdot \Sigma_{(n,m) \in M} 1.0 - \mathrm{Sim}(n,m)}{|\mathrm{subn}|}$$

The graph edit similarity induced by the mapping M is:

$$1.0 - \frac{\mathrm{wskipn} \cdot \mathrm{fskipn} + \mathrm{wskipe} \cdot \mathrm{fskipe} + \mathrm{wsubn} \cdot \mathrm{fsubn}}{\mathrm{wskipn} + \mathrm{wskipe} + \mathrm{wsubn}}$$

The graph edit similarity of two graphs is the maximal possible similarity induced by a mapping between these graphs.

For example, using the weights $\mathrm{wsubn} = 1.0$, $\mathrm{wskipn} = 0.1$ and $\mathrm{wskipe} = 0.3$, the graph edit similarity that is induced by the mapping that maps 'Order' to 'Order', and 'Verification invoice' to 'Verify invoice' in figure 1 is: $1.0 - \frac{0.1 \cdot 0.33 + 0.3 \cdot 1.0 + 1.0 \cdot 0.7}{0.1 + 0.3 + 1.0} \approx 0.73$. This is also the maximal possible similarity induced by a mapping and, hence, this is the graph edit similarity of the two graphs.

3   Algorithms

To compute the graph edit similarity of two process graphs, we must find the mapping that induces the maximal similarity. We could construct all possible mappings and return the one with maximal similarity. However, this approach has factorial complexity. Accordingly, this section presents four possible heuristic algorithms to address this problem.

3.1   Greedy Algorithm

We first propose a greedy algorithm (Algorithm 1) that incrementally constructs a mapping between a pair of process graphs. The algorithm starts by marking all possible pairs of nodes from the two graphs as open pairs. (For all algorithms we assume that pairs of nodes that cannot substitute each other, as defined in definition 2, are not considered.) In each iteration, the algorithm selects an open pair that most increases the similarity induced by the mapping, and adds this pair to the mapping.[2] The selected pair consists of two nodes. Since each node can only be mapped once, the algorithm removes from the set of open pairs, all pairs in which one of the selected nodes appears. The algorithm iterates until there is no open pair left that can increase the similarity induced by the mapping.

The algorithm is in $O(n^3)$ where n is the number of nodes of the largest graph. Indeed, in the first iteration we consider up to n^2 open pairs, in the second iteration $(n-1)^2$ open pairs, etc. And $\Sigma_{i=1}^{n} i = n(n+1)(2n+1)/6$.[3] Also, the algorithm has a quadratic space complexity (the set of open pairs). Unfortunately, the algorithm may lead to a suboptimal mapping, because it selects an open pair that most increases the similarity induced by the mapping at a particular time, but in doing so, it may discard open pairs that would increase the similarity induced by mapping at a later iteration.

For example, in figure 1 the open pair ('Order', 'Order') is chosen in the first iteration, because adding this pair to the mapping increases the similarity score most. All open pairs in which 'Order' appears are then removed from the set of open pairs. In the second iteration, the open pair ('Verification invoice', 'Verify invoice') is chosen. All open pairs in which either 'Verification invoice' or 'Verify invoice' appears are removed. This leaves no open pairs and the algorithm returns the mapping { ('Order', 'Order'), ('Verification invoice', 'Verify invoice')}.

3.2   Exhaustive Algorithm with Pruning

The second algorithm (Algorithm 2) recursively explores all possible mappings, but when the recursion tree reaches a certain size, the algorithm prunes it to keep only the mappings with the highest similarity. In the extreme case, the algorithm is thus exponential, but the pruning parameters will control its complexity.

---

[2] The similarity induced by a mapping is given by function s as per definition 4.

[3] Computing the graph edit similarity induced by a mapping can be done in constant time (amortized), because when we add a pair we already know the graph edit similarity induced by the existing mapping.

Algorithm 1. Greedy algorithm

---

input: two business process graphs $G_1 = (N_1, E_1, \lambda_1)$, $G_2 = (N_2, E_2, \lambda_2)$

init

  openpairs $\Leftarrow N_1 \times N_2$

  map $\Leftarrow \emptyset$

begin

  while *exists* $(n, m) \in$ openpairs, *such that* $s(\text{map} \cup \{(n, m)\}) > s(\text{map})$ *and*

  there does not exist another pair $(o, p) \in$ openpairs, *such that*

  $s(\text{map} \cup \{(o, p)\}) > s(\text{map} \cup \{(n, m)\})$ **do**

    map $\Leftarrow$ map $\cup \{(n, m)\}$

    openpairs $\Leftarrow \{(o, p) \in \text{openpairs} | o \neq n, p \neq m\}$

  end

  return $s(\text{map})$

end

The algorithm starts by initializing the set of unfinished mappings to an empty mapping, with all nodes from the two graphs mapped as 'free' to be mapped. It repeatedly prunes the set of unfinished mappings and performs a step in which finished mappings are added to the set of finished mappings and unfinished mappings are extended with an additional pair of nodes. It repeats this until there are no more unfinished mappings. It then returns the finished mapping with the highest similarity score.

The pruning function (shown separately) tests if the set of unfinished mappings has reached the size 'pruneat' (a parameter of the algorithm). If it has, it returns a set of mappings (of size 'pruneto') with the highest similarity score.

The recursion step is also shown in a separate function. The recursion step takes each unfinished mapping. If the unfinished mapping has no nodes that are free to be mapped, the mapping is added to the set of finished mappings. Otherwise, the algorithm takes each possible combination of pairs of free nodes and creates a new unfinished mapping in which that pair is added to the existing unfinished mapping (and the nodes from the pair are removed from the sets of free nodes). It includes pairs in which free nodes are not mapped (i.e. they are removed from the sets of free nodes, but not added to the unfinished mapping).

For example, in figure 1 the set of unfinished mappings is initialized to $\{(\emptyset, \{O, V\}, \{O, R, V, S\})\}$ (using the first letter of node labels as identifier). In the first step, the algorithm takes this unfinished mapping and, since neither $\{O, V\}$ nor $\{O, R, V, S\}$ is empty, it generates a mapping for each combination of a node from $\{O, V\}$ and a node from $\{O, R, V, S\}$, i.e. $(\{(O, O)\}, \{V\}, \{R, V, S\})$, $(\{(O, R)\}, \{V\}, \{O, V, S\})$, $(\{(O, V)\}, \{V\}, \{O, R, S\})$, …. It also generates one mapping for each possible removal of a node from one of the two sets, generating: $(\emptyset, \{V\}, \{O, R, V, S\})$, $(\emptyset, \{O\}, \{O, R, V, S\})$, $(\emptyset, \{O, V\}, \{R, V, S\})$, …. The generated mappings form the new set of unfinished mappings. In the next step the generation of new unfinished mappings is repeated for each of these mappings.

This example illustrates that the set of unfinished mappings increases exponentially. Pruning will keep the size of the set within acceptable bounds. Suppose

---

Algorithm 2. Exhaustive algorithm with pruning

---

input: two business process graphs $G_1 = (N_1, E_1, \lambda_1)$, $G_2 = (N_2, E_2, \lambda_2)$

function prune(unfinished)
begin
  if |unfinished| < pruneat **then**
    return unfinished
  else
    return a set pruned, such that pruned $\subseteq$ unfinished, |pruned| = pruneto and
    $\forall p \in$ pruned : $\neg \exists u \in$ unfinished : $s(\text{first}(u)) > s(\text{first}(p))$
  end
end

function step(unfinished)
begin
  newunfinished $\Leftarrow \emptyset$
  foreach $(\text{map}, \text{free}_1, \text{free}_2) \in$ unfinished **do**
    if $(\text{free}_1 = \emptyset) \vee (\text{free}_1 = \emptyset)$ **then**
      finished $\Leftarrow$ finished $\cup$ map
    else
      newunfinished $\Leftarrow$ newunfinished $\cup$
        $\{(\text{map} \cup \{(f_1, f_2)\}, \text{free}_1 - \{f_1\}, \text{free}_2 - \{f_2\}) | f_1 \in \text{free}_1, f_2 \in \text{free}_2\} \cup$
        $\{(\text{map}, \text{free}_1 - \{f_1\}, \text{free}_2) | f_1 \in \text{free}_1\} \cup$
        $\{(\text{map}, \text{free}_1, \text{free}_2 - \{f_2\}) | f_2 \in \text{free}_2\}$
    end
  end
  return newunfinished
end

init
  unfinished $\Leftarrow \{(\emptyset, N_1, N_2)\}$
  finished $\Leftarrow \emptyset$
begin
  repeat
    unfinished $\Leftarrow$ prune(unfinished)
    unfinished $\Leftarrow$ step(unfinished)
  until unfinished $= \emptyset$
  return $s(\text{map})$, *such that* map $\in$ finished *and* $s(\text{map})$ *is maximal*
end

---

that 'prune at' is set to 2 and 'prune to' is set to 1, then the set of unfinished mappings will be pruned after the first step, because the set will have reached a size of 2. It will be pruned back to a set the set $\{(\{(O, O)\}, \{V\}, \{R, V, S\})\}$ of size 1, because this mapping has the highest similarity score.

3.3 Process Heuristic Algorithm

The third algorithm is a variation of the exhaustive algorithm. It also builds a recursion tree of possible mappings, but it starts by mapping the source nodes of the business process graphs, then mapping nodes that immediately follow the source nodes, etc. Since it is plausible that nodes closer to the start of a process should be mapped to nodes closer to the start of the other process (and conversely), this should yield a higher-quality pruning. Indeed, the algorithm is more likely to prune mappings with node pairs that are further apart in terms of their distance to the starts of their processes.

Algorithm 3 shows only the initialization of the algorithm and the 'step' function. The 'prune' function and the algorithm itself are the same as for the exhaustive algorithm 2. The algorithm starts by initializing the set of unfinished mappings to an empty mapping with all nodes marked as 'free' and all source nodes marked as 'current'. With each 'step' the algorithm takes an unfinished mapping. If the unfinished mapping has no 'current' nodes, the mapping is added to the set of finished mappings. Otherwise, the algorithm takes each possible combination of pairs of 'current' nodes and creates a new unfinished mapping in which that pair is added. The nodes from the pair are removed from the sets of free nodes. The current nodes are set to include the post-sets of the nodes from the pairs. Only free nodes are included in the sets of current nodes. Pairs in which 'current' nodes are not mapped are also included. The algorithm assumes that process graphs always have source nodes, an assumption that is valid for common process modeling notations (e.g. EPC, BPMN, BPEL).

For example, in figure 1 the set of unfinished mappings is initialized to $\{(\emptyset, \{O, V\}, \{O, R, V, S\}), \{O\}, \{O\}\}$ (using the first letter of the labels to identify each node). In the first step, the algorithm will take this unfinished mapping and, because neither set of current nodes ($\{O\}$ nor $\{O\}$) is empty. From this mapping, it generates one mapping in which the current nodes are mapped, generating

---

Algorithm 3. Process heuristic algorithm

---

    function step(unfinished)
    begin
      newunfinished $\Leftarrow \emptyset$
      foreach $(\mathsf{map}, \mathsf{free}_1, \mathsf{free}_2, \mathsf{curr}_1, \mathsf{curr}_2) \in$ unfinished **do**
        if $(\mathsf{curr}_1 = \emptyset) \vee (\mathsf{curr}_1 = \emptyset)$ **then**
          finished $\Leftarrow$ finished $\cup$ map
        else
          newunfinished $\Leftarrow$ newunfinished $\cup$
            $\{(\mathsf{map} \cup \{(c_1, c_2)\}, \mathsf{free}_1 - \{c_1\}, \mathsf{free}_2 - \{c_2\}, (\mathsf{curr}_1 \cup c_1\bullet) \cap (\mathsf{free}_1 -$
            $\{c_1\}), (\mathsf{curr}_2 \cup c_2\bullet) \cap (\mathsf{free}_2 - \{c_2\}))|c_1 \in \mathsf{curr}_1, c_2 \in \mathsf{curr}_2\}\cup$
            $\{(\mathsf{map}, \mathsf{free}_1 - \{c_1\}, \mathsf{free}_2, (\mathsf{curr}_1 \cup c_1\bullet) \cap (\mathsf{free}_1 - \{c_1\}), \mathsf{curr}_2)|c_1 \in \mathsf{curr}_1\} \cup$
            $\{(\mathsf{map}, \mathsf{free}_1, \mathsf{free}_2 - \{c_2\}, \mathsf{curr}_1, (\mathsf{curr}_2 \cup c_2\bullet) \cap (\mathsf{free}_2 - \{c_2\}))|c_2 \in \mathsf{curr}_2\}$
      end
    end
    return newunfinished
    end

    init
      unfinished $\Leftarrow \{(\emptyset, N_1, N_2, \{n|n \in N_1, \bullet n = \emptyset\}, \{n|n \in N_2, \bullet n = \emptyset\})\}$
      finished $\Leftarrow \emptyset$

$\{(\{(O, O)\}, \{V\}, \{R, V, S\}), \{V\}, \{R\}\}$. It also generates mappings for each possible removal of a current node, generating $\{(\emptyset, \{V\}, \{O, R, V, S\}), \{V\}, \{O\}\}$
and $\{(\emptyset, \{O, V\}, \{R, V, S\}), \{O\}, \{R\}\}$. The generated mappings form the new
set of unfinished mappings. This example illustrates that the set of unfinished
mappings explodes less rapidly for this algorithm than for the exhaustive algorithm. It also illustrates that mappings of nodes closer to the start of the process
are explored first.

3.4 A-Star Algorithm

The fourth algorithm (Algorithm 4) is based on the well-known A-star heuristic
search, which has been applied to the problem of graph matching in [14]. In each
step, the algorithm selects the existing partial mapping map with the maximal
graph edit similarity. The algorithm then takes a node n_1 from graph G_1 that
has not yet been mapped, and creates a mapping between this node and every
node n_2 of G_2 such that n_2 does not already appear in map. Let us say that m
such nodes n_2 exist. The algorithm then creates m new mappings, by adding
(n_1, n_2) to map. In addition, one mapping is created where (n_1, ϵ) is added to
map (ϵ is a "dummy" node). This latter pair represents the case where node n_1
has been deleted. This step is repeated until all nodes from G_1 are mapped. It
can be proven that the result is an optimal mapping.

The number of steps performed by the algorithm is bounded by $O(n^2 m)$
where n and m are the number of nodes in G_1 and G_2. However, $O(m^n)$ partial
mappings need to be maintained during the search [14]. To reduce the memory
requirements, we modified the algorithm so as to avoid mapping nodes with very
different labels. If the string-edit similarity between two node labels is less than
a cut-off value, we do not consider the possibility of mapping these nodes.

For example, if we consider the models in figure 1 and a cut-off value of
0.6, two mappings, $\{('Order', 'Order')\}$ and $\{('Order', \epsilon)\}$, are created in the
first iteration. Since other candidate node pairs have a string-edit similarity
smaller than the cut-off value, no mapping is created for them. In the second
iteration, the algorithm selects the mapping $\{('Order', 'Order')\}$ and creates
two new mappings $\{('Order', 'Order'), ('Verification invoice', 'Verify invoice')\}$
and $\{('Order', 'Order'), ('Verification invoice', \epsilon)\}$. The algorithm stops in the
third iteration with a complete mapping $\{('Order', 'Order'), ('Verification of
invoice', 'Verify invoice')\}$ and with nodes 'Receive goods' and 'Store goods'
being considered as insertions. Thus, the algorithm discards the two partial
mappings $\{('Order', \epsilon)\}, \{('Order', 'Order'), ('Verification of invoice', \epsilon)\}$.

4 Evaluation

In this section, we present an experimental evaluation of the algorithms discussed
above in terms of quality of retrieval results and in terms of execution time.

4.1 Experimental Setup

We derived an experimental dataset from the SAP reference model. This is
a collection of 604 business process models (described as EPCs) capturing

---

Algorithm 4. A-star algorithm

---

input: two business process graphs $G_1 = (N_1, E_1, \lambda_1)$, $G_2 = (N_2, E_2, \lambda_2)$
init
   open $\Leftarrow \{\{(n_1, n_2)\} | n_2 \in N_2 \cup \{\epsilon\}, Sim(n_1, n_2) > \mathsf{ledcutoff} \vee n_2 = \epsilon\}$, for some
   $n_1 \in N_1$
begin
  while open $\neq \emptyset$ **do**
    select map $\in$ open, such that $s(\mathsf{map})$ is maximal
    open $\Leftarrow$ open $-$ {map}
    if $\mathrm{dom}(\mathsf{map}) = N_1$ **then**
      return $s(\mathsf{map})$
    else
      select $n_1 \in N_1$, such that $n_1 \notin \mathrm{dom}(\mathsf{map})$
      foreach $n_2 \in N_2 \cup \{\epsilon\}$, such that either $n_2 \notin \mathrm{cod}(\mathsf{map})$ and
      $Sim(n_1, n_2) > \mathsf{ledcutoff}$ or $n_2 = \epsilon$ **do**
        map$'$ $\Leftarrow$ map $\cup \{(n_1, n_2)\}$
        open $\Leftarrow$ open $\cup$ {map$'$}
      end
    end
  end
end

---

business processes supported by the SAP enterprise system. We randomly extracted 100 business process models from this collection and tagged them as "document models". On average each model contained 21.6 nodes with a minimum of 3 and a maximum 130 nodes. The average size of node labels was 3.8 words. From the 100 document models we randomly extracted 10 models. These models became the "search query models". We modified some of these models to investigate the effect of certain types of changes (for example taking a subgraph) on the performance of the algorithms. We did not observe any noteworthy effects. Therefore, we will only present overall averaged results.

Next, we manually compared each of the 1000 pairs (search model, document model) and ranked their degree of similarity on a 1-7 Likert scale. This manual comparison was done by three process modeling experts, including the first author of this paper. For a given search model sq, we sorted the 100 pairs (sq, document model) in descending order according to the human expert score. Finally, for each algorithm, we applied it to each pair ("search query model", "document model") and sorted the results (for each of the 10 queries) in descending order according to the similarity score retrieved by the algorithm. The resulting sorted lists were used to calculate the average precision.

The algorithms depend on several parameters:

- wskipn, wsubn and wskipe which denote the weight given to node deletion, node substitution and edge deletion (see Definition 4).
- ledcutoff (label edit cut-off): a number between zero and one representing the minimum similarity that two nodes must have so that we can consider

their substitution. For example, if the cut-off is 0.5 "Pay Invoice" and "Pay Allowance" will not be mapped since their similarity is 0.3.

– pruneat is the maximum allowed size of the recursion tree. When the recursion tree reaches this level, it is pruned, down to a size of pruneto.

In the experiments, we considered multiple variants of each algorithm corresponding to different parameter settings. The implementation of the proposed algorithms (and several others) can be found in the "Graph Matching Analysis Plug-in" of the ProM process mining and analysis framework[4].

4.2 Results

Table 1 shows the mean average precision and the average execution times of the similarity search techniques under study. Average precision is a measure commonly used to evaluate the quality of search techniques that return ranked lists of results [3]. It is the average of the precision scores at each point where a relevant document appears in the ranked list. Given a ranked list of results of size n, the average precision is $\Sigma_{j=1}^{n}(precision[j] \times rel[j])/R$, where R is the number of relevant documents, $rel[j]$ is one if the document of rank j in the list is relevant, zero otherwise, and $precision[j] = \Sigma_{k=1}^{j}rel[k]/j$ (i.e. the precision at rank j). Intuitively, average precision is higher when relevant documents appear earlier in the ranked list. The mean average precision of a search technique over a given set of queries is the mean of the average precision of the technique over each of the queries.

As explained above, each algorithm has a number of parameters. The mean average precisions reported in the table correspond to the scores obtained for the best possible settings of each algorithm. All four algorithms depend on parameters wskipn, wsubn and wskipe explained above. We varied each of these parameters from 0 to 1 in increments of 0.1 and ran the experiments with all possible combinations of parameter values in this range. By analyzing the mean average precisions obtained for every combination of parameter values, we noticed that the "Greedy", "Exhaustive" and "Heuristic" algorithms give their best results for settings such that $2 \times$ (wskipn + wskipe) $\sim$ wsubn. One can notice that the optimal parameter settings for these three algorithms (Table 1) closely satisfy this condition.

The exhaustive and the process heuristic algorithm rely on parameters pruneat and pruneto to determine when should pruning occur and to what extent. We tested different values of pruneat (50, 100, 200, etc.) and different ratios pruneat/pruneto (0.1, 0.2, etc.). We found that a value of pruneat = 100 is sufficient. Larger values do not improve the results significantly, but they degrade performance. Similarly we found that a ratio pruneat/pruneto = 0.1 is sufficient, larger ratios do not significantly improve the outcome. Accordingly, we settled for pruneat = 100 and pruneto = 10.

The A-star algorithm relies on a parameter ledcutoff. Again, we experimented with different values of this parameter and found that a value of 0.5 yields

---

[4] http://prom.sourceforge.net

Table 1. Summary of results

Algorithm	wskipn	wsubn	wskipe	Mean avg. precision	Execution time
Greedy	0.1	0.9	0.4	0.84	3.8 sec.
Exhaustive	0.1	0.8	0.2	0.82	53.7 sec.
Process Heuristic	0.1	0.8	0.2	0.83	14.2 sec.
A-star	0.2	0.1	0.7	0.86	15.7 sec.

optimal results among those that we were able to test. We could not experiment with values significantly below 0.5, because if the threshold is too low, the memory requirements of the A-star algorithm grow substantially and the performance degrades to the point of making the technique impractical. This is the reason why this parameter is important for the A-star algorithm, whereas the other algorithms rely on pruning. A side-effect of using the ledcutoff parameter is that the algorithm favours insertions and deletions over substitutions. To compensate for this effect, the values of wskipn and wskipe need to be set higher than wsubn, in other words, deletions/insertions need to be given higher weight than substitutions. For the A-star algorithm, we noticed that all settings of wskipn, wskipe and wsubn that satisfy this condition given high average precisions.

The A-star algorithm slightly outperforms the others in terms of mean average precision. Looking closer, we noticed that A-star outperforms all other techniques in 6 out of 10 queries and yields equal results in a seventh query. It slightly underperforms the others in queries 6, 8 and 10.

Table 1 also displays the average execution time of 5 runs of each algorithm. For these measurements, we used the parameter settings giving the highest mean average precision. In each run, we executed all 10 queries, i.e. 1000 pairwise process model comparisons in total. All tests were conducted on a laptop with a dual core Intel processor, 2.4 GHz, 4 GB memory, running Mac OSX and SUN Java Virtual Machine version 1.6 (with 512MB of allocated memory).

Not surprisingly, the greedy algorithm is considerably faster than all others. Its execution time per search query is less than half a second. The A* and the process heuristic algorithms have comparable execution times – around 1.5 seconds per query. The exhaustive algorithm is significantly slower.

5 Related Work

To the best of our knowledge there exist eight other initiatives that address algorithms for measuring the similarity between business process models or similar models [1,7,10,11,12,15,16,21]. Of these initiatives five present algorithms to measure the similarity between business process models [7,10,11,12,15], two present algorithms to measure the similarity between state machines [16,21] and one presents algorithms to measure the similarity between a business process and a set of execution traces [1]. Our algorithms are the only ones that are validated for use in similarity search. Nejati et al. [16] validate their algorithms,

Table 2. Comparison of related work

Paper	Similarity of	Validated	Basis for similarity
This paper	process models	for similarity search	edit distance
Nejati et al. [16]	state machines	for merging state machines	bi-similarity
Wombacher [21]	state machines	for correlation with human judgement	process conformance language construction
Li et al. [10]	process models	no	change patterns
Minor et al. [15]	process models	no	edit distance
Lu and Sadiq [11]	process models	no	features
Madhusudan et al. [12]	process models	no	similarity flooding
Van der Aalst et al. [1]	process model and execution traces	no	process conformance
Ehrig et al. [7]	process models	no	semantic similarity
Grigori et al. [8]	service protocols	for similarity search	edit distance (A*)

but for suitability as a technique for merging state machines. Wombacher [21] validates the correlation of the similarity scores found by his technique with similarity scores assigned according to human judgement. We have done a similar validation in previous work [19]. The different initiatives have very different bases for computing the similarity. Nejati et al. [16] use a combination of label similarity, comparison of the depth of a state-machine fragment in a hierarchical state-machine and bi-similarity of the fragment. Wombacher [21] evaluates three algorithms; one is based on conformance of a set of execution traces (first generated from a process model) to a business process, similar to the work by Van der Aalst et al. [1]; the other two are based on comparison of the language that is represented by a state machine. Li et al. [10] compare process models by 'counting' the number of high-level change operations needed to transform one process into another. This can be seen as a specialized case of edit distance, using a specific set of transformation operations. Like this paper Minor et al. [15] use graph edit distance as a basis for comparing process models. Lu and Sadiq [11] measure the presence or absence of 'features' in process models as a basis for comparison. Madhusudan et al. [12] use an algorithm known as 'similarity flooding' [13]. Ehrig et al. [7] use a combination of structural properties of process models and similarity of labels of tasks, based on the distance of words in those labels in terms of whether they are, for example, synonyms (which we called 'semantic similarity' in previous work [5]). Table 2 summarizes the related work on business process model comparison.

The algorithms that we studied are based on graph edit distance [4]. However, the actual distance metric we used is different from traditional graph edit distance metrics. Our metric considers the ratio between the actual graph edit distance and the maximum possible distance. In addition, we added various parameters to the algorithms and fine-tuned these parameters for the computation of similarity of business process models. Of the algorithms that we tested, the greedy algorithm and the exhaustive algorithm with pruning are general algorithms to solve recursive problems. The process heuristic algorithm is similar to Neuhaus and Bunke's planar graph matching algorithm [17]. The main difference is that their algorithm starts with a random pair of graph nodes for comparison, while we assume that business process models have source nodes and sink nodes and we start by mapping source nodes. The A-star algorithm that we present is due to Messmer [14]. We adapted it to exclude mappings of node pairs that are deemed improbable based on the string edit distance of their labels. This algorithm was also applied in [8] for similarity search of service protocol specifications captured in BPEL and WSCL. The authors showed that the algorithm performs well on a small collection of service protocols (5 protocols and variants).

6   Conclusion

Among the four process similarity search techniques presented in this paper, the greedy and the A-star ones offer the most interesting tradeoffs. The A-star algorithm offers a slightly better mean average precision but is significantly slower. Still, the execution times of the A-star algorithm can be acceptable for repositories of a few hundred models. The other two techniques, based on an exhaustive search with pruning, offer a less attractive quality/scalability tradeoff.

The graph matching algorithms studied in this paper attempt to establish 1-to-1 correspondences between nodes in the compared process models (i.e. a node in a process model is related to at most one node in the other process model). One can think of variants of these algorithms that would calculate 1-to-N or N-to-M correspondences, e.g. algorithms that would consider the possibility of a node being split into multiple ones or multiple nodes being merged into one. Such graph matching algorithms have been considered in other application domains [2]. We plan to investigate such variants in future work.

This paper focuses on similarity of business processes with respect to tasks and control-flow relations between tasks. Other aspects of business processes can be considered when determining similarity, e.g. data and resources. Also, process models can be annotated with information that helps to determine the similarity more precisely, such as ontological information [7] and textual documentation. Exploiting such additional information is an avenue for future work.

Acknowledgments. This research was supported by the European Regional Development Fund through the Estonian Centre of Excellence in Computer Science.

References

1. van der Aalst, W.M.P., de Medeiros, A.K.A., Weijters, A.J.M.M.T.: Process equivalence: Comparing two process models based on observed behavior. In: Dustdar, S., Fiadeiro, J.L., Sheth, A.P. (eds.) BPM 2006. LNCS, vol. 4102, pp. 129–144. Springer, Heidelberg (2006)
2. Ambauen, R., Fischer, S., Bunke, H.: Graph edit distance with node splitting and merging, and its application to diatom identification. In: Hancock, E.R., Vento, M. (eds.) GbRPR 2003. LNCS, vol. 2726, pp. 259–264. Springer, Heidelberg (2003)

3. Buckley, C., Voorhees, E.M.: Evaluating evaluation measure stability. In: Proc. of the ACM SIGIR Conference, pp. 33–40 (2000)
4. Bunke, H.: On a relation between graph edit distance and maximum common subgraph. Pattern Recognition Letters 18(8), 689–694 (1997)
5. Dijkman, R.M., Dumas, M., van Dongen, B.F., Käärik, R., Mendling, J.: Similarity of business process models: Metrics and evaluation. Working Paper 269, BETA Research School, Eindhoven, The Netherlands (2009)
6. Documentair structuurplan, http://www.model-dsp.nl/ (accessed: Feburary 20, 2009)
7. Ehrig, M., Koschmider, A., Oberweis, A.: Measuring similarity between semantic business process models. In: Proc. of APCCM 2007, pp. 71–80 (2007)
8. Grigori, D., Corrales, J.C., Bouzeghoub, M.: Behavioral matchmaking for service retrieval: Application to conversation protocols. Inf. Syst. 33(7-8), 681–698 (2008)
9. Levenshtein, I.: Binary code capable of correcting deletions, insertions and reversals. Cybernetics and Control Theory 10(8), 707–710 (1966)
10. Li, C., Reichert, M.U., Wombacher, A.: On measuring process model similarity based on high-level change operations. Technical Report TR-CTIT-07-89, CTIT, Enschede, The Netherlands (2007)
11. Lu, R., Sadiq, S.K.: On the discovery of preferred work practice through business process variants. In: Parent, C., Schewe, K.-D., Storey, V.C., Thalheim, B. (eds.) ER 2007. LNCS, vol. 4801, pp. 165–180. Springer, Heidelberg (2007)
12. Madhusudan, T., Zhao, L., Marshall, B.: A case-based reasoning framework for workflow model management. Data Knowl. Eng. 50(1), 87–115 (2004)
13. Melnik, S., Garcia-Molina, H., Rahm, E.: Similarity flooding: A versatile graph matching algorithm and its application to schema matching. In: Proc. of ICDE 2002, pp. 117–128 (2002)
14. Messmer, B.: Efficient Graph Matching Algorithms for Preprocessed Model Graphs. PhD thesis, University of Bern, Switzerland (1995)
15. Minor, M., Tartakovski, A., Bergmann, R.: Representation and structure-based similarity assessment for agile workflows. In: Weber, R.O., Richter, M.M. (eds.) ICCBR 2007. LNCS (LNAI), vol. 4626, pp. 224–238. Springer, Heidelberg (2007)
16. Nejati, S., Sabetzadeh, M., Chechik, M., Easterbrook, S., Zave, P.: Matching and merging of statecharts specifications. In: Proc. of ICSE 2007, pp. 54–63 (2007)
17. Neuhaus, M., Bunke, H.: An error-tolerant approximate matching algorithm for attributed planar graphs and its application to fingerprint classification. In: Fred, A., Caelli, T.M., Duin, R.P.W., Campilho, A.C., de Ridder, D. (eds.) SSPR&SPR 2004. LNCS, vol. 3138, pp. 180–189. Springer, Heidelberg (2004)
18. Rosemann, M.: Potential pitfalls of process modeling: part a. Business Process Management Journal 12(2), 249–254 (2006)
19. van Dongen, B.F., Dijkman, R., Mendling, J.: Measuring similarity between business process models. In: Bellahsène, Z., Léonard, M. (eds.) CAiSE 2008. LNCS, vol. 5074, pp. 450–464. Springer, Heidelberg (2008)
20. Weske, M.: Business Process Management: Concepts, Languages, Architectures. Springer, Berlin (2007)
21. Wombacher, A.: Evaluation of technical measures for workflow similarity based on a pilot study. In: Meersman, R., Tari, Z. (eds.) OTM 2006. LNCS, vol. 4275, pp. 255–272. Springer, Heidelberg (2006)

Controllability in
Temporal Conceptual Workflow Schemata

Carlo Combi and Roberto Posenato

Dipartimento di Informatica, Università degli Studi di Verona
strada le Grazie 15, 37134 Verona Italy
{carlo.combi,roberto.posenato}@univr.it

Abstract. Workflow technology has emerged as one of the leading technologies in modelling, redesigning, and executing business processes. Currently available workflow management systems (*WfMS*) and research prototypes offer a very limited support for the definition, detection, and management of temporal constraints over business processes. In this paper, we propose a new advanced workflow conceptual model for expressing time constraints in business processes and, in particular, we introduce and discuss the concept of *controllability* for workflow schemata and its evaluation at process design time. Controllability refers to the capability of executing a workflow for any possible duration of tasks. Since in several situations durations of tasks cannot be decided by *WfMSs*, even tough the minimum and the maximum durations for each task are known, checking controllability is stronger than verifying the consistency of the workflow temporal constraints.

1 Introduction

A workflow management system (*WfMS*) fully takes over the responsibility for the coordinated execution of tasks of a business process. Organisations use *WfMSs* to streamline, automate, and manage business processes that depend on information systems and human resources (e.g., provisioning telephone services, processing insurance claims, and handling bank loan applications) [1,2]. *WfMSs* provide tools to support the modelling of business processes at a conceptual level, to coordinate the execution of the component activities according to the model, to monitor the execution progress and to report various statistics about business processes and resources involved in their enactment. Many business processes have restrictions such as a limited duration of subprocesses, terms of delivery, dates of re-submission, or activity deadlines. Generally, time violations increase the cost of a business process because they lead to some form of exception handling [3]. Therefore, a *WfMS* should provide the process manager with the necessary information about a process, its time restrictions, and its actual time requirements. In addition, the process manager needs tools to anticipate time problems, to avoid consequent time constraint violations, and to take decisions about the relative priorities of processes according to some predefined temporal constraints [4].

Currently, existing *WfMSs* offer only a limited support for modelling and managing time constraints associated to processes and their activities [5]. This support takes

U. Dayal et al. (Eds.): BPM 2009, LNCS 5701, pp. 64–80, 2009.
© Springer-Verlag Berlin Heidelberg 2009

place through monitoring activity deadlines: however, the consistency of these deadlines and the side effects of missing them are not considered [4]. In the last decade some proposals have been presented to manage different temporal aspects of workflows as temporal constraints and deadlines both at design-time and at run-time [4,6]. At the best our knowledge, the concept of *controllability* has not yet been deeply considered for temporal workflows: controllability refers to the capability of executing a workflow for any possible duration of tasks. Since in several situations durations of tasks cannot be decided by *WfMSs*, even tough the minimum and the maximum durations for each task are known, checking controllability is stronger than verifying the consistency of the workflow temporal constraints [3,4,5,6]: in other words, each task duration cannot be imposed or decided by the *WfMS* that can only schedule all the tasks assuming that task durations respect their allowed ranges; therefore, it is necessary to know if some schedule is possible before to start the execution.

According to this scenario, in this paper we propose a framework for the conceptual design of workflows, which takes into account the modelling and the management of temporal aspects in *WfMSs*. Throughout the paper, we will use a motivating example taken from the domain of healthcare to introduce and discuss our proposal. In particular, we focus on the following specific features:

– *Temporal conceptual modelling of workflows:* we present a workflow conceptual model capturing the temporal aspects of main workflow activities (e.g., minimum and maximum durations of activities, delays and temporal constraints between them, deadlines and other temporal constraints). The conceptual model we propose is a block-structured one: in this way we are allowed to focus on temporal aspects without considering possible inconsistencies in the workflow schema. Moreover, further constructs as cycles or compound tasks are not considered at the current stage, as they require further analysis based on the results we will present in this paper.
– *Checking the controllability of temporal workflow schemata:* we propose a general method to determine whether, given a workflow schema based on our conceptual model, there may exist workflow executions where temporal constraints are satisfiable for any given duration of tasks. More precisely, in this paper we will focus on controllability for basic workflow patterns, i.e., *sequential and parallel patterns.*

The paper is structured as follows. In Sect. 2 we consider proposals from the literature having some relevance for workflow design with temporal features. In Sect. 3 we propose an extension of the Temporal Workflow model presented by Combi et al. [7] and provide a motivating example taken from the cardiology domain. In Sect. 4 we show how to check the controllability in common workflow patterns of a temporal workflow. Finally, in Sect. 5 we provide some discussion on the algorithmic aspects of controllability for workflow schemata and sketch some concluding remarks.

2   Related Work

In this section we consider the main proposals in the literature dealing with temporal constraints in workflow systems.

Eder et al. in [3,4,8] specify the Timed Workflow Graph (*TWG*) to represent temporal properties of tasks named "activity nodes". *TWG* is a directed acyclic graph (*DAG*), in which nodes are activities and oriented edges are control flows. *TWG* provides a notation indicating different kinds of execution flow, such as parallel or alternative ones, but doesn't provide the possibility of explicitly defining delays between activities: for instance, if an industrial process includes several short tasks, the incapability of explicitly modelling delays between executions can lead to incorrect time evaluations. *TWG* provides temporal constraints representing upper and lower bounds for ending the execution of a task with respect to either to the start of the workflow or to the end of another previous task.

Marjanovic et al. define in [6] a conceptual model that classifies temporal aspects of workflow schemata as: "basic temporal constraint", "limited duration constraint", "deadline constraint" and "interdependent temporal constraint". A basic temporal constraint limits the expected duration of one single task. A limited duration constraint is an upper bound for the duration of the workflow execution. An interdependent temporal constraint limits the time distance between two tasks in a workflow model. Marjanovic et al.'s model does not permit any delay interval between consecutive activities: the completion instant of a task corresponds to the starting instant of the subsequent task. Moreover, they represent the kinds of flow by using connector nodes, which differ from task nodes as they do not require agents. However, the connector nodes have no temporal features, such as minimum and maximum duration, and all the connector nodes in the model correspond to instantaneous activities.

The model by Bettini et al. [5] is quite different from the previous ones. First of all, nodes of a workflow graph are not tasks but correspond to temporal instants. Every task is represented by two nodes: the starting instant and the ending one. Every edge in the workflow graph represents the temporal distance between two nodes: any edge label is an interval representing the allowed time distances between the connected nodes. If a pair of nodes represents the starting and the ending instants of a task, the label on the connecting edge represents the allowed durations of the task. In the same way, if a pair of nodes represents the ending instant of a task and the starting instant of another task, the connecting label edge represents the allowed delay between the first task execution ending instant and the second task execution starting instant. By this formalism, workflow edges can be used to limit the time distance between all the nodes of a workflow schema: this approach is able to represent very complex processes. On the other side, it makes no difference between edges representing execution flows and edges limiting the time distance between nodes. In [5] a polynomial algorithm ($O(n^4)$ where n is the number of nodes) is provided to check the existence of a *free schedule*: a schedule is free when it is possible to statically fix, before the beginning of the execution, the start times of all tasks without constraining their durations, and satisfying all the constraints.

A problem similar to the workflow controllability check has been studied in the AI area of temporal constraint networks for planning [9,10]. Starting from the consideration that a temporal planner has to manage the likely uncertainty about the duration of processes, Vidal et al. propose an extension of the Simple Temporal Network, the Simple Temporal Network With Uncertainty (*STNU*), where the edges, i.e., constraints, are divided into two classes, the *contingent links* and *requirement links*. Contingent links represent processes of uncertain duration, where finish timepoints (i.e., STNU nodes) are decided by Nature within the limits imposed by the bounds defined on the contingent links. Requirement links represent all the other processes whose finish timepoints are controlled by the agents that execute processes. Morris et al. show that in the framework of *STNU* the checking controllability algorithm is polynomial, i.e., $O(n^4)$, w.r.t. the number of *STNU* nodes [10].

3   The Conceptual Temporal Workflow Model

Workflow modelling is an effective technique for understanding, automating, and documenting business processes. A conceptual workflow model produces high-level specifications of workflows that are independent from the workflow management software.

Several workflow models have been developed based on different modelling concepts (e.g., Petri's Net variants, precedence graph models, precedence graphs with control nodes, state charts, control structure based models, and so on) and on different representation models (programming language style text based models, simple graphical flow models, structured graphs, and so on) [4,11,12].

The conceptual model presented in this paper focuses on formalising all the temporal aspects related to single atomic activities (*tasks*) and to their temporally adaptive activation sequences. By using our model, temporal properties of workflow schemata can be deeply modelled.

As for the basic atemporal concepts, our temporal conceptual model is based on the atemporal one by Casati et al. [11], that we find quite general, it is not influenced by any particular commercial *WfMS*, and it is one of the models closest to the recommendations from the *WfMC* [1]. The graphical notation we introduce for the constructs of the conceptual model is a straightforward extension of the widely adopted Business Process Modelling Notation (*BPMN*) [2] to consider temporal aspects.

3.1   The Temporal Conceptual Model

Conceptual models allow designers to represent workflow schemata (*process models*) which capture the behaviour of processes describing the activities and their execution flow. A workflow schema defines the tasks to be performed, their order of execution, and assignment criteria to agents [11].

In our model a workflow schema is a directed graph, called *Workflow Graph*. Nodes correspond to activities and edges represent control flows that define the task dependencies that a *WfMS* has to consider when managing the order of execution of tasks.

There are two different activity types: *task* and *connector*. Tasks represent the elementary work units that collectively achieve the process goal. A task can be initial (Start), final (End) or intermediate. Connectors are the elementary work units executed by the *WfMS* to achieve a correct and ordered sequence for task execution. A path between two nodes represents an order of execution among the set of nodes of the path and it is called *flow*.

In the following, we present the syntactic properties of all the workflow graph components. We restrict ourselves to the main constructs; several other components, e.g., cycles, supertasks and multitasks, require a depth analysis with regard to their temporal behaviour and this is beyond the scope of this paper.

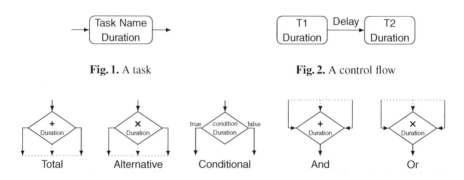

Fig. 1. A task   Fig. 2. A control flow

Fig. 3. *split* connectors   Fig. 4. *join* connectors

Tasks. A task is the basic modelling object in workflow schemata and represents the atomic unit of work to be executed. Tasks have many properties, represented by attributes. Task Name and execution Duration are mandatory attributes. Duration specifies the allowed temporal spans of the activity and its effective value cannot be set by the *WfMS*. More details on temporal aspects are in Sect. 3.3. Every task has one incoming edge and one outgoing edge (see Fig. 1).

Control Flows. Control flow is an oriented edge that connects two activities: the former activity must be finished before starting the execution of the latter one (see Fig. 2). Every edge has a temporal property, Delay, that denotes the allowed times that can be set by the *WfMS* for its internal activities for possibly delaying the execution of workflows according to the given temporal constraints. More details are given in Sect. 3.3.

Connectors. Connectors represent internal activities executed by the *WfMS* to achieve a correct and coordinated execution of tasks. Differently from tasks, connectors are directly executed by the *WfMS* and do not need to be assigned to any agent; the mandatory Duration attribute of a connector specifies the temporal spans allowed to the *WfMS* for executing the connector activity and the effective duration of the connector can be decided by the *WfMS*. As in the *WfMC* Reference Model [1], in our model we have two connector types: *split* and *join*. As depicted in Fig. 3, *split* connectors are nodes with one incoming edge and two or more outgoing edges: after the execution of the predecessor, (possibly) several successors have to be considered for the execution. The set of nodes that can start their execution is given by the features of each split connector. A *split* connector can be: *Total*, *Alternative* or *Conditional*. *Join* connectors are nodes with two or more incoming edges and one outgoing edge only, as shown in Fig. 4: *join* connectors merge more flows into one single flow. A *join* connector can be either *And* or *Or*.

Start and End. Each workflow can contain exactly one Start and one or more Ends, graphically represented by a circle with one ingoing/outgoing edge respectively. They have no temporal property.

    In the following we assume to deal with block-structured workflow graphs: checking this property of workflow graphs has been deeply studied in the literature and an effective algorithm, based on Petri Nets, has been proposed by van der Aalst et al. [13].

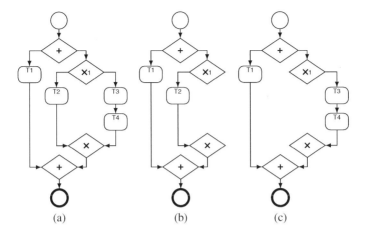

Fig. 5. Workflow graph (a) and its two *wf-paths* (b) and (c)

3.2 Workflow Paths (*wf-paths*)

Due to conditional and alternative flows, not all the cases (i.e., executions) of one workflow schema perform exactly the same set of tasks. We group workflow cases into *workflow paths* (*wf-paths*) in accordance with the activities actually executed. As in [6], a *wf-path* refers to a set of workflow cases that contain exactly the same activities (i.e., for each alternative flow in the workflow graph, the same successor node is chosen; for each conditional connector the same successor activity is selected). Therefore, a *wf-path* can be regarded as a workflow subgraph in which every alternative or conditional connectors have exactly one successor.

Figure 5 shows two *wf-paths* (b) and (c) of a workflow graph (a): the alternative connector ✕1 leads to the creation of two possible flows. Thus, each of these subgraphs represents the subset of workflow activities that are all executed for a particular *wf-path*. If a workflow schema has no alternative or conditional connectors, only one *wf-path* occurs.

3.3 Modelling Time and Temporal Aspects

Our model considers instants and durations as elementary temporal types [14]. Instants are points on the time domain, while durations are lengths on the time domain. Intervals are derived types and can be defined as the time span between starting and ending instants. Instants are represented through timestamps: each timestamp is defined at a given granularity. In this paper we adopt the approach proposed by Goralwalla et al. [14] for modelling granularities: a (calendric) *granularity* is a unit of measurement for spans of time. For example, the granularity of days (day) stands for a duration of 24 hours. More generally, a granularity is a special kind of, possibly varying, duration that can be used as a unit of time. Such granularities may thus be used as a unit of measure for expressing durations and also for specifying time points; in the last case, granularities are used for expressing the distance of a time point from a *reference time point*, chosen as origin of the time axis. In the following, without loss of generality we adopt the granularities of the Gregorian calendar, i.e., year, month, week, day, hour, minute, as units of measure for expressing durations.

In Sect. 3.1 we proposed two temporal properties: Duration for nodes and Delay for edges. Moreover, we proposed to consider the task Duration as a constraint on the effective duration of the task that cannot be modified by the *WfMS* while the connector Duration as a constraint on the effective duration that can be adjusted by the *WfMS*. As regards Delay, even if it could be simulated by a dummy task and this substitution could simplify the workflow graph structure, we prefer to maintain the Delay concept to underline that it represents a time span that could be spent by the *WfMS* to adaptively coordinate the task executions instead of a time span that is available to a dummy agent.

In the real world, workflow designers cannot always use precise values for durations of activities and edges. For example, tasks can be performed by human agents, and generally the duration of a task cannot be precisely known at workflow design time. Moreover, delays could be set by the *WfMS* to correctly manage the overall temporal constraints the workflow execution has to satisfy. Therefore, allowed durations/delays should be expressed by using ranges like "expected duration ± time tolerance". Our conceptual model describes all the durations/delays by an attribute having the form [MinDuration, MaxDuration] Granularity where $0 \leq$ MinDuration $\leq$ MaxDuration $\leq \infty$, as depicted in Fig. 6. Range bounds represent the minimum and maximum allowed durations. The minimum duration can represent either an estimate of the required time to carry out the activity or a constraint to the necessary time to carry out it. The maximum duration represents a *deadline* to the time to carry out the activity. Since each workflow component has to have the Duration attribute, if the workflow designer does not set a duration, we assume that there is no explicit temporal constraint and therefore we set the attribute value to $[1, +\infty]$ MinGranularity, where MinGranularity is the finest granularity managed by the *WfMS*. We do not admit $[0, 0]$ MinGranularity to underline that no activity can be executed without time consumption. A user can always set [0,n] Granularity as a duration attribute to specify that an activity can last 0 Granularity, if Granularity is not the minimal granularity used by the *WfMS* to measure the time.

Temporal Constraints. Besides the basic temporal constraints, expressed through durations and delays as previously described, our conceptual model allows one to express several other kinds of temporal constraints: from the business perspective they are defined by laws and regulations, business policies, common practises, as well as mutual agreements and expectations related to efficiency of business practise [6]. In general, temporal constraints are complex enough to be captured separately as an aspect of workflow modelling rather than being covered by the workflow execution properties [3]. Differently from the duration attributes, temporal constraints are not mandatory for each workflow component: they model additional temporal properties and must be controlled by the *WfMS*. In order to underline the difference between a temporal constraint derived from a duration and a temporal constraint explicitly stated, we denote these last ones as *relative* constraints. A *relative constraint* limits the time distance (duration) between the starting/ending instants of two non-consecutive workflow activities. Our model provides one type of relative constraint, expressed according to the following pattern:

$$I_F[\text{MinDuration}, \text{MaxDuration}]I_S \text{ Granularity},$$

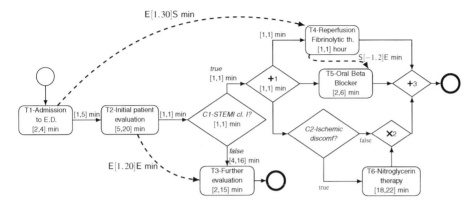

Fig. 6. Workflow graph example of patient admission to an hospital. The dashed edges represent relative constraints.

where (i) I_F marks which instant of the first activity to use ($I_F = S_{<activity>}$ for the starting execution instant or $I_F = E_{<activity>}$ for the ending one; the subscript can be omitted if it is clear form the context.); (ii) I_S marks the instant for the second activity in the same way; (iii) [MinDuration, MaxDuration] Granularity represents the allowed range for the time distance between the two instants I_F and I_S.

A finite positive MaxDuration value models a *deadline* as defined in other workflow models [4,6], since it corresponds to the maximum global allowable execution time for the activities that are present on possible flows between the two activities of the constraint. On the other hand, a finite positive MinDuration represents the minimum execution time that has to be spent from I_F before proceeding after I_S: if the global time spent to execute all activities between I_F and I_S is less than MinDuration, then the WfMS has to dynamically manage a suitable exception (like to sleep, for example) depending on the specific applications. Negative MinDuration and MaxDuration can be interpreted similarly, assuming that in this case I_S precedes I_F. We assume therefore that $-\infty \leq$ MinDuration $\leq$ MaxDuration $\leq \infty$.

3.4 A Motivating Example from Healthcare

As an example of a real workflow schema, in Fig. 6 we propose an excerpt from the guideline to the diagnosis and treatment of *ST-segment Elevation Myocardial Infarction (STEMI)*, published by the American College of Cardiology/American Heart Association in 2004 [15], represented as a temporal workflow.

The case starts as the patient is admitted to the Emergency Department (E.D.) (task T1) that has not to require more than four minutes. After the admission, the patient is examined by a physician (task T2) who can take a time between five and twenty minutes to make the examination. If the diagnosis is a STEMI occurrence (connector C1), then a well-know set of therapy and diagnosis activities has to be fired (*true* flow). Otherwise, a further patient evaluation has to be done (*false* flow). Since the guideline considers only STEMI patients, we have decided to close the *false* flow by a generic task (task T3) to represent a further evaluation. The *true* flow is composed by three parallel flows starting from total connector +1. The uppermost flow refers to the main therapeutic action in presence of a myocardial infarction: reperfusion is obtained through a fibrinolytic therapy (task T4). The central flow refers to the complementary therapeutic action consisting of the assumption of beta blocker drugs (task T5). The lowest flow contains the (possible) activities related to therapies for ischemic discomfort: after the evaluation of the presence of ischemic discomfort (conditional connector C2), a nitroglycerin therapy is provided (task T6).

After all these therapeutic actions, the workflow ends. It is possible to observe that there are different temporal constraints for tasks (if durations are not specified, then they are set to the default value of $[1, +\infty]$ min since we assume that the minimum granularity managed by the *WfMS* is that of minutes). The intertask constraint $E_{T1}[1, 30]S_{T4}$ min between the end of task T1 and the beginning of task T4 represents the most important recommendation from the guideline to successfully apply the fibrinolytic therapy to patients.

Relative constraints are conceptually more expressive than the deadline constructs used in other models. In fact, relative constraints can model other temporal bindings, as depicted by the T2-T3 relative constraint of Fig. 6. This constraint fixes to 20 minutes the maximum time distance between the end of T2 and the end of T3 and it has to be evaluated in the C1-false flow. Relative constraints cannot be set for activities belonging to mutually exclusive flows. For instance, in Fig. 6, relative constraints cannot connect T3 to T4 or T3 to T5.

Let us consider possible issues when executing a workflow instance according to the given temporal constraints. A first problem is related to the existence of a suitable duration/delay for each activity/edge within the allowed range satisfying all the specified relative constraints, because the durations of tasks and delays of edges are not independent. For example, considering the *wf-path* T1-T2-T3, if the delay of edge C1-T3 is set to 16, the duration of task T3 must be set to 2 in order to satisfy the relative constraint $E_{T2}[1, 20]E_{T3}$ min and cannot be arbitrarily chosen in the range $[2, 15]$.

A second stronger issue is related to the existence of a suitable duration/delay for each connector/edge such that the overall workflow satisfies the relative constraints without fixing task durations (i.e., it is possible to choose the connector durations or edge delays without knowing the duration of the following tasks). This is interesting because often task durations cannot be set by the *WfMS*. For example, considering the *wf-path* T1-T2-T4||T5, if the delay of edge T1-T2 is set to 5, the duration of task T2 can be arbitrarily chosen in the allowed range still satisfying the relative constraint $E_{T1}[1, 30]S_{T4}$ min. As this property holds also for the other allowed values of the delay of edge T1-T2, we call this *wf-path controllable*. It is worth noting that the *wf-path* T1-T2-T3 is not controllable as there are no allowed delay values for C1-T3 that guarantee the satisfaction of the relative constraint $E_{T2}[1, 20]E_{T3}$ for any allowed duration of task T3.

Controllability arises another kind of constraint between activities. Let us consider for the *wf-path* T1-T2-T4||T5 the constraint $S_{T4}[-1, 2]E_{T5}$ min, describing the fact that reperfusion (T4) neither can start more than 2 minutes before nor can start more than 1 minute after the end of oral therapy (T5). As we are not able to control the duration of T5, the start of reperfusion (T4) must be explicitly related to both the end and the start of T5. To satisfy the relative constraint $S_{T4}[-1,2]E_{T5}$ min, T4 could start either after the end of T5 or 4 minutes after the start of T5, even if this last one has not yet ended: in next section we will discuss how it works.

4   Controllability of Workflows

The successful completion of a process often depends on the correctness of temporal aspects modelled at design time. If relative constraints are such that any of them cannot be satisfied, the process cannot be performed successfully. Therefore, preliminary temporal evaluations are needed to state whether the specified relative constraints can be satisfied by any case.

In general we say that a workflow schema is *controllable* if the *WfMS* is able to perform any *wf-path* satisfying all relative constraints, all delays, all connector durations without any settings about the (allowed) task durations involved in the *wf-path*. These preliminary temporal evaluations are exponential in the number of the alternative connectors and conditional connectors because a workflow schema may represent many *wf-paths* and therefore the evaluations have to be done for each flow separately.

In this paper we focus on how it is possible to check the controllability of a single *wf-path*; we will discuss how to deal with a workflow schema at the end of the paper.

In Sect. 3.2 we defined a *wf-path* as a workflow subgraph in which every alternative or conditional connector has exactly one successor. Analysing the structure of workflow schemata it is straightforward to verify that (i) if a workflow schema does not contain any total connector, then each *wf-path* is represented as one graph-path *(sequential path)* and (ii) if the workflow schema contains at least one total connector, then at least one *wf-path* is represented as two or more graph-paths *(parallel paths)*.

The problem of controllability checking arises when there is at least one relative constraint that involves two or more tasks. If relative constraints involve only connectors, then there is no a controllability problem because the possible duration assignments are independent of any task duration and the problem is more simple [16].

In the following we show how to check the controllability in *sequential paths* and in *parallel paths* starting from simple *patterns* for them; more structured patterns can be reduced to these simple ones. Hereinafter, we assume that all the temporal constraints (activity durations, edge delays and relative constraints) have been mapped into equivalent constraints at the finest granularity [14]. For sake of simplicity, we will omit the specification of granularity in the considered constraints.

4.1   Controllability on Sequential Paths

Let us consider three simple sequential patterns each containing a relative constraint as in Fig. 7, where in sub-figure (a) the constraint is between the start instants of the two tasks, in sub-figure (b) it is between the end ones and in sub-figure (c) it is between the start instant of the first task and the end one of the second. We do not consider a relative constraint of the form $E_{T1}[p,q]S_{T2}$ because it is represented by the edge connecting the two tasks. In all patterns the range $[p,q]$ has to have p and q non negative because any negative value would be meaningless.

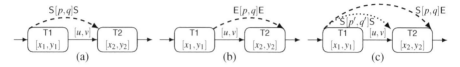

Fig. 7. Three sequential patterns with a relative constraint. In (c) the relative constraint $S[p',q']S$ (dotted) is induced by $S[p,q]E$.

In the pattern of Fig. 7-(a) the composition of the task duration and of the delays has to comply with the relative constraint. Task duration cannot be modified, therefore the *WfMS* can only decide the duration of delay, after the task T1 is executed. In order to verify whether it is possible to guarantee that the relative constraint can be satisfied for every possible T1 duration, it is sufficient to verify whether the range $[p - y_1, q - x_1] \subseteq [u,v]$. If so, the range of values that is permitted by the designer allows the *WfMS* to control the pattern. Otherwise, the range $[p - y_1, q - x_1]$ is either empty or it contains some values that are possible as delay values but that are not permitted by the designer: in the first case the relative constraint is inconsistent with the task duration, while in the second one the pattern is not controllable. In detail, if $[u',v'] = [p - y_1, q - x_1] \subseteq [u,v]$ is not empty, then at run-time the *WfMS* has to pick the delay value in the new range $[u',v']$ according to the T1 duration. The range $[p - y_1, q - x_1]$ is determined observing that the delay has to be minimum when T1 lasts its maximum allowed time and has to be maximum when T1 lasts its minimum time so that the sum of times results to be in $[p,q]$ range.

As an example, if T1 duration is $[6,8]$, the edge delay is $[1,11]$ and the relative constraint is $[10,12]$, then the new delay range is $[2,6] \subset [1,11]$. Therefore the pattern is controllable and the new delay range is $[2,6]$.

The same approach can be adopted when there is a relative constraint involving only connectors and edges.

The pattern of Fig. 7-(b) is similar to the case (a), but the task duration is still unknown when the *WfMS* has to decide the duration of the delay. In order to guarantee that the relative constraint can be satisfied for every possible T2 duration, it is necessary to impose a stronger condition on the delay range. The range $[p - y_2, q - x_2]$ is not sufficient because it is possible that a value chosen by the *WfMS* would be not suitable with the effective duration of T2. For example, assuming that T2 duration is the same of T1 and all the other values are as above (considering the new restricted range $[u,v] = [2,6]$), if the *WfMS* chooses to set the delay to 2 and the following T2 requires time 6, the total time is 8, lower than the allowed bound 10. Therefore, edge delay values have to be calculated as values that can be used for any T2 duration value: the minimum restricted valid range is $[p - x_2, q - y_2] \subseteq [p - y_2, q - x_2]$. The pattern is controllable if $[p - x_2, q - y_2]$ has the same relation with $[u,v]$ as for the (a) pattern. In the example above, the new delay range is $[4,4] \subset [2,6]$, therefore the pattern is controllable.

The same pattern of Fig. 7-(b) is depicted in the healthcare-related example in Fig. 6: the constraint on *wf-path* T1-T2-T4||T5 induces a relative constraint between the end of T1 and the end of T2 with range $[6,25]$ min. It can be derived by considering all the delays, durations and relative constraints of tasks and edges between T1 and T4. A

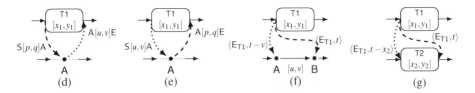

(d)   (e)   (f)   (g)

Fig. 8. Four parallel patterns with a relative constraint. The dotted edges are induced relative constraints by the composition of the given relative constraint and the T1 duration. For sake of simplicity, we put A in the labels of relative constraints, as A could represent either a starting or an ending instant of an activity. In (f) and (g) the relative constraints are *wait* constraints.

general method to deriving all the possible temporal constraints is given by reducing the *wf-path* to a Simple Temporal Problem (*STP*) which is known to be solved in $O(n^3)$ using an *all-pairs shortest path* algorithm as the Floyd-Warshall one [16]. As already mentioned, this *wf-path* is controllable. Indeed, the range $[p,q]$ is $[6,25]$ and the range $[x_2,y_2]$ is $[5,20]$; thus, the minimum restricted valid range for $[u,v]$ is $[6-5,25-20] = [1,5]$ as the range for the edge T1-T2 delay in Fig. 6.

On the other hand, the *wf-path* T1-T2-T3 is not controllable. Indeed, the induced constraint range between the end of C1 and the end of T3 results to be $[3,18]$. This constraint requires that the delay range of the edge C1-T3 should be $[3-2,18-15] = [1,3]$ to guarantee the controllability. Instead, the delay range in Fig. 6 is $[4,16]$ and therefore the *wf-path* is not controllable.

Finally, we analyse the pattern where the relative constraint has the $S_{T1}[p,q]E_{T2}$ form as in Fig. 7-(c). The controllability check can be done in two steps. In the first step, a new relative constraint between S_{T1} and S_{T2} is defined with temporal range determined applying the rule of pattern (b). If the induced relative constraint is not empty, in the second step, the controllability of the T1 duration together with the edge delay w.r.t. the new constraint is verified applying the pattern (a). If all steps are successfully performed, the pattern is controllable.

As an example, considering $[6,8]$ as duration range for both T1 and T2, $[2,6]$ as delay range and $S_{T1}[17,19]E_{T2}$ as relative constraint, we obtain that the induced relative constraint is $S_{T1}[11,11]S_{T2}$ and that this new constraint involves $[3,5]$ as delay range. Since $[3,5] \subset [2,6]$, the pattern is controllable and $[3,5]$ becomes the new delay range.

Any sequential path is a generalisation of the pattern (c).

4.2   Controllability on Parallel Paths

Let us consider four simple parallel patterns each containing a relative constraint as in Fig. 8, where in sub-figure (d) the constraint is between the start instant of task T1 and a generic instant A on a parallel flow (either start or end instant of an activity), in (e) it is between a generic instant A and the end instant of task T1 on a parallel flow, in (f) the relative constraint has a special label and is between the start of T1 and the instant B that can be either the end point of an edge or the end instant of a connector and, in (g) there is a pattern similar to (f) but with a task instead of points A and B. In the following, we will determine new constraints and check the controllability of the pattern w.r.t. T1 duration.

In the pattern of Fig. 8-(d) the composition of duration of T1 and of the relative constraint results in the derived constraint $A[u,v]E_{T1} = A[x_1 - q, y_1 - p]E_{T1}$ as in sequential pattern (a), although here the direction of $A[u,v]E_{T1}$ is reversed. This pattern is useful to propagate the original constraint for the overall evaluation of the *wf-path*.

The pattern of Fig. 8-(e) is the most interesting one. If we consider the T1 duration and the relative constraint, it is possible to determine a new constraint between the start instant of T1 and the instant A, that represents the most interesting information for evaluating the controllability of the pattern. If $q < 0$ then A has to happen after the end instant of T1. The controllability is sure because the duration of T1 is known: it is sufficient to choose a suitable value in the range $[x_1 - q, y_1 - p]$ (depending on the T1 duration) as delay between the start of T1 and A.

If $p \geq 0$, A has to happen before (or at) the end of T1. It is necessary to guarantee that the constraint holds whatever T1 duration is. In similar way as done for the (b) sequential pattern (here the derived constraint has opposite orientation w.r.t. the corresponding delay edge in pattern (b)), it is sufficient to fix the constraint between the start of T1 and A to be $[y_1 - q, x_1 - p]$ to have the controllability.

If $p < 0$ and $q \geq 0$, then A may occur before or after the end of T1. In this case it is not possible to set an unique range to guarantee the controllability; it is necessary to set a new constraint (*wait* constraint) between the start of T1 and A that is conditioned by the end of T1. The *wait* constraint has the special label $\langle E_{T1}, y_1 - q \rangle$ that means: A could occur either when (1) "T1 has ended (and within $|p|$ time units)" or when (2) "$y_1 - q$ time units have elapsed since the start of T1 and T1 has not yet finished" (if A does not occur when condition (2) holds, it will happen that the following end of T1 will trigger the condition (1)). Indeed, if A could occur before $y_1 - q$ time units, there were a constraint violation when T1 lasts y_1 time units. Sometimes the *wait* constraint can be simplified: if $(y_1 - q) \leq x_1$, then a lower bound can be set because the condition (2) is always verified before the end of T1 could happen: so the constraint can be represented as $[y_1 - q, E_{T1} + |p|]$.

The constraint on *wf-path* T1-T2-T4||T5 of the healthcare-related example in Fig. 6 between T4 and T5 is an instance of the pattern of Fig. 8-(e) where the range $[p,q]$ is $[-1,2]$ min and the range $[x_1,y_1]$ is $[2,6]$; thus, the derived wait constraint between the start of T4 and the start of T5 is $\langle E_{T5}, y_1 - q \rangle = \langle E_{T5}, 4 \rangle$.

In the pattern of Fig. 8-(f) we show how to propagate a possible *wait* relative constraint (possibly originated in previous steps of the analysis: see pattern (e)) to the link that shares the same endpoint with the relative constraint. The instants A and B can be either the endpoints of a edge or the start and the end instants of a connector (not a task!), respectively. In the former case, $[u,v]$ represents the delay range while in the latter one the duration of the connector. The relative constraint between S_{T1} and B has a *wait* condition $\langle E_{T1}, t \rangle$: B can happen either after that E_{T1} is occurred or t time units after S_{T1} are elapsed. In this case it is necessary to propagate the *wait* condition to the instant A in order to guarantee the controllability of the whole pattern. The propagation of the *wait* constraint S_{T1} carries another *wait* constraint between S_{T1} and A with condition $\langle E_{T1}, t - v \rangle$: if A could occur before time $t - v$, B could occur before t and this determines a constraint violation if T1 were still running.

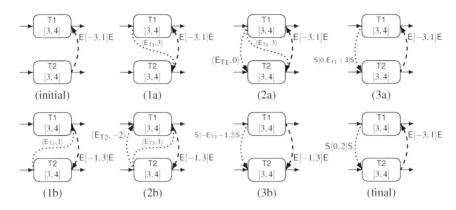

Fig. 9. Example of controllability check of a simple parallel *wf-path* T1-T2 (initial). (1a) is the application of pattern (e) w.r.t. T1 duration. (2a) is the result of pattern (f) application and (3a) is the simplification of the wait. (1b)–(3b) represent the applications of patterns w.r.t. T2 duration. (final) is the combination of (3a) and (3b).

The pattern in Fig. 8-(g) seems to be similar to Fig. 8-(f): the only difference is that instants A and B are now the start and the end instants of a task, respectively. It is worth to note that this difference requires a more strict constraint propagation: in a similar way as done for pattern (b), it is simple to show that applying the argument made for pattern (f) is not sufficient because it is not possible to decide when E_{T1} occurs. Therefore it is necessary to set another *wait* constraint between S_{T1} and S_{T2} with condition $\langle E_{T1}, t - x_1 \rangle$.

As a brief example, let us consider the parallel pattern of Fig. 9-(initial): two parallel tasks with a relative constraint on their end instants. In general, all possible constraint propagations have to be done to determine the ranges of delays, connector durations and relative constraints to guarantee the controllability. Here, it is interesting to know which is the more general relative constraint on start instants of tasks, induced by the given constraint, that guarantees the controllability. In Fig. 9-(1a)–(3a), we report the sequence of propagation that starts from T1 duration and in Fig. 9-(1b)–(3b) the sequence of propagation that starts from T2 duration. Starting from T1 duration, the new induced constraint results to be $S_{T1}[0, E_{T1} + 3]S_{T2}$ while starting from T2 duration it results to be $S_{T1}[-(E_{T2} + 1), 2]S_{T2}$. Since $-(E_{T2} + 1) < 0$ and $E_{T1} + 3 > 2$, the composition of these two new constraints yields the final constraint $S_{T1}[0, 2]S_{T2}$ as depicted in Fig. 9-(final).

5   Discussion and Conclusions

As suggested by considering parallel patterns in the previous section, in order to verify the controllability of a *wf-path*, the constraints propagation has to be made w.r.t. all possible combinations of durations and delays applying the six basic patterns (a,b,d–g) appropriately. The evaluation terminates when either a constraint violation is detected or no new range restrictions are determined. In the former case, the *wf-path* is uncontrollable and the workflow schema too. On the contrary, if all *wf-paths* of a schema are controllable, then we say that the workflow schema is controllable.

It is possible to show that the algorithm proposed by Morris and Muscettola [10] could be extended to deal with our temporal workflow model: task durations are represented as contingent links, while edges and temporal constraints are requirement links. The controllability of each *wf-path* can be checked in $O(n^4)$ time w.r.t. the number n of activities present on the *wf-path* [17]. Intuitively, the controllability is checked executing the following actions: (1) the *wf-path* is reduced to a corresponding *STP* where constraints between nodes can be the standard ones or the *contingent* ones; a *contingent* constraint cannot be squeezed; (2) the STP is solved applying alternatively an *all-pairs shortest paths* algorithm and an ad-hoc technique that propagates the contingent constraints in a similar way to the approach described in this paper, until a final state is found. If the STP admits a solution, then the original *wf-path* is controllable and the ranges induced by the STP solution are the new ranges for the connectors and edges (task durations cannot be squeezed!). Otherwise, the *wf-path* is not controllable.

It's worth to note that the number of *wf-paths* can be exponential w.r.t. the graph order because in a workflow schema there could be an arbitrary sequence of alternative or conditional operators. Despite of the exponential number of *wf-paths*, the controllability of a workflow schema can be evaluated in $O(n^4)$ on the graph corresponding to a workflow schema obtained from the original one by substituting all *Conditional* and *Alternative* connectors with *Total* ones and by substituting all the *Or* connectors with the *And* ones. In this way all the possible constraints are considered together and, therefore, the resulting controllability is checked against an over-constrained workflow schema. All the partial workflow schemata corresponding to single *wf-paths* are controllable if the above over-constrained schema is controllable. Moreover, delays and connector ranges determined by considering the over-constrained schema allow the *WfMS* to choose a duration for them without preventing the execution of any *wf-path* that contains the activities already done. Moving from design-time to run-time, controllability can be reconsidered during the execution of a workflow schema by (1) considering the actual values for duration of activities and for delays already done and (2) by removing from the schema all *wf-paths* that have not been executed. This could produce a less constrained workflow schema, i.e., with wider temporal ranges for not-yet-executed delays and connectors.

The concept of workflow controllability seems to be closed to that of free schedule [5]. A free schedule corresponds to a controllable *wf-path*, while it deserves further investigations to verify whether any controllable *wf-path* corresponds to one or several free schedules: indeed it seems that a controllable *wf-path* could correspond even to a schedule where the starting point of tasks cannot be completely set when a workflow execution starts.

As for the application of our temporal conceptual workflow model, since it is always important to evaluate the goodness of a model on real-life cases, we are cooperating with the *YAWL* [18] development group in order to develop *YAWL* extensions that manage temporal aspects of *YAWL* workflow schemata both at design-time and run-time. A prototype of such extended *YAWL* has been successfully used to manage some simple (till now) healthcare processes [7].

In conclusion, in this paper we have proposed a new advanced workflow conceptual model for expressing time constraints and we have introduced the concept of controllability for workflow schemata that are block-structured and do not contain cycles or compound-tasks. Currently, we are investigating on (1) extending these results to a model that includes both cycles and compound-tasks and on (2) the possibility to extend our approach to unstructured workflow models.

References

1. Workflow Management Coalition, Hollingsworth, D.: The workflow reference model (1995), http://www.wfmc.org/standards/framework.htm
2. Object Management Group (OMG): Business process definition metamodel (bpdm), beta 1 (2007), http://www.omg.org/cgi-bin/doc?dtc/2007-07-01
3. Eder, J., Panagos, E., Rabinovich, M.I.: Time constraints in workflow systems. In: Jarke, M., Oberweis, A. (eds.) CAiSE 1999. LNCS, vol. 1626, pp. 286–300. Springer, Heidelberg (1999)
4. Eder, J., Panagos, E.: Managing time in workflow systems. In: Workflow Handbook 2001. Workflow Management Coalition (WfMC), pp. 109–132 (2000)
5. Bettini, C., Wang, X.S., Jajodia, S.: Temporal reasoning in workflow systems. Distributed and Parallel Databases 11, 269–306 (2002)
6. Marjanovic, O., Orlowska, M.E.: On modeling and verification of temporal constraints in production workflows. Knowl. Inf. Syst. 1, 157–192 (1999)
7. Combi, C., Gozzi, M., Juárez, J.M., Oliboni, B., Pozzi, G.: Conceptual modeling of temporal clinical workflows. In: TIME, pp. 70–81. IEEE Computer Society, Los Alamitos (2007)
8. Ede, J., Gruber, W., Panagos, E.: Temporal modeling of workflows with conditional execution paths. In: Ibrahim, M., Küng, J., Revell, N. (eds.) DEXA 2000. LNCS, vol. 1873, pp. 243–253. Springer, Heidelberg (2000)
9. Vidal, T., Fargier, H.: Handling contingency in temporal constraint networks: from consistency to controllabilities. J. Exp. Theor. Artif. Intell. 11, 23–45 (1999)
10. Morris, P.H., Muscettola, N.: Temporal dynamic controllability revisited. In: Veloso, M.M., Kambhampati, S. (eds.) AAAI, pp. 1193–1198. AAAI Press / The MIT Press (2005)
11. Casati, F., Ceri, S., Pernici, B., Pozzi, G.: Conceptual modelling of workflows. In: Papazoglou, M.P. (ed.) ER 1995 and OOER 1995. LNCS, vol. 1021, pp. 341–354. Springer, Heidelberg (1995)
12. Mangan, P.J., Sadiq, S.W.: A constraint specification approach to building flexible workflows. Journal of Research and Practice in Information Technology 35, 21–39 (2003)
13. van der Aalst, W.M.P., Hirnschall, A., Verbeek, H.M.W(E.): An alternative way to analyze workflow graphs. In: Pidduck, A.B., Mylopoulos, J., Woo, C.C., Ozsu, M.T. (eds.) CAiSE 2002. LNCS, vol. 2348, pp. 535–552. Springer, Heidelberg (2002)
14. Goralwalla, I.A., Leontiev, Y., Özsu, M.T., Szafron, D., Combi, C.: Temporal granularity: Completing the puzzle. J. Intell. Inf. Syst. 16, 41–63 (2001)
15. Antman, E.M., et al.: ACC/AHA guidelines for the management of patients with ST-elevation myocardial infarction. Circulation 110, 588–636 (2004)
16. Dechter, R., Meiri, I., Pearl, J.: Temporal constraint networks. Artif. Intell. 49, 61–95 (1991)
17. Morris, P.: A structural characterization of temporal dynamic controllability. In: Benhamou, F. (ed.) CP 2006. LNCS, vol. 4204, pp. 375–389. Springer, Heidelberg (2006)
18. van der Aalst, W.M.P., Aldred, L., Dumas, M., ter Hofstede, A.H.M.: Design and implementation of the YAWL system. In: Persson, A., Stirna, J. (eds.) CAiSE 2004. LNCS, vol. 3084, pp. 142–159. Springer, Heidelberg (2004)

Towards Algorithmic Generation of Business Processes: From Business Step Dependencies to Process Algebra Expressions

Márcio K. Oikawa[1], João E. Ferreira[1], Simon Malkowski[2], and Calton Pu[2]

[1] Institute of Mathematics and Statistics
University of São Paulo
{koikawa,jef}@ime.usp.br
[2] Center for Experimental Research in Computer Systems
Georgia Institute of Technology
{zmon,calton}@cc.gatech.edu

Abstract. Recently, a lot of work has been done on formalization of business process specification, in particular, using Petri nets and process algebra. However, these efforts usually do not explicitly address complex business process development, which necessitates the specification, coordination, and synchronization of a large number of business steps. It is imperative that these atomic tasks are associated correctly and monitored for countless dependencies. Moreover, as these business processes grow, they become critically reliant on a large number of split and merge points, which additionally increases modeling complexity. Therefore, one of the central challenges in complex business process modeling is the composition of dependent business steps. We address this challenge and introduce a formally correct method for automated composition of algebraic expressions in complex business process modeling based on acyclic directed graph reductions. We show that our method generates an equivalent algebraic expression from an appropriate acyclic directed graph if the graph is well-formed and series-parallel. Additionally, we encapsulate the reductions in an algorithm that transforms business step dependencies described by users into digraphs, recognizes structural conflicts, identifies Wheatstone bridges, and finally generates algebraic expressions.

Keywords: business process modeling, directed acyclic graphs, series-parallel reductions, and process algebra.

1 Introduction

Today's large enterprise applications typically include powerful workflow engines as a central module in their information systems. Additionally, the paradigms of web services and mobile computing require mapping flows of work among humans, systems, or both as parallel control-flows that interact and communicate among each other. Despite numerous approaches to formal control of these workflows, there still exists a wide gap between complex business process modeling and workflow languages based on formal representation (e.g., graph-oriented

U. Dayal et al. (Eds.): BPM 2009, LNCS 5701, pp. 80–96, 2009.

models [1], Petri nets [2], and process algebras [3]) [4]. The reason for this gap lays in the difficult task of exhaustively capturing and modeling all dependencies between all business steps (i.e., atomic activities) in complex business processes. This task becomes even more difficult as the number of steps grows to stricter business process requirements caused by the rapid evolution of markets, more complex technologies, and shrinking time-to-market for new products. Although some previous efforts have addressed business process modeling with precise mapping of object relations and process dependencies, none of them has offered explicit support for automated generation of business process representation.

This paper focuses on complex business process modeling, where automated generation and accurate representation of business step dependencies becomes a critical factor. We have applied graph transformation techniques to define algorithmic derivations, which preserve algebraic properties and take advantage of the topological characteristics of series-parallel digraphs. The input for our algorithm are step dependencies described by users. Consecutively, the algorithm generates acyclic directed graphs (DAGs), checks potential structural conflicts, and implements a formal method for polynomial-time reduction of DAGs to build algebraic expressions. In our methodology inconsistencies between corresponding split and merge points identify structural problems such as deadlocks and lacks of synchronization. Another output of our algorithm is the identification of Wheatstone bridges, which characterize regions in the graph where the generation of algebraic expressions may be difficult or even unfeasible. Wheatstone bridge circuits are traditionally known from measuring instruments in electric circuits. Recently, they have been employed as structural mechanism for reducibility in PERT (Project Evaluation and Review Technique) stochastic networks [5].

The main goal of this work is to automatically generate equivalent algebraic representations of DAG-based structures, which represent internal dependencies, branching, as well as synchronizing rules in business processes. The main contribution is the definition of a formal reduction system that recognizes a class of DAGs (well-formed series-parallel DAGs) for what is possible to quickly define an equivalent ACP algebraic expression. In the case of forming-rule violations, our algorithm identifies regions of structural conflicts. These conflicts represent topological structures, which cannot be expressed algebraically and have to be translated into process algebra under relaxation of equivalence requirements. The scope of this paper are DAG-based models; hence, internal process behaviors that are not observable in DAGs (e.g., multiple instances, canceling patterns [6], and mobile behavior [7]) are not addressed and remain potential future work topics.

The rest of this paper is organized as follows. After an overview of related work in Section 2, we introduce some important concepts of the graph transformation system, which has formed the basis of our work, in Section 3. In Section 4, we define expression digraphs, build a reduction system, and present the algorithmic generation of business processes and algebraic expressions from business step dependencies. Section 5 discuss some aspects about Wheatstone bridges. We conclude the paper in Section 6.

2   Related Work

To this day, no definite consensus has been reached on the best way of representing business process models, and there exists a considerable body of work addressing the definition of a general purpose language for workflows and business processes. For instance, XPDL, defined by the Workflow Management Coalition (WfMC) [1], is a popular standard based on graph-oriented modeling. Petri net based models [2] are another approach that provides a consistent framework for derivation of workflow properties. Similarly, the JBoss community offers the Java Process Definition Language (JPDL) [8], which allows process definition as combination of declaratively specified process graphs with sets of Java classes. Recently, process algebras [9] have been advocated by many business process management efforts. Especially within the areas of service-oriented architectures [10] and collaborative information systems [11, 12, 13], process algebras have been successfully applied to represent and control the life cycles of complex orders. Another important process algebra contribution is the Bigraphical Reactive System [14], a meta model for global ubiquitous computing that generalizes the process algebra approach for mobile environments [15]. Naturally, the question which of these formalisms is most suitable remains subject to a controversial discussion [16]. In our work, we have chosen the Algebra of Communicating Processes (ACP) [17] as representative algebraic formalism. However, this paper does not necessarily advocate a single formalism since there exist unidirectional mappings of ACP to other formalisms such as pi-calculus or Petri nets.

From a workflow modeling perspective, many different approaches have been devised that aid in the formulation of workflows in corporate information system environments. For example, workflow patterns [18] are an important initiative for identification of comprehensive workflow functionality and provide the basis for an in-depth comparison of many workflow management systems. Recently, the elimination of modeling errors through explicit support for pattern selection in user-determined contexts has been studied [19]. Moreover, data-driven approaches [20, 21, 22] define a new paradigm for the integration of complex data and process structures. Similarly, the concept of business artifacts [23, 24] facilitates business process modeling through operational modeling. Conceptual schemas, which represent rules and processes shared among complex collaborating information systems, are another approach to business process modeling [25]. However, all of these methodologies and tools neglect the need of automation that results from growing complexity. Their formulations on how to generate models often remain partly implicit, and consequently, significant aspects of the modeling that are required for automation are unformalized.

In general, automated generation of algebraic expressions for (complex) business processes from dependency sets remains a difficult problem [16], and it is comparable to intractable problems on arbitrary DAGs [38]. Our work addresses this problem, identifying a class of digraphs for what an algebraic expression can be produced efficiently. However, there are some contributions that explore the automatic manipulation of dependencies. In contrast to our work, their sole goal is process characterization and control. For instance, in dataflow programming [26] dependencies are modeled explicitly to guide activity scheduling. The dependencies are optimized for the generation of minimal dependency sets, which guarantees high concurrency and reduced maintenance cost for process execution.

Our approach is also related to work on graph transformation systems [27]. In such systems, transformation rules are applied to specific sets of graphs, allowing the observation of certain properties and behaviors. We take advantage of this concept and use graph transformation systems to define a set of graph topological transformations, which preserve algebraic properties and, at same time, compose representative expressions for them.

3   Graph Reduction

In the following, we introduce some concepts of graphs and graph transformation systems, which form the foundation of our approach. Interested readers should refer to the cited sources (e.g., [29, 30]) for a more comprehensive discussion. Readers familiar with these formalisms can skip to the next section.

Directed Acyclic Graphs. A *directed graph* (or *digraph* for short) is a pair $D = (V, A)$ of *vertices* and *arcs*, such that $A \subseteq V \times V$ [28]. If more than one digraph is considered, we refer to the sets of vertices and arcs as $V(D)$ and $A(D)$, respectively. Given two vertices u and v, an arc uv represents one relation, in which v depends on u, and it is graphically represented by an arrow from u to v. In this case, u is referred to be the *head* of uv, and v is its *tail*. A vertex u is called *source* if it does not depend on any other vertex, and it is called *sink* if no vertex is dependent on it. A *path* in D is a sequence of vertices $v_1 v_2 \ldots v_{k-1} v_k$, such that $v_i v_{i+1} \in A(D)$, and all vertices v_i are distinct, for $i = 1, 2, \ldots, k - 1$. A cycle is a path where $v_1 = v_k$. A *DAG* is defined as a digraph without cycles.

A *line digraph* of D is a digraph $Q = L(D)$, such that $V(Q) = A(D)$ and $A(Q) = \{uv : u, v \in V(Q)$, the head of u coincides with the tail of $v\}$. A *series-parallel digraph* (or *sp-digraph* for short) is a special type of DAG, recursively defined as follows [29, 30].

- A digraph formed by two vertices joined by a single arc is a sp-digraph, called *trivial*;
- Let D_1 and D_2 be sp-digraphs, and s_i and t_i be their source and sink vertices for $i = 1, 2$. Then the result of either of the following compositions is also a sp-digraph:
  - parallel composition: identify s_1 with s_2 and t_1 with t_2.
  - sequential composition: identify t_1 with s_2.

A sample sp-digraph is shown in Figure 1(a). An important property of series-parallel DAGs is that they are a class of digraphs, for which there exists an efficient identification algorithm based on topological reductions [31, 32]. Such reductions are either sequential or parallel, and examples for each are provided in Figure 1(b). Applying these reductions iteratively to a DAG G, one eventually obtains a reduced (i.e., minimal) DAG G' with one of the following two properties.

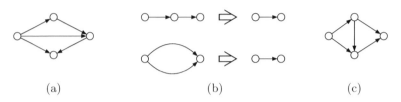

Fig. 1. Sp-digraphs: (a) an example of a sp-digraph; (b) sequential reduction (top) and parallel reduction (bottom); (c) Wheatstone bridge

– G' is a trivial sp-digraph;
– G' has a Wheatstone bridge inside its structure.

G is a sp-digraph if and only if G' satisfies the first property. Otherwise, the second property holds for G'. The topology of a Wheatstone bridge [30] is shown in Figure 1(c).

Graph Transformation System. A *GTS* consists of a set of transformation rules that can be applied to certain graphs to build other graphs. The main purpose of a GTS is to formalize rules for a closed class of graphs. This allows to analyze specific graph properties by examining the outcome of these transformations [33]. In particular, we are interested in two properties, confluence [34] and termination, which are presented after the following basic concepts.

Definition 1. *(Labeled graph [35]) Let Ω_V and Ω_A be alphabets for vertex and arc labels. A labeled graph is a tuple $G = (V, A, l_v, l_a)$, such that V is the vertex set, A is the arc set, and the functions $l_v : V \to \Omega_V$ and $l_a : A \to \Omega_A$ are labeling functions for vertices and arcs.*

A *morphism graph* $f : G \to G'$ is a pair $f =< f_v : V(G) \to V(G'), f_a : A(G) \to A(G') >$ of functions, which preserve sources, targets, and labels. A morphism f is an *isomorphism* if both f_v and f_a are bijective. Two isomorphic graphs G and G' are represented as $G \cong G'$.

Definition 2. *(Production [35, 34]) A production $p = (L \xleftarrow{l} K \xrightarrow{r} R)$ consists of two injective graph morphisms l and r as well as three finite graphs L, K, and R, called left hand side, interface, and right hand side, respectively.*

A production defines a transformation rule, that transforms L into R. Digraph K is the interface between L and R and represents all elements in L that will be preserved in R. The transformation consists of deleting the elements $L - K$ from digraph L, and adding all elements in $R - K$. Operation $L - K$ renders all elements of L that are not in K, and the same analogy applies to $R - K$.

Definition 3. *(Pushout [35,34]) Given a set of graphs X, Y, Z, W, W' and two morphisms $y : X \to Y$ and $z : X \to Z$, a tuple $< W, f : Z \to W, g : Y \to W >$ is called pushout of $< z, y >$ if the following properties are observed.*

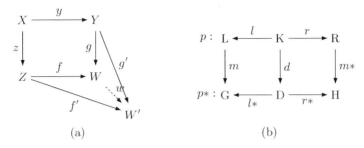

Fig. 2. Pushout examples: (a) graphical representation; (b) double pushout

- *Commutativity condition:* $g \circ y = f \circ z$;
- *Universal property: for all graphs W' and morphisms $g' : Y \to W'$ and $f' : Z \to W'$, there exists a unique morphism $w : W \to W'$ such that $w \circ g = g'$ and $w \circ f = f'$.*

A graphical representation of a pushout is provided in Figure 2(a).

Definition 4. *(Direct Derivation [35, 34]) Let G and H be digraphs, $p = (L \xleftarrow{l} K \xrightarrow{r} R)$ a production, and $m : L \to G$ an injective morphism. Then G directly derives H through p and m, denoted by $G \overset{p,m}{\Longrightarrow} H$, if two pushouts exist such that the notation in Figure 2(b) applies. In this case D is called context graph, and we sometimes write $G \overset{p}{\Longrightarrow} H$ if m is not relevant.*

Intuitively, the left pushout corresponds to the identification of elements L in G and the deletion of elements $m(L)$ from G. The right pushout corresponds to the addition of elements $m * (R) - d(K)$ to $m(G)$. A *derivation* from G to H is a sequence of direct derivations $G = G_0 \Longrightarrow \ldots \Longrightarrow G_n = H$ for some $n \geq 0$ and may be denoted by $G \overset{*}{\Longrightarrow} H$.

Independence and Confluence. Two direct derivations $G \overset{p_1}{\Longrightarrow} G_1$ and $G \overset{p_2}{\Longrightarrow} G_2$ commute if there is a graph H such that $G_1 \overset{p_2}{\Longrightarrow} H$ and $G_2 \overset{p_1}{\Longrightarrow} H$. In the double-pushout approach, this is the case if the occurrences of two steps do not overlap in deleted nodes and edges. A graph transformation system is *confluent* if for each pair of derivations $G \Longrightarrow G_1$ and $G \Longrightarrow G_2$ there exists a graph H such that $G_1 \overset{*}{\Longrightarrow} H$ and $G_2 \overset{*}{\Longrightarrow} H$ [36, 34]. Confluence implies that every graph can be transformed into at most one irreducible graph. One of the possible methods for proving confluence is to show that (i) a transformation is terminating and (ii) all overlapping direct derivations (i.e., *critical pairs*) converge to isomorphic graphs (i.e., *joinability*). This method is known as Newman's Lemma [34].

Let G, G_1, G_2 and H be digraphs, and let p_1 and p_2 be productions, such that $G \overset{p_1}{\Longrightarrow} G_1$ and $G \overset{p_2}{\Longrightarrow} G_2$. The pair $G_1 \overset{p_1}{\Longleftarrow} G \overset{p_2}{\Longrightarrow} G_2$ is called *critical* if the direct derivations $G \overset{p_1}{\Longrightarrow} G_1$ and $G \overset{p_2}{\Longrightarrow} G_2$ are not *independent*. The GTS is confluent if there exists some H_1 and H_2, such that $G_1 \overset{*}{\Longrightarrow} H_1 \cong H_2 \overset{*}{\Longleftarrow} G_2$.

Termination. A graph transformation system is called *terminating* if infinite derivations $G1 \Rightarrow G2 \Rightarrow G3 \Rightarrow \ldots$ are impossible [36]. Simple sufficient conditions for termination are that each rule reduces the size of a graph or the number of occurrences of a fixed subset of labels.

4   Algorithmic Generation of Business Process

Our methodology requires adding semantic values to arcs and vertices by assigning algebraic elements to their labels. Therefore, we define the concept of expression digraphs, which combines elements of classical digraphs and algebraic expressions, in the following. Building upon this foundation, we present our algorithmic procedure for automated generation of algebraic expressions.

4.1   Expression Digraph

Definition 5. *(Expression graph) An expression digraph is a digraph $D=(V,A, l_v,l_a)$, such that V is a vertex set, A is an arc set, and the functions $l_v : V \to \{\cdot,+,\|,s,t\}$ and $l_a : A \to \mathcal{T}(\Sigma) \cup \{\lambda\}$ are labeling functions for vertices and arcs, respectively.*

An expression graph is a labeled graph, in which the labeling functions l_v and l_a assign algebraic features to vertices and arcs. Function l_v maps each non-terminal vertex to an algebraic composition operator (i.e., sequential, alternative, or parallel). Terminal vertices are mapped to s (source) or t (sink) and do not have any specific algebraic behavior. Function l_a maps each arc either to an algebraic term or to λ (see explanation below). By joining all algebraic terms inside a single arc, it is possible to obtain a global algebraic expression, which is called $term(D)$. Symbol λ has to be defined in order to cover specific cases, in which an arc does not describe the execution of an algebraic term. Therefore, λ represents an empty or neutral action without functional participation in the execution of the process. The behavior of λ is shown in Figure 3 for an arbitrary algebraic term x.

$$x \cdot \lambda = \lambda \cdot x = x$$
$$x \parallel \lambda = \lambda \parallel x = x$$

Fig. 3. Properties for action λ

In essence, an expression graph is a data structure that makes explicit all process terms, their dependencies, and their internal compositional relationships. Consequently, each vertex represents more than just a composition between terms; it also represents an execution state of the business process. Vertices with out-degree greater than one are called *branching vertices* and represent split points of the business process. Vertices with in-degree greater than one are similarly called *synchronizing vertices* and represent merge points of the business process. The execution options for branching and synchronizing vertices are illustrated in Figure 4.

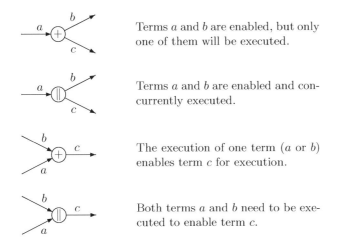

Terms a and b are enabled, but only one of them will be executed.

Terms a and b are enabled and concurrently executed.

The execution of one term (a or b) enables term c for execution.

Both terms a and b need to be executed to enable term c.

Fig. 4. Graphical representations for branching and synchronizing vertices

Well-formed structures have to use the same operator in each pair of branching and corresponding synchronization vertices. Given the algebraic terms a and b, Figure 5(a) shows a well-formed alternative structure. Well-formed parallel structures are constructed analogously by using parallel operators in both vertices. In all cases when the business processes does not satisfy this characteristic, a structural conflict (i.e., *deadlock* or *lack of synchronization*) is unavoidable. Deadlocks are parallel synchronizations of terms triggered by alternative branching (Figure 5(b)). Notice the parallel synchronization for terms a and b after an alternative trigger. Lacks of synchronization are unintentional multiple executions of one or more terms after an alternative synchronization (Figure 5(c)). In the example it is easy to see that terms a and b are concurrently triggered and not synchronized.

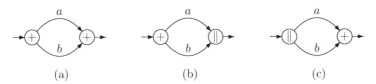

(a)   (b)   (c)

Fig. 5. Branching and synchronizing in expression graphs: (a) well-formed alternative composition; (b) deadlock; (c) lack of synchronization

4.2 Reduction System for Generation of Algebraic Expressions

A reduction system is a GTS, in which all productions reduce the graph dimension. It is common to refer to these productions as *reductions*. This subsection introduces reduction system $\mathcal{G}$ for expression digraphs. $\mathcal{G}$ is composed of the three productions shown in Figure 6. We call reductions (a), (b), and (c) alternative reduction, parallel reduction, and sequential reduction, respectively. The

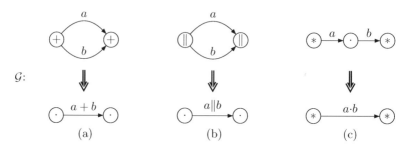

Fig. 6. Reduction system $\mathcal{G}$: (a) alternative reduction; (b) parallel reduction; (c) sequential reduction, where character '*' indicates either an alternative $(+)$ or a parallel $(\|)$ operator

purpose of $\mathcal{G}$ is to replace more complex graph regions with single arcs while preserving the semantics through equivalent algebraic expressions in the new arc labels.

Lemma 1. *Let D and H be expression DAGs and r a reduction in $\mathcal{G}$ such that $D \overset{r}{\Rightarrow} H$. Then $term(D)$ is equivalent to $term(H)$.*

Proof. Since we consider algebraic expressions, the equivalence criterion is bissimilarity. Thus, the proof of this lemma may be restricted to proving bissimilarity between $term(D)$ and $term(H)$. For this purpose, we show that the generation of H does not modify the behavior of the process represented by D. Next, we present a constructive proof for alternative reductions. The same approach can be analogously applied to the other reductions as well.

Recall alternative reductions from Figure 6(a). According to the definition of expression digraphs in Section 4.1, the original digraph represents the alternative execution of terms a and b. Since the execution trees of all possible instances of D and H are isomorphic, $term(D)$ is bissimilar to $term(H)$. ☐

Once we guarantee the equivalence preservation in $\mathcal{G}$, we proceed to prove that any sequence of transformations leads to the same result (i.e., $\mathcal{G}$ is convergent). This can done by showing that $\mathcal{G}$ is *confluent* and *terminating*. For proving confluence, it sufficient to show that all critical pairs converge. In fact, $\mathcal{G}$ has only one critical pair, that converges. It occurs where two overlapping sequential reductions share a single vertex of the digraph. With some observations, we should confirm that the expressions produced from them are bissimilar, so that $\mathcal{G}$ is *confluent*. Regarding the terminating, it is sufficient to show that any transformation reduces number of elements of digraph.

The results of Lemma 1 guarantee the preservation of algebraic properties of arc labels after the application of all reductions. Together with confluence and terminating properties, $\mathcal{G}$ shows that any sequence of reductions leads to the same final expression DAG. A direct consequence of these results is the generation of a single algebraic expression for each expression DAG—regardless of the transformation sequence. This result is the core foundation for our algorithm, which is described next.

4.3   Algorithm

This subsection explains how to use expression digraphs for generating businesses processes represented in algebraic expressions from a set of predefined dependencies.

Preparing input information. As previously discussed, our starting point is a set of direct dependencies between activities in a business process. Therefore, we assume our initial input to be a set of dependencies, such as shown in Figure 7. In this phase vertices represent activities of the business process, which is common in many software tools and intuitive for non-expert users.

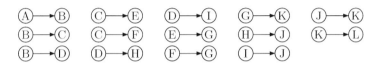

Fig. 7. Initial set of dependencies for the business process

Figure 7 shows some activities that participate in more than one dependency. For example, activities B, C, and D have two successors each, while activities G, J, and K have two predecessors each. Therefore, it is necessary to provide additional information concerning the branching and synchronizing. A refinement of the previous set of dependencies leads to the identification of branching and synchronizing vertices, as listed in Figure 8. For each branching and synchronizing vertex, it is necessary to define the corresponding composition rule (i.e., alternative or parallel).

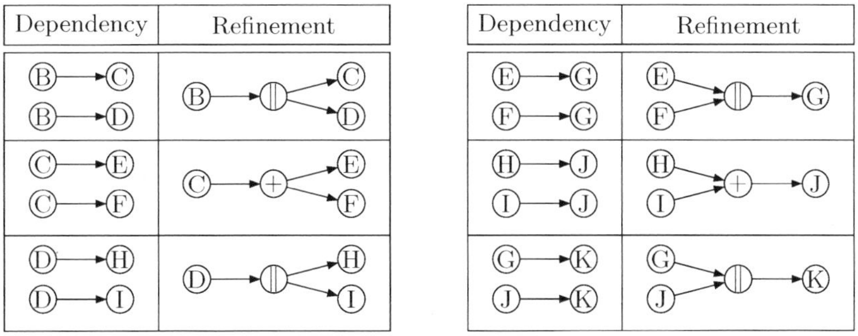

Dependency	Refinement
B →C B →D	B →◫< C / D
C →E C →F	C →⊕< E / F
D →H D →I	D →◫< H / I

Dependency	Refinement
E →G F →G	E / F >◫→ G
H →J I →J	H / I >⊕→ J
G →K J →K	G / J >◫→ K

Fig. 8. Refinement of dependencies by definition of branching and synchronizing rules

After this step the graph-based representation of the current specification can be built. The resulting digraph is shown in Figure 9 (i). Note that this is not an expression graph yet. For generating an expression digraph, we have to build a line digraph first. The activity labels are placed on all arcs, and all non-terminal vertices are mapped to algebraic composition operators. Following this steps, we create an expression graph (Figure 9(a)). The latter serves as input for the algorithm. One of its requirements is that the expression digraph is acyclic and has just one connected component. These two properties can be tested in polynomial time by classical graph algorithms. The presence of cycles in this structure prohibits the use of the algorithm, and identification of more than one connected component indicates a possible failure of the original project.

Executing the algorithm. At its core Algorithm 1 relies on the successive application of reductions over an expression DAG. These reductions are applied according to GTS $\mathcal{G}$, illustrated in Figure 6. For each reduction, arcs and algebraic terms are joined, generating ampler sets of algebraic terms. During some reductions, the algorithm can also identify structural conflicts by comparing the coherence among vertices. The execution terminates when no more reductions are applicable, returning a reduced expression DAG. For well-formed sp-expression DAGs, the returned DAG is isomorphic to the trivial sp-digraph and has the complete algebraic expression as its arc label. Algebraic expressions are not generated if any of the following conditions is true:

– The expression DAG is not well-formed, which indicates the presence of structural conflicts. Algorithm 1 returns a DAG, in which these conflicts are identified;
– The expression DAG is not series-parallel, which indicates the presence of one or more Wheatstone bridges. Algorithm 1 returns a reduced DAG with explicit Wheatstone bridges.

Algorithm 1 uses the following functions:

– $\mathrm{pred}(v)$: returns a set with all of the predecessors of vertex v;
– $\mathrm{suc}(v)$: returns a set with all of the successors of vertex v;
– $l_v(v)$: returns the algebraic operator related to the vertex v;
– $l_a(uv)$: returns the algebraic term related to the arc uv;
– $\|uv\|$: returns the number of arcs between vertices u and v.

In general, sequential reduction replaces a sequence of two terms through a composition of them. This step is included in lines 5–14 where a loop changes a pair of consecutive arcs into a single one and updates the labels with an algebraic sequential composition. The example in Figure 9 (b) shows the application of many sequential reductions on the initial expression DAG (Figure 9 (a)).

After the sequential reductions, the expression DAG may have pairs of vertices with redundant arcs, susceptible to parallel and alternative reductions. Similarly to sequential reductions, parallel and alternative reductions simplify a DAG by removing redundant arcs and updating arc labels to compositions of algebraic terms (Figure 9 (c)). The latter step is carried out by the loop in lines 16–34. This step is also responsible for the identification of structural conflicts, because the correspondent branching and synchronizing vertices are compared in lines 19–21.

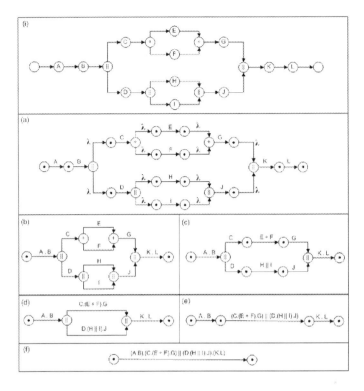

Fig. 9. Graphic representation of our algorithm: the dependency graph (i), the initial expression DAG (a), and its reductions (b)–(f)

The aforementioned procedure is repeated inside the loop in lines 4-35, which exits when no more reduction is applicable. This condition can be easily tested by comparing the previous and current number of arcs, for instance. Exemplary loop cycles and the terminating condition are illustrated in Figure 9 by sub-figures (d), (e), and (f).

Considering loops at lines 4–35 and 16–34, this algorithm takes $O(|V|.(|V| + |A|))$ time, since the outer loop investigates all vertices and inner loops trace vertices and arcs. In fact, as we have multi-digraphs, $|A| \geq |V|$, so the algorithm takes time $O(|V|.|A|)$. Despite of this time consumption, some contributions suggests that is possible to achieve a better time consumption, around $O(|A| + |V|)$, using strategies of identification of series-parallel graphs. These algorithms consider the application of appropriate auxiliary data structures [29, 37]. So, we believe that the time consumption should be improved using some customized data structures for digraphs.

The final result of the algorithm is a reduced expression DAG. If the original expression DAG is well-formed and series-parallel, the reduced graph will be a trivial series-parallel expression DAG, and the correspondent algebraic expression will be placed on its sole arc label.

However, obtaining algebraic expression in some cases might be problematic. If the original expression DAG contains conflicts, the algorithm can be used to return an expression DAG with the explicit indication of the conflict point. This may be done by inserting some information about the vertices during the execution of lines 19–21. An important case to be considered separately occurs when the original DAG is not series-parallel. In this case the reduced DAG will contain at least one Wheatstone bridge. Generation of algebraic expressions from Wheatstone bridges causes difficulties, because of a lack of explicit correspondence between pairs of split and merge points. Instead of being synchronized, split points are shared with potentially many merge points and vice-versa.

5   Wheatstone Bridges

Wheatstone bridges indicate structures on which series-parallel reductions do not work. The topology of Wheatstone bridges (Figure 1(c)) may be found in any non sp-digragh (e.g., Figure 10(a)). When considering the semantics of split and merge rules, the analysis shows that many of these configurations contain structural conflicts. This can be illustrated with a short example. Consider the simple Wheatstone bridge in Figure 10(b). All of the arcs have an algebraic sub-term in ACP (a, b, c, d and e). This configuration contains a deadlock, which occurs (over v_3) when term b is executed, and a lack of synchronization, which occurs (over v_4) when term a is executed. In fact, it is possible to observe that almost all combinations of these vertices have some conflict, which propagates through the business process instance. The only two completely conflict-free cases are when all of the rules are the same, either alternative or parallel. In these two cases, it is possible to apply separate algorithms to generate local ACP expressions and re-apply the Algorithm 1. For alternative vertices, the algebraic expression for Wheatstone bridges can be built in polynomial time, unlike the parallel vertices, which may require exponential time.

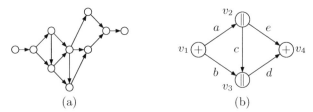

(a)   (b)

Fig. 10. Wheatstone bridges: (a) An arbitrary digraph with Wheatstone bridges. (b) An isolate Wheatstone bridge with conflicts.

The presence of Wheatstone bridges within business process models complicates certain aspects, such as execution control and system scalability. Execution control concerns the efficient management of all possible instances of a business process. Practical experience shows that many configurations of Wheatstone

Algorithm 1. Algorithm for generation of algebraic expressions and identification of structural conflicts and Whetstone bridges

Require: An expression digraph D
1:
2: *has_conflict* ← **false**
3:
4: **while** D can be reduced **do**
5:     **for all** non-terminal $v \in V(D)$ **do**
6:         **if** $d^-(v) = d^+(v) = 1$ **then**
7:             $u \leftarrow \text{pred}(v);$
8:             $w \leftarrow \text{suc}(v);$
9:             $A(D) \leftarrow A(D) \cup \{(u,w)\}$
10:            $l_a(uw) \leftarrow l_a(uv) \cdot l_a(vw)$
11:            $A(D) \leftarrow A(D)\backslash\{(u,v),(v,w)\}$
12:            $V(D) \leftarrow V(D)\backslash\{v\}$
13:        **end if**
14:    **end for**
15:
16:    **for all** $v \in V(D)$ and $\neg has\_conflict$ **do**
17:        **for all** $w \in V(D)$ e $w \in \text{suc}(v)$ **do**
18:            **if** $\|vw\| > 1$ **then**
19:                **if** $l_v(v) \neq l_v(w)$ **then**
20:                    *has_conflict* ← **true**
21:                **end if**
22:                {Let $a_1, a_2, ..., a_k$ be the arcs $v \rightarrow w$ in $A(D)$.}
23:                $l'_a(vw) \leftarrow l_a(a_1)$
24:                $A(D) \leftarrow A(D)\backslash\{a_1\}$
25:                **for all** $2 \leq i \leq k$ **do**
26:                    $l'_a(vw) \leftarrow l'_a(vw) \ l_v(v) \ l_a(a_i)$
27:                    $A(D) \leftarrow A(D)\backslash\{a_i\}$
28:                **end for**
29:                $A(D) \leftarrow A(D) \cup \{vw\}$
30:                $l_a(vw) \leftarrow l'_a(vw)$
31:                $l_v(v) \leftarrow l_v(w) \leftarrow \cdot$
32:            **end if**
33:        **end for**
34:    **end for**
35: **end while**
36: **return** D

bridges contain conflicts, and their actions may propagated to the rest of business process. Regarding the scalability, even for Wheatstone bridges without conflicts, the addition of new future dependencies (involving their internal vertices) will likely result in conflicts, unless these dependencies preserve the previous rule. For Wheatstone bridges with conflicts, new additions will preserve or even increase the conflict influence in the entire model.

Solely during the modeling phase, there exists the opportunity of planning the control flow of the business process to avoid future problems with Wheatstone bridges. When starting from an existing business process model, it should be guaranteed that all new sub-processes will not generate new conflicts. Therefore, Algorithm 1 can be used as a supporting tool for modeling and reviewing of business processes. The algorithm identifies conflicts and Wheatstone bridges and provides a correct algebraic expression for series-parallel regions.

6   Conclusion

This paper presented a technique for reducing acyclic digraphs to equivalent ACP algebraic expressions with explicit handling of structural conflicts. Our method required the extension of the classical definition of acyclic digraphs by assigning labels to vertices and arcs. With the help of these constructs, we have formalized the step-by-step generation of equivalent algebraic terms through systematic digraph reduction. Based on these findings we developed an algorithm, which is able to transform user specified business step dependencies into algebraic business process specifications. Additionally, our algorithm identifies conflicts in business process models through the comparison of corresponding split and merge points during the reduction phase. Our method directly benefits tasks that involve complex business process modeling and allows the automated generation of business step dependencies in business processes. Since our solution automatically identifies regions with deadlocks and lacks of synchronization, it can also be used for evaluating models (e.g., constructed through process mining). Our future and ongoing work include the adaptation of the algorithm for the treatment of cycles and indirect dependencies. Our ultimate goal is to enable the automated generation of algebraic expressions involving communication and multiple instances in business processes.

Acknowledgement

This work has been supported by grant# 06/00375-0, from FAPESP (São Paulo State Research Foundation). Additional support is provided by grant# 482139/2007-2 from CNPq (Brazilian National Research Council).

References

1. Workflow Management Coalition (2008), http://www.wfmc.org
2. Murata, T.: Petri nets: Properties, analysis and applications. Proceedings of the IEEE 77(4), 541–580 (1989)
3. Baeten, J.C.M., Weijland, W.P.: Process algebra. Cambridge University Press, New York (1990)
4. Bergstra, J., Ponse, A., Smolka, S.: Handbook of Process Algebra. Elsevier, Amsterdam (2001)
5. Ghomi, S.F., Rabbani, M.: A new structural mechanism for reducibility of stochastic PERT networks. European Journal of Operational Research (2003)

6. Russell, N., Hofstede, A.H.M.T., Mulyar, N.: Workflow control-flow patterns: A revised view. Technical report (2006)
7. Smith, H., Fingar, P.: Workflow is just a pi process (2003), http://www.fairdene.com/picalculus/workflow-is-just-a-pi-process.pdf
8. jBPM (2008), http://www.packtpub.com/article/jboss-jbpm-concepts-jpdl-jbpm-process-definition-language
9. Puhlmann, F.: Soundness verification of business processes specified in the pi-calculus. On the Move to Meaningful Internet Systems, 6–23 (2007)
10. Woodley, T., Gagnon, S.: BPM and SOA: Synergies and challenges. In: Ngu, A.H.H., Kitsuregawa, M., Neuhold, E.J., Chung, J.-Y., Sheng, Q.Z. (eds.) WISE 2005. LNCS, vol. 3806, pp. 679–688. Springer, Heidelberg (2005)
11. Ferreira, J.E., Takai, O.K., Pu, C.: Integration of collaborative information system in internet applications using riverfish architecture. In: CollaborateCom (2005)
12. Ferreira, J.E., Takai, O.K., Braghetto, K.R., Pu, C.: Large scale order processing through navigation plan concept. In: IEEE SCC, pp. 297–300 (2006)
13. Braghetto, K.R., Ferreira, J.E., Pu, C.: Using control-flow patterns for specifying business processes in cooperative environments. In: Cho, Y., Wainwright, R.L., Haddad, H., Shin, S.Y., Koo, Y.W. (eds.) SAC, pp. 1234–1241. ACM, New York (2007)
14. Gualtieri, A., Dell'Armi, T., Leone, N.: Process representation and reasoning using a logic formalism with object-oriented features. In: BPM Workshops, pp. 153–163 (2006)
15. Milner, R.: Communicating and Mobile Systems: the Pi-Calculus. Cambridge University Press, Cambridge (1999)
16. van der Aalst, W.M.P.: Pi calculus versus petri nets: let us eat humble pie rather than further inflate the pi hype. BPM Trends 3, 1–11 (2005)
17. Fokkink, W.: Introduction to Process Algebra. Springer, New York (2000)
18. van der Aalst, W.M.P., ter Hofstede, A.H.M., Kiepuszewski, B., Barros, A.P.: Workflow patterns. Distributed and Parallel Databases 14, 5–51 (2003)
19. Gschwind, T., Koehler, J., Wong, J.: In: Applying Patterns during Business Process Modeling, pp. 4–19 (2008)
20. Müller, D., Reichert, M., Herbst, J.: Data-driven modeling and coordinaiton of large process structures. On the Move to Meaningful Internet Systems (2007)
21. Müller, D., Reichert, M., Herbst, J.: A new paradigm for the enactment and dynamic adaptation of data-driven process structures. In: Advanced Information Systems Engineering, pp. 48–63 (2008)
22. Müller, D., Reichert, M., Herbst, J., Köntges, D., Neubert, A.: Corepro sim: A tool for modeling, simulating and adapting data-driven process structures. In: Dumas, M., Reichert, M., Shan, M.-C. (eds.) BPM 2008. LNCS, vol. 5240, pp. 394–397. Springer, Heidelberg (2008)
23. Nigam, A., Caswell, N.S.: Business artifacts: an approach to operation specificaiton. IBM Journal 42, 428–445 (2003)
24. Bhattacharya, K., Caswell, N.S., Kumaran, S., Nigam, A., Wu, F.Y.: Artifact-centered operational modeling: Lessons from artifact-centered operational modeling: Lessons from customer engagements. IBM Journal 46, 703–721 (2007)
25. Zuliane, D., Oikawa, M.K., Malkowski, S., Alcazar, J.P., Ferreira, J.E.: The riverfish approach to business process modeling: Linking business steps to control-flow patterns. In: CollaborateCom (2008)
26. Wu, Q., Pu, C., Sahai, A., Barga, R.S.: Categorization and optimization of synchronization dependencies in business processes. In: ICDE, pp. 306–315 (2007)

27. Gadducci, F.: Graph rewriting for the π-calculus. Mathematical. Structures in Comp. Sci. 17, 407–437 (2007)
28. Bang-Jensen, J., Gutin, G.: Digraphs: Theory, Algorithms and Applications, 2nd edn. Springer, Heidelberg (2002)
29. Duffin, R.J.: Topology of series-parallel networks. J. Mathematical Analysis and Applications, 303–318 (1965)
30. Valdes, J.: Parsing flowcharts and series-parallel graphs. Technical report 1978 (1978)
31. Valdes, J., Tarjan, R., Lawler, E.: The recognition of series parallel digraphs. SIAM J. Comput., 298–313 (1982)
32. Schoenmakers, B.: A new algorithm for the recognition of series parallel graphs. CWI Report CS-R9504 (1995)
33. Ohlebusch, E.: Advanced Topics in Term Rewriting. Springer, Heidelberg (2002)
34. Plump, D.: Confluence of graph transformation revisited. In: Processes, Terms and Cycles, pp. 280–308 (2005)
35. Corradini, A., Montanari, U., Rossi, F., Ehrig, H., Heckel, R., Loewe, M.: Algebraic approaches to graph transformation, part i: Basic concepts and double pushout approach. Technical Report TR-96-17, Corso Italia 40, 56125 Pisa, Italy (1996)
36. Andries, M., Engels, G., Habel, A., Hoffman, B., Kreowski, H.-J., Kuske, S., Plump, D., Schürr, A., Taentzer, G.: Graph transformation for specification and programming. Science of Computer Programming 34, 1–54 (1999)
37. Korenblit, M., Levit, V.E.: On algebraic expressions of series-parallel and Fibonacci graphs. In: Calude, C.S., Dinneen, M.J., Vajnovszki, V. (eds.) DMTCS 2003. LNCS, vol. 2731. Springer, Heidelberg (2003)
38. Naummann, V.: Measuring the distance to series-parallelity by path expressions. In: Mayr, E.W., Schmidt, G., Tinhofer, G. (eds.) WG 1994. LNCS, vol. 903, pp. 269–281. Springer, Heidelberg (1995)

Extending BPM Environments of Your Choice with Performance Related Decision Support

Mathias Fritzsche[1], Michael Picht[2], Wasif Gilani[1], Ivor Spence[3], John Brown[3], and Peter Kilpatrick[3]

[1] SAP Research CEC Belfast, United Kingdom
mathias.fritzsche@sap.com, wasif.gilani@sap.com
[2] SAP Product & Technology Unit Suite Core, Germany
michael.picht@sap.com
[3] Queen's University Belfast, United Kingdom
i.spence@qub.ac.uk, tj.Brown@qub.ac.uk, p.kilpatrick@qub.ac.uk

Abstract. What-if Simulations have been identified as one solution for business performance related decision support. Such support is especially useful in cases where it can be automatically generated out of Business Process Management (BPM) Environments from the existing business process models and performance parameters monitored from the executed business process instances. Currently, some of the available BPM Environments offer basic-level performance prediction capabilities. However, these functionalities are normally too limited to be generally useful for performance related decision support at business process level. In this paper, an approach is presented which allows the non-intrusive integration of sophisticated tooling for what-if simulations, analytic performance prediction tools, process optimizations or a combination of such solutions into already existing BPM environments. The approach abstracts from process modelling techniques which enable automatic decision support spanning processes across numerous BPM Environments. For instance, this enables end-to-end decision support for composite processes modelled with the Business Process Modelling Notation (BPMN) on top of existing Enterprise Resource Planning (ERP) processes modelled with proprietary languages.

1 Introduction

Business processes are the foundation of any enterprise. Their efficiency has an important effect on the profitability and hence on the success of a company regardless of its size or domain. Therefore, the goal of any enterprise is to continuously optimize business process execution and adapt it to changes within the market environment or the company itself. Enterprise application vendors aim to support this by the notion of "closed loop of continuous process optimization" (see Figure 1). In this paper, all tooling related to this loop is bundled under the term BPM Environment. One phase within this loop is the business processs configuration and business process composition (see CONFIGURE and COMPOSE in Figure 1). This phase enables business analysts to use tools like

U. Dayal et al. (Eds.): BPM 2009, LNCS 5701, pp. 97–112, 2009.

Fig. 1. Decision Support integrated in a closed loop of continuous process optimization

NetWeaver BPM [1], JCOM [2] or EMC Documentum Process Suite [3] to compose business logic, e.g. for Composite Applications, on top of services provided by configured back-end processes, such as the ERP processes offered by SAP Business Suite or Business ByDesign [4]. One example of such an extension is provided in one of our previous works [5]. Additionally, the business execution (see EXECUTE) needs to be supported by BPM Environments as well.

Tooling for the analysis of the business process history is provided by some BPM environments (see ANALYSE) in order to enable business process monitoring and analysis. This Analysis step provides performance data for already executed business processes, and offers functionalities and UI capabilities to the users such as sales unit managers, etc., to monitor and analyse the historic process performance data. This monitoring is then interpreted by users into decisions, such as organisational changes or modifications of the business process itself, meant to improve the future business performance.

However, decisions deduced from monitoring and analysis tooling are not sufficient in case of a high degree of complexity in resource intensive processes (e.g. layered use of resources, complex workflows, etc.) or in the statistical distribution of the monitored performance data or of plan data. The monitoring and analysis based decision making process, therefore, might not always be helpful to completely eliminate performance issues caused by a suboptimal scheduling of resource, under-utilization, bottle-necks, etc.

Thus, business performance related decision support is needed to deal with such cases (see DECIDE). For such support, performance analysis models normally need to be manually built in order to deal with complex resource scheduling problems, for instance, via what-if simulation. This task is time consuming, expensive and requires simulation related skills. A user further needs to have the required performance modelling expertise and the necessary skills to be able to interpret simulation results properly. The same applies for modelling in order to solve an optimization problem. Therefore, it is more appropriate to integrate directly such decision support into the existing business process modelling tools as part of BPM Environments. Integrated decision support is provided by a number of BPM Environments, like the EMC Documentum Process Suite [3], but based on basic level process simulation capabilities. Sophisticated business performance decision support, such as simulation of resource sharing scenarios among different departments or integration of optimization engines is missing in most environments [6]. Finally, non of the existing BPM Engines enables sophisticated end-to-end decision support spanning Composite Application as well as back-end processes [5].

In this paper we define an architecture which enables non-intrusive integration of sophisticated performance related decision support into existing BPM Environments and describe our industrial experiences with applications of this architecture for processes provided by existing ERP software, processes modelled with NetWeaver BPM and processes of the JCOM [2] environment.

The paper is structured as followed: The next section describes different kinds of questions answered by performance related decision support for business processes. Section 3 motivates the need for the integration of performance related decision support into existing BPM environments in a way which abstracts concrete business process modelling. Section 4 describes the proposed architecture which is then evaluated in Section 5. Section 6 provides an overview of the related work. Finally, Section 7 concludes the paper.

2 Background: Performance Related Decision Support for BPM Environments

We experienced that support is needed within complex business processes to investigate questions related to distribution of resources, working times, throughput and utilization. This is especially required in cases where there is a high degree of complexity in resource intensive processes (e.g. layered use of resources, shared use of resources, complex workflows, etc.) or in the statistical distribution of the history data or the plan data. Our business performance related decision support addresses the following type of resource related questions:

1. Can available staff cope with the future business growth?
2. Would a change of business conditions (e.g. change in the lower boundary of sales order approval request) improve the business performance?
3. Would a redistribution of resources between departments help to achieve overall performance targets?
4. How many employees are needed at which point in time?
5. Where is the predicted bottle-neck of the process?

Questions 1-3 can be answered via discrete event simulations [7]. For such a prediction, the control flow oriented business behaviour model, e.g. a Business Process Modelling Notation (BPMN) conforming model, needs to be combined with business process instance data indicating the resource related behaviour data of business process instances over a period of time. This time period can span historic process instances as well as future ones. Examples of resource related behaviour are the time needed to execute a BPMN Activity by one employee or the working time contingent of this employee.

Thus, *Process Model Data* (control flow related behaviour) needs to be combined with *Process Instance Data* (resource related behaviour). In the literature, numerous transformations can be found [8,9] where process models, such as UML Activity Diagrams are combined with process instance data in order to generate input for a discrete event simulation tool. On the other hand, numerous transformations can be found [10,11] as well where the same data is used to generate models analysed analytically. Analytical performance analysis tools produce results normally significantly faster than simulations but they are using mathematical assumptions which make the results less accurate. Also, they can normally only be used up to a certain size of input model due to the state space explosion problem. For instance, [10] transforms UML Activity Diagrams to Layered Queuing Networks (LQN) which can be analytically solved with the LQN solver (LQNS)[12]. LQN solver, as compared to other analytic approaches, especially considers layered use of resources, for instance, in case one resource needs to wait for another one in a rendezvous like communication scheme [13] in order to process activities. Moreover, if business performance objectives, constraints and requirements are modelled as well, it might be useful to integrate an optimization engine, such as that provided via the tool AnyLogic [14], in order to automatically execute a number of what-if simulations in order to propose an optimal solution. Such an approach can be used to answer question 4. Furthermore, additional computations can be done to answer requests like question 5, which are based on the results from the analytical or simulation based analysis.

3   Motivation

Some BPM Environments already provide basic what-if simulation capabilities [6]. Others turn to specialists to undertake simulation studies, and those specialists often prefer more sophisticated simulation tools [6]. Sophisticated simulation tools, such as the AnyLogic simulation tool [14], enable simulation of resource sharing scenarios among different departments. This functionality is normally not supported by the integrated simulation capabilities of existing BPM tools, such as the EMC Documentum Process Suite [3].

Furthermore, most BPM tools with simulation capabilities do not offer decision support functionality besides simulation functionality. However, a combination of different kinds of performance analysis techniques can help to turn what-if simulation results into information that business domain experts better understand. This is required as business domain experts are typically not performance modelling experts [5]. For instance, the automated combination of what-if simulations with an optimization engine might be useful in order to provide suggestions about, for example, how a perfect resource scheduling should look like. Also a combination of a simulation or a analytic performance prediction tool with a bottle-neck analysis based on the prediction results might be useful for a user to know which process step should be improved to prevent a bottleneck in the future. Moreover, none of the existing BPM Environments enable decision support spanning processes across numerous BPM Environments. This would enable end-to-end decision support for composite processes modelled with BPMN based modelling tools (e.g. NetWeaver BPM) on top of existing back-end processes (provided by existing ERP software) modelled with proprietary modelling tools. Concluding, an approach is required which enables the integration of sophisticated performance related decision support into a number of existing BPM Environments, which don't have this functionality yet or only offer basic level functionality. This integration especially needs to abstract from the business process modelling tool.

4   Proposed Architecture

Our proposed architecture for such an integrated performance related decision support in shown in Figure 2. This architecture refines the so called Model-Driven Performance Engineering architecture (MDPE) [15,16] which was originally designed for rather hardware resource related performance decisions. However, performance is a concept which cross cuts numerous domains. Thus, a performance modelling approach such as MDPE can be used for the BPM domain as well.

In the following paragraphs we define the various actors involved in this architecture which are bundeled within the *MDPE Workbench*:

The *Decision Support Integrator* extends the BPM Environment of choice with performance decision support functionality. A *Performance Modelling Actor* is a part of the Decision Support Integrator. It abstracts sophisticated *Performance Analysis Tools* on the one hand and a BPM Environment of choice on the other hand. Furthermore, the Performance Modelling Actor requires *Performance Parameter* as input. Examples for such parameters are how many sales order requests have occurred/will occur per day in the previous/next 12 months (see *History Data* and *Plan Data* in Figure 2). A more detailed description of the Performance Modelling is given in Subsection 4.1. A *Decision Support Calculator*, another part of the Decision Support Integrator, enables us to interconnect a number of different *Performance Analysis Tools* and to use these to generate a *Decision Support Result* as described in detail in Subsection 4.3.

The *Instance Data Manager* described by Subsection 4.2 is needed to provide access to the *Process Instance Data* for all actors of the Decision Support Integrator, and to enable editing and analysis of input and output data for the decision support in a language which a business domain expert understands.

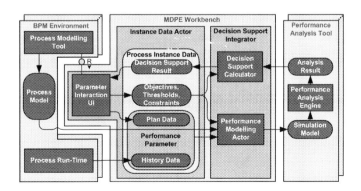

Fig. 2. Proposed Architecture as Block Diagram [17]

4.1   Performance Modelling Actor

The Performance Modelling Actor (see Figure 4) provides an abstraction layer for sophisticated Performance Analysis Tools including the Process Runtime on the one hand and Process Modelling Tools on the other hand.

The MDPE approach uses Tool Independent Performance Model (TIPM) which has been designed based on the Core Scenario Model (CSM) [18] by the TU Dresden, SAP Research and the simulation tool provider XJTech as a generic performance analysis model representation. Each TIPM is transformed to at least one Tool Specific Performance Model (TSPM) as shown in Figure 4. A TSPM is specific for a given Performance Analysis Tool, such as the discrete event simulation engine AnyLogic [14]. Compared to that, the TIPM is an inter-mediate language between Performance Analysis Tools and Process Modelling Tools. Thus, the TIPM helps to apply performance related decision support for a number of BPM Environments. A description of the TIPM meta-model follows.

TIPM based Abstraction. A TIPM combines the behavioural information from the Process Models with Process Instance Data. The behavioural information is represented in the meta-model of the TIPM (see Figure 3) with the meta-elements *Step* and *PathConnections* which are part of a *Scenario*. An example for such a Scenario in the business process domain is "Sales Order Processing" for a certain sales office in Philadelphia. Resources can be shared among multiple Scenarios, such as the case that the Marketing department with 10 employees is shared between the *Sales Order Processing of a sales office in Philadelphia* and the *Sales Order Processing of a sales office in Chicago*.

Performance Parameters (see Figure 2) need to be collected by an automated parameter importer out of the Process Runtime in the case of History Data (see in Figure 4) as proposed by Rozinat et al. [19] or defined as Plan Data. Performance Parameters are used to populate the following fields in the TIPM (see Figure 3):

– *Resource.multiplicity (called* Capacity *in this paper):* This metric indicates how many units are available in a pool of resources e.g. 10 employees in the Philadelphia sales office.

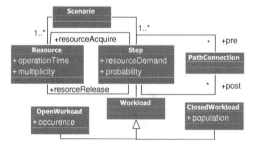

Fig. 3. Simplified Performance Analysis Model called TIPM

- *Resource.operationTime:* This metric indicates how much work can be done by one resource unit in a period of time. For instance, it specifies the resource efficiency of an employee.
- *Step.resourceDemand:* Indicates the net resource consumption of a Step, e.g. how much net working time is needed in order to create a Sales Order.
- *Step.probability:* Indicates the probability that a step is reached from the previous step.
- *Resource Link:* Is the reference between the Step and the Resource (see ResourceRelease and ResourceAcquire in Figure 3). It specifies, for instance, which process steps in Sales Order Processing have to be executed by the Marketing department.
- *Workload:* Specifies the occurrence or the population of arriving requests either in case of an OpenWorkload (occurrence), or a ClosedWorkload (population). An example of a open workload is the number of arriving sales requests per day in a business process for sales order processing; a closed workload example would be number of consultants starting a business trip immediately upon return from the previous one.

In the following subsection a description is provided of how the TIPM interconnects process modelling tools within BPM Environments with the Performance Analysis Tools.

Modular Model Transformations. As shown in Figure 4, the TIPM induces the need for a model automated transformation chain in order to first transform Process Models and the Performance Parameters to a TIPM and then to transform the TIPM to one or more TSPMs. The transformations are implemented within so called source- and target adapters as shown in Figure 4. These transformations are modularized into numerous transformation steps as described in [20]. This, for instance, enables separation of the structural concern of the *TIPM2AnyLogic_Sim* transformation (see Figure 4) from the concern of the actual XML representation of an AnyLogic simulation model. This decoupling further enables a high degree of reusability, as we are able to reuse some transformation steps for a number of source- and target adapters [20].

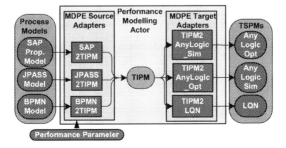

Fig. 4. Performance Modelling Actor as Block Diagram[17]

Figure 4 shows the source- and target adapters that we implemented. It can be seen that we are able to extend three different BPM Environments. Each of these environments is based on different modelling languages. We, therefore, have to support as input for the Performance Modelling Actor: SAP proprietary models, employed for back-end processes delivered by Business Suite or Business ByDesign; JPASS models for the extension of the JCOM environment; and BPMN models to extend the NetWeaver BPM environment employed for composite processes.

We added three different target adapters to our workbench. Thus, three different performance analysis methodologies can be used from the three different BPM environments which shows the high degree of extensibility of our solution enabled by the TIPM. One target adapter contains the transformations between TIPM and the simulation tool AnyLogic that we currently use as discrete event simulation engine (see *TIPM2AnyLogic_Sim* in Figure 4). Another is used for AnyLogic optimization experiments (see *TIPM2AnyLogic_Opt* in Figure 4). Moreover, we are considering analytic performance analysis. Therefore, the current MDPE implementation also supports the transformation of the TIPM to Layered Queuing Networks (see *TIPM2LQN* in Figure 4) in order to be used as input for the LQNS tool [12].

Each transformation in the chain has not only the direct transformation result as output but also, as a by-product, a *Trace Model* which stores the information about which model element(s) *a* is transformed to which model element(s) *b*. In [21] we described how this Trace Model is achieved as a by-product without additional effort from the developer of a transformation via the so-called Higher Order Transformations. The use of these trace models is described in the following subsection.

4.2   Instance Data Actor

As shown in Figure 4, the transformations within the source adapters combine Performance Parameters with the behaviour modelled within Process Models in order to generate a TIPM. Most of the Performance Parameters need to be extracted as History Data out of a business process history log provided by a Business Process Runtime (see Figure 2). However, Plan Data can be defined or modified by the user. Additionally, the user needs to specify the Target Values, Objectives and Constraints and to understand the Decision Support Results (see Figure 2). Thus, it is required that the user can set and view this Process Instance Data (see Figure 2) based on the Process Models and by using a vocabulary of his/her business process domain.

The following two subsections, therefore, describe how these Process Instance Data (see Figure 2) are represented and managed through the automated model transformation chain introduced by the abstraction provided by the TIPM; and the high degree of modularity for the implementation of the TIPM related model transformations.

Management of Process Instance Data. Decision Support Results (see Figure 2), such as a simulation based prediction that a threshold will not pass in the future, are set based on a TSPM, but need to be visualized based on the original Process Models. We therefore use the trace models generated as by-products of the transformations within source-and target adapters to navigate backward through the automated transformation chain, from the TSPM model elements to the model elements of the original Process Model. However, we have to deal with a high number of source and target adapters, and therefore a high number of model transformations, trace models and intermediate models of the different model transformation chains. Thus, a systematic solution was required to represent the linkage between source and target models of the different model transformations and the related trace models. This linkage is stored, as proposed by Bèzivin [22], into a so-called *megamodel*, which is a specialized model to represent relationship between modelling artefacts. A specialized version of such a megamodel [23] together with the trace models enables us to modularize the transformation chains between Process Models and TSPMs into as many transformation steps as one wants, as shown in [24].

Representation of Process Instance Data. The representation of the Process Instance Data has to be done in a way that the meta-model of the Process Model is not polluted. This pollution leads to contradicting the separation of concerns principle [25]. Additionally, it is not always possible to have access to the meta-model of the Process Model [24]. Thus, an approach such as UML profiles was not sufficient for our case.

In our approach all Process Instance Data is defined within separate annotation models [24] which are conforming to annotation meta-models. Therefore, for the definition of, for instance, Performance Parameters, Decision Support Results and the Objectives, we had to define a number of separate annotation meta-models [24]. Each of these meta-models is specific for the business domain which enables the user of our architecture to view and edit the different annotation models via a specific Parameter Interaction UI (see Figure 5) [24] in vocabulary he/she understands.

As described in [24], our annotation meta-models are refining the weaving meta-model provided by the ATLAS group [26]. This meta-model enables the definition of links to other models [27,28]. Thus, due to the fact that our annotation meta-models are based on the weaving meta-model, our approach enables annotation of additional information to Process Model of the BPM Environment without polluting them.

Figure 5 shows the application of our annotation models for two different process modelling tools: The BPMN based NetWeaver BPM editor and an editor for back-end processes based on a SAP proprietary modelling language. The "Start Process" node selected in the back-end process editor (see right side of Figure 5) and the annotated workload for the sales office "Chicago" is visualized as "Process Instance Occurrence" in the Parameter Interaction UI (see left side of Figure 5). Moreover, the bottom right of the figure shows a planned "Occurrence" of a process instance, between 01.10.2009 and 31.12.2009, which is 2 tasks per day.

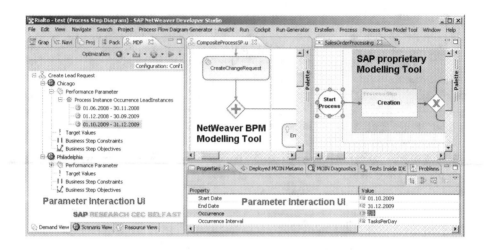

Fig. 5. Integration of the Parameter Interaction UI into two modelling tools: The NetWeaver BPM editor (middle) and an editor for back-end processes (right)

The Parameter Interaction UI therefore encapsulates the functionality to enrich the Process Modelling Tool of the BPM Environment with capabilities to visualize the annotated Process Instance Data based on the Process Model and edit some of this data. The current implementation of this annotation editor is Eclipse framework specific which restricts the application of our implementation to BPM Environments, using Eclipse based Process Modelling Tools. The main concepts can however be applied to any Process Modelling Tool.

For the non-intrusive integration of annotation models into a process modelling tool it is necessary to notify about the currently selected graphical model element to the Parameter Interaction UI. Therefore, it was required to implement a minor extension (less than 100 lines of code) for the SAP proprietary counterpart of the Eclipse Graphical Editor Framework (GEF) [29], to call the Parameter Interaction UI if the selected process flow model element is changed. This extension can be reused for numerous modelling editors. The JPASS tool is however not based on a graphical framework like GEF. We therefore additionally developed a minor extension for the JPASS tool. Hence, the only place where, in few cases, an Eclipse based process model editor needs to be modified in order to extend it with performance related decision support, is to notify about the currently selected graphical model element to the Parameter Interaction UI.

4.3 Decision Support Calculator

In the previous subsection we described how the Process Instance Data is represented as annotation models, and how we interconnect different Process Modelling Tools with multiple Performance Analysis Tools. The current subsection describes how the Decision Support Results (see Figure 2), e.g. if thresholds will be met in the future, are calculated based on the output of one or more

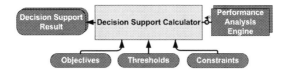

Fig. 6. Decision Support Calculator as Block Diagram [17]

Performance Analysis Tools and user provided Objectives, Constraints and Thresholds. This calculation is done by the *Decision Support Calculator*, which is depicted by Figure 6. This actor combines Performance Analysis Results with its own logic in order to output a Decision Support Result.

Based on the currently available Performance Analysis Tools for our current proposed architecture, a combination of different kinds of decision support is enabled:

- A *Threshold Checker* either executes what-if simulations by calling the Any-Logic discrete event simulation tool based on the history and plan data, or triggers analytical predictions from the LQNS tool based on average calculations from these data. The performance analysis results are compared with the user provided thresholds.
- An *Optimization Engine* executes automatically a number of what-if simulations by calling the AnyLogic Optimization Engine in order to fulfil user provided threshold but also to have the best possible result with regard to user provided Objectives. The possible configurations for the what-if simulations are restricted via the user provided Constraints.

The final Decision Support Results are also represented as annotation models which are used to enrich the original Process as described in the previous subsection. Hence, a user is, for instance, able to see which activities in a BPMN process will not fulfil certain thresholds in case of future business growth.

5 Experiences Gained

From the architectural point of view, the high degree of modularity within the proposed architecture enabled us to gain advantages in terms of extensibility and reusability. Extensibility is demonstrated when we recently extended our solution with the LQNS tool. This additional performance analysis methodology is usable for the users of all the BPM Environments which we have already extended. Reusability has been demonstrated also: The JPASS modelling tool (within the JCOM BPM Environment), has been recently extended with our architecture. The effort of writing the required transformations took less than one week of development effort. However, the effort of integrating the first Process Modelling Tool in our architecture took us around six weeks of development effort. This is due to the fact that all TIPM to TSPM transformations could be reused for the JPASS integration including some transformation steps provided by the already existing Process Model to TIPM transformation.

The cost for this high degree of reusability and extendibility, introduced with our architecture, is additional memory consumption and performance footprint-which is however, for current applications of our architecture, not yet critical.

In order to gain experiences with our tooling from the functionality point of view, we applied it for the SAP demo company called Akron Heating. Akron Heating does not exist in reality but the business processes and data of this company are maintained within SAP just as the data of a real one, for experimental purposes. Below we discuss an example combination of different what-if questions and an process optimization based on Akron Heating in order to demonstrate that the current combination of Performance Analysis Tools provided by our architecture is applicable for industrial usage.

Akron has three sales offices in the US (Denver, New York and Philadelphia). To leverage business in the USA Midwest area the head of the company decided to set up a new sales office in Chicago in September 2008. Based on the monitoring and analysis tooling, the head of sales discovers that one of the processes has not been executed with the expected efficiency across all sales units. Thus, based on a process model he/she is able to investigate the source of this poor business process performance based on predefined thresholds of Key Performance Indicators (KPI) which are e.g. indicating the historic end-to-end processing times. Based on a drilldown of these processing times, he/she finds out that especially the historic performance of one process step executed in the Chicago office is not sufficient.

Since we extended a SAP proprietary modelling editor for back-end processes with our tooling, the user is now able to investigate the impact of a number of potential changes in the process execution with a combination of different automatically generated what-if simulations (discrete-event simulations) and process optimizations. For the what-if simulation, two Scenarios are annotated (see left side of Figure 5) to the process model and transformed to the TIPM (see 4.1): one for the process executed in Philadelphia and one for the Chicago business process. It follows a description of a four step performance analysis:

- In a first step, the user does a what-if simulation in order to predict the outcome of training by reducing the annotated planned working time consumption (called Resource Demand in the TIPM) of one process step. A discrete event simulation based prediction which utilizes the AnyLogic tool shows, based on the process model, that if the training of the department in Chicago made the employees as efficient as the employees in other departments, all processing targets would be met.
- In a second what-if question the user wants to investigate, again via a discrete event simulation, if the staff in the Chicago office can handle the future business growth by increasing the planned Process Instance Occurrence (called OpenWorkload in the TIPM). The result of the simulation demonstrates that a business growth would lead to a resource problem within the department in Chicago.
- A third what-if simulation shows how our decision support tooling can help to identify if staff of other departments can compensate this resource problem in the case that some resources are shared among the departments. This is done

by modifying annotated responsibilities of the Philadelphia staff for marketing related process steps of the Chicago office (responsibilities are represented as Resource Links in the TIPM). In the TIPM the Philadelphia resource can be linked between the marketing related Steps of the Chicago process. This is possible as the staff of the two departments is represented in the TIPM independent of the formerly mentioned Scenarios.

- In the fourth step, the application of the AnyLogic optimization engine shows, via Optimization Assessment, what is the optimal sharing of resources among the departments; e.g. how many working hours have to be provided for the Chicago process related tasks by Philadelphia staff.

Additionally, we could have predicted the impact of changing a business condition, such as the lower boundary of an approval request. Also, in case the analysed process gets extended with a NetWeaver BPM process, our tool is still able to support the head of sales at Akron heating.

Furthermore, we are able to combine simulations with a bottle-neck analysis in order to indicate future bottle-necks. We anticipate to also gain industrial experiences with this additional Performance Analysis Engine.

Concluding, the possibility to combine different sophisticated decision methodologies with a number of process modelling tools provided by different BPM Environments has been identified as very beneficial. However, we identified the need for an automated History Data import for our solution which we have not implemented yet. Thus, History Data is currently annotated manually which is too time consuming for industrial application. An automated History Data import would calculate the historic Probabilities and Occurrences for a specific process, e.g. by counting the number of executed process instances. Furthermore, the working time consumptions can be calculated based on the process step durations. Additionally, resource Capacities and Resource Links can be calculated by interpreting those resources as part of the Capacity which has been used in the past. This importer also needs to provide a way to systematically deal with uncertainties in the History Data, for instance, due to a high variance or too few executed process instances. This should especially enable users to provide assumptions for cases where the confidence in the historic data is too low. Moreover, an integration of additional data sources, such as Human Resource (HR) data from the organizational management is required for future versions of our tooling. Especially, the allocation of persons to projects or organisational units needs to be accessed from HR data in order to calculate capacities. We, therefore, also require a mechanism to enable import of such additional data sources.

6   Related Work

From the application point of view, the closest related work to our knowledge is that concerned with BPM Environment such as EMC Documentum Process Suite [3] which provides simulation capabilities which are normally simplistic [6]. Our approach enables one to benefit from the know-how and functionality contained in a sophisticated performance decision support system, which enables, for example, sophisticated model simulations, optimizations and static analysis including a combination of them. The closest work to our knowledge from the architectural point of view is the PUMA architecture [30] which is based on a Core Scenario Model (CSM) [18], similar to the TIPM. However, the PUMA approach cannot be applied to BPM Environments as it is modelling Performance Parameters as UML Profiles, which are applicable only when UML models are employed as Process Models. Our approach is based on annotation models and provides a significantly higher degree of flexibility. Our approach can therefore be used to annotate any kind of Process Model, for instance, BPMN models used for NetWeaver BPM, or numerous SAP proprietary models used in existing ERP solution. Furthermore, we are able to support visualization of the Process Instance Data, such as Plan Data and the Decision Support Results, in the language a domain expert can understand and based on the original Process Models. Finally, our approach considers multiple views, namely: Objectives, Constraints and Requirements, as proposed in [16]. This enables better decision support than that provided by the PUMA approach, as we can, for instance, automatically propose optimal solutions.

7   Conclusions and Future Work

In this paper we proposed a generic architecture which enables extension of existing BPM Environments, having basic-level or no decision support, with capabilities for sophisticated performance related decision support. In case the proposed solution is applied, this decision support is executed via a mouse click. It is especially useful for resource scheduling questions, which arise particularly in the case of highly complex resource intensive processes (e.g. layered use of resources, complex workflows, etc.) or where the statistical distributions of the history data are complex. Thus, our approach helps to improve understanding of resource usages within complex business processes.

Due to the integration of sophisticated decision support tooling, existing BPM Environments can benefit from the know-how and functionality contained in such tools, which enables, for example, sophisticated model simulations, optimizations and static analysis. Furthermore, our architecture enables the integration and combination of multiple sophisticated decision support tools in an efficient way and without polluting original models with additional information for performance analysis, which is sometimes not possible [24]. Our architecture further enables to integrate sophisticated decision support tooling in such a way that it is straightforward to be used by business domain experts using the BPM Environments at runtime and design time of a business process.

Additionally, we abstract the BPM Environment itself which enables us to apply our decision support for end-to-end processes which are possibly managed with a number of BPM Environments.

We anticipate to extend our approach with a graphical indication of uncertainties in the historic performance data, which is used as input for the automatically generated performance analysis models. Such uncertainties, for instance, historic resource demands with a high variance, etc., will be presented to the user to allow input of available assumptions.

Disclaimer

References

1. Snabe, J.H., Rosenber, A., Molle, C., Scavillo, M.: Business Process Management: The SAP Roadmap (2008)
2. JCOM (2008), http://www.jcom1.com/
3. Associates, B.S.: The BPMS Report: EMC Documentum Process Suite 6.0 (2007)
4. SAP AG (2009),
   http://www.sap.com/solutions/sme/businessbydesign/index.epx
5. Fritzsche, M., Gilani, W., Fritzsche, C., Spence, I.T.A., Kilpatrick, P., Brown, J.: Towards utilizing model-driven engineering of composite applications for business performance analysis. In: Schieferdecker, I., Hartman, A. (eds.) ECMDA-FA 2008. LNCS, vol. 5095, pp. 369–380. Springer, Heidelberg (2008)
6. Harmon, P., Wolf, C.: The state of business process management (2008)
7. Banks, J., Carson, J.S., Nelson, B., Nicol, D.: Discrete-Event Simulation. Prentice-Hall, Englewood Cliffs (2005)
8. Rozinat, A., Wynn, M.T., van der Aalst, W.M.P., ter Hofstede, A.H.M., Fidge, C.J.: Workflow Simulation for Operational Decision Support Using Design, Historic and State Information. In: Dumas, M., Reichert, M., Shan, M.-C. (eds.) BPM 2008. LNCS, vol. 5240, pp. 196–211. Springer, Heidelberg (2008)
9. Balsamo, S., Marzolla, M.: A simulation-based approach to software performance modeling. In: ESEC/FSE-11th. ACM, New York (2003)
10. D'Ambrogio, A.: A model transformation framework for the automated building of performance models from uml models. In: WOSP 2005. ACM Press, New York (2005)
11. Bertolino, A., Marchetti, E., Mirandola, R.: Real-time UML-based performance engineering to aid manager's decisions in multi-project planning. In: WOSP 2002. ACM Press, New York (2002)
12. Franks, R.G.: DISSERTATION: Performance Analysis of Distributed Server Systems (1999)
13. Woodside, C.M., Neilson, J.E., Petriu, D.C., Majumdar, S.: The Stochastic Rendezvous Network Model for Performance of Synchronous Client-Server-like Distributed Software. IEEE, Los Alamitos (1995)
14. XJ Technologies: AnyLogic — multi-paradigm simulation software, http://www.xjtek.com/anylogic/

15. Fritzsche, M., Johannes, J.: Putting Performance Engineering into Model-Driven Engineering: Model-Driven Performance Engineering. In: Giese, H. (ed.) MODELS 2008. LNCS, vol. 5002, pp. 164–175. Springer, Heidelberg (2008)
16. Fritzsche, M., Gilani, W., Spence, I., Brown, T.J., Kilpatrick, P., Bashroush, R.: Towards performance related decision support for model driven engineering of enterprise soa applications. In: 15th ECBS 2008, vol. 0. IEEE, Los Alamitos (2008)
17. Knöpfel, A., Gröne, B., Tabeling, P.: Fundamental Modeling Concepts: Effective Communication of IT Systems. John Wiley & Sons, Chichester (2006)
18. Petriu, D.B., Woodside, M.: An intermediate metamodel with scenarios and resources for generating performance models from UML designs. Software and Systems Modeling 6(2) (2007)
19. Rozinat, A., Wynn, M., Aalst, W., Hofstede, A., Fidge, C.: Workflow Simulation for Operational Decision Support Using Design, Historic and State Information. In: Dumas, M., Reichert, M., Shan, M.-C. (eds.) BPM 2008. LNCS, vol. 5240, pp. 196–211. Springer, Heidelberg (2008)
20. Fritzsche, M., Jouault, F., Lämmel, R., Gilani, W.: Model Transformation Chains to integrated Performance related Decision Support into BPM Tool Chains. In: Invited submission fro the post- proceedings of the GTTSE 2009. LNCS. Springer, Heidelberg (2009)
21. Fritzsche, M., Johannes, J., Zschaler, S., Zherebtsov, A., Terekhov, A.: Application of tracing techniques in model-driven performance engineering. In: 4th ECMDA Traceability Workshop, ECMDA-TW (2008)
22. Bézivin, J., Jouault, F., Rosenthal, P., Valduriez, P.: Modeling in the Large and Modeling in the Small. In: Aßmann, U., Aksit, M., Rensink, A. (eds.) MDAFA 2003. LNCS, vol. 3599, pp. 33–46. Springer, Heidelberg (2005)
23. Barbero, F., Jouault, J.: Model Driven Management of Complex Systems: Implementing the Macroscope's Vision. In: 15th ECBS 2008. IEEE, Los Alamitos (2008)
24. Fritzsche, M., Johannes, J., et al.: Systematic usage of embedded modelling languages in model transformation chains. In: SLE 2008. LNCS, vol. 5701. Springer, Heidelberg (2009)
25. Mehr, F., Schreier, U.: Modelling of Message Security Concerns with UML. In: 9th ICEIS (2007)
26. Fabro, M.D.D., Bézivin, J., Valduriez, P.: Weaving Models with the Eclipse AMW plugin. In: Eclipse Modeling Symposium, Eclipse Summit Europe (2006)
27. Vara1, J.M., Castro1, M.V.D., Fabro, M.D.D., Marcos, E.: Using Weaving Models to automate Model-Driven Web Engineering proposals. In: ZOCO 2008/ JISBD (2008)
28. Voelter, M., Groher, I., Kolb, B.: Mechanisms for Expressing Variability in Models and MDD Tool Chains. In: MDSD in Embedded Systems (2007)
29. Eclipse, C.: Eclipse graphical editing framework (gef) – version 3.4.2 (2009), http://www.eclipse.org/gef
30. Woodside, M., Petriu, D.C., Petriu, D.B., Shen, H., Israr, T., Merseguer, J.: Performance by unified model analysis (PUMA). In: WOSP 2005. ACM Press, New York (2005)

Business Process-Based Resource Importance Determination

Stefan Fenz[1], Andreas Ekelhart[2], and Thomas Neubauer[2]

[1] Institute of Software Technology and Interactive Systems
Vienna University of Technology, A-1040 Vienna, Austria
fenz@ifs.tuwien.ac.at
http://www.ifs.tuwien.ac.at
[2] Secure Business Austria, A-1040 Vienna, Austria
{ekelhart,neubauer}@securityresearch.at
http://www.sba-research.org

Abstract. Information security risk management (ISRM) heavily depends on realistic impact values representing the resources' importance in the overall organizational context. Although a variety of ISRM approaches have been proposed, well-founded methods that provide an answer to the following question are still missing: How can business processes be used to determine resources' importance in the overall organizational context? We answer this question by measuring the actual importance level of resources based on business processes. Therefore, this paper presents our novel business process-based resource importance determination method which provides ISRM with an efficient and powerful tool for deriving realistic resource importance figures solely from existing business processes. The conducted evaluation has shown that the calculation results of the developed method comply to the results gained in traditional workshop-based assessments.

Classification: Static process analysis.

1 Introduction

As almost every business decision is based on data, reliable information technology (IT) is a prerequisite for business continuity and therefore crucial for the entire economy [1,2]. The importance of information technology brought up the urgent need to ensure its continuous and reliable operation and to protect the processed and stored information respectively. Recent research has shown the impact of security breaches on the market value of organizations. According to [3] organizations lost on average approximately 2.1% of their market value within two days surrounding security breaches. The interconnectedness of the global economic system enables information security threats such as computer viruses to proliferate in a very fast way. Due to the rising economic relevance of IT risks, organizations should strive for adequately managing these risks.

Information security risk management (ISRM) is a process which allows IT managers to balance the operational and economic costs of protective measures

U. Dayal et al. (Eds.): BPM 2009, LNCS 5701, pp. 113–127, 2009.

and achieve gains in mission capability by protecting the IT systems and data that support their organizations' mission [4]. The two main phases of this process are Risk Assessment, which focuses on risk identification and evaluation, and Risk Mitigation, which refers to prioritizing, implementing, and maintaining the appropriate risk-reducing measures. Continual evaluation and assessment are necessary to keep the required level of security and thus are cornerstones in successful risk management. As we focus in this work on the Impact Analysis, which is part of the Risk Assessment process, we will briefly state the theoretical groundwork. In the information security context risk is defined as a function of the probability of a given threat-source exercising a particular potential vulnerability, and the resulting impact of that adverse event on the organization [4]. According to NIST 800-30 the level of impact is determined by the potential mission impacts and in turn produces a value for affected IT assets and resources. This description points out the information necessary for a successful impact analysis, keen understanding and knowledge of the processes performed, and secondly, system and data criticality values of connected resources (*importance* to an organization). The importance indicates the organizational impact if the considered resource is not longer able to conduct its designated tasks (we focus on the availability aspect). Even though a great deal of research has been conducted and manifold ISRM approaches evolved in the past 30 years, gathering this data is still mostly a manual and work intensive process, relying on interviews and questionnaires with system and information owners. The following problems are connected with the determination of the importance of an organization's resources:

- Business processes are subject to constant change. While flexible workflow design is a key factor in keeping pace with modern market trends [5,6,7], it poses a major challenge for ISRM [8,9]. Changing or newly introducing business processes requires a reevaluation of the current risk situation. Resources could be used in a dangerous new context or new activities could introduce critical vulnerabilities. Considering time consuming risk assessments, companies often refrain from continuous risk evaluation.

- Detailed and correct knowledge about business processes and attached resources is required, otherwise gained risk values will be incorrect. A consistent and up-to-date documentation of processes and connected resources is often not available and time consuming and error prone to create.

- While system and information owners should have a grounded knowledge of the processes and resources in their domain, resources can be used by various processes. Aggregating the resource importance from the process- to the organization-wide level is, again, time consuming and error prone.

- Even if there are well defined rating criteria, due to the involvement of various system and information owners (e.g., multiple departments) an objective rating cannot be guaranteed.

Determining the resource importance, based on business processes, is an elemental and reoccurring step in ISRM. With regard to the identified problems our research aims at answering the following question:

- How can business processes be used to determine resources' importance in the overall organizational context?

First, we elaborate on the research question by analyzing existing approaches in the field of business process analysis (cf. Section 2). Second, we aim at developing concepts to determine the organization-wide importance of resources based on business processes and the corresponding activities (cf. Section 4). Third, the gathered research results are prototypical implemented (artifact-building) and evaluated by comparing its output to a traditional workshop-based assessment (cf. Section 5 and 6).

2   Existing Approaches

This section provides an overview how existing approaches address ISRM with focus on the resource importance determination. Today, a collection of information security risk management methods, standards and best-practice guidelines, such as CRAMM [10], NIST SP 800-30 [4], CORAS [11], OCTAVE [12], EBIOS [13], and recently ISO 27005 [14] exist. High level standards such as NIST SP 800-30 and ISO 27005 address the step of determining the importance of an organization's resources by recommending the collection of information on business processes and system/data criticality and sensitivity. Should there be no existing documentation available, such as business impact analysis (BIA) reports, interviews should be conducted with system and information owners to determine the impact level of IT systems and data in case of loss or degradation of confidentiality, integrity, and/or availability. The magnitude of impact can be assessed quantitative or qualitative.

The reference model for process-oriented IT risk management by Sackmann [8,15] connects Business Processes, IT Applications and Infrastructure, Vulnerabilities, and Threats to model IT security relevant risks and their effects on each layer. Therefore, Sackmann's reference model can be used for modeling threat consequences on business processes. The main problem of the approach is that it is not possible to describe how the modeled business processes and IT applications/infrastructure interrelate in detail. IT applications and infrastructure are assigned on the process- and not on the activity-level. Therefore, it is not possible to determine realistic importance values of the required IT applications/infrastructure, leading to biased risk values for the business process.

Another approach described in [16] uses the Tropos Goal Risk framework in the context of business continuity management. Business objects are annotated with utility values for the organization. Those high level goals can be achieved by tasks which again can depend upon resources. Negative events affect resources and thereby threaten the business goals. Utility values for goals are assigned manually by business owners in advance. Resource utility can be calculated by summing up the values generated by a resource. While this approach offers a possibility to determine resource utility the following open challenges remain: (i) no standardized business process modeling language has been used, (ii) path possibilities have not been taken into consideration, and (iii) multiple usage of a resource in one process is not addressed.

In 2003 van der Aalst et. al. [17] point out that for information-intensive products, such as insurances, loans, permits, and many other services, the relationship with the supporting workflow process is often neglected. They primary goal was to support users in designing efficient and effective workflows based on product information rather than on subjective interpretations of managers, consultants, and IT experts. Their work is insofar important for our research as they strive to provide a methodology to automatically calculate the value of process elements. A Product/Data Model with nodes representing end-products, raw materials, purchased products, and subassemblies, is the basis for their calculations. In this tree-like structure various paths lead to the top element. Costs and required throughput times of child elements define the parent's characteristics. After node characteristics (costs, flow time, probability, and constraints) have been quantitatively defined, it is possible to provide insights, such as cost or flow time, on paths to reach the top level product.

Important to mention is the approach in [18] which explicitly focuses on business process-oriented resource evaluation. To improve accuracy of risk analysis results, they argue that resources have different values according to their business contribution, department utilization and user position, and are not sufficiently defined by purchase costs or maintenance expenses. Delphi teams apply weights for the 'business process-oriented classification factors' for each resource and thereby the resource value is calculated. While this approach offers categories and a methodology to evaluate resource values it still depends on Delphi teams to analyze business processes and resources, and to assign values accordingly to their cognition and experience.

Our paper makes a first step towards addressing the shortcomings of existing approaches and provides a business process-based resource importance determination method. Based on the organization's business processes, their overall organizational importance, and the resources required by their activities, the proposed method automatically determines the organization-wide importance of the involved resources. The advantages of the proposed solution are: (i) the necessary input data is restricted to machine-interpretable business process representations including required resources and the importance of the business process, and (ii) assuming that the required input data is already available our approach provides ISRM with fast results regarding resource importance, which are based on the business processes' structure and resource involvement.

3   Preliminaries

In this paper we use Petri nets to model business processes (cf. [19,20,21]). For the purpose of this paper, places represent the current state and causal dependencies of the business process whereas transitions represent the activities involved in the considered business process. According to [20] we use the building blocks AND-split, AND-join, OR-split, and OR-join to model sequential, conditional, and parallel routing. Sequential routing deals with casual relationships between activities. Compare *A1 - P2 - A2* in Figure 1 for an example. Parallel routing uses AND-split and AND-join to model parallel activities (see the AND-split at *A4* and the AND-join at *A12* in Figure 1). Conditional routing is modeled by OR-split and OR-join building blocks to allow for routing which may vary between cases [20]. Place *P3* and *P15* in Figure 1 show a typical OR-split and OR-join. With regard to the stated research questions it is required that these typical business process building blocks are supported by our contribution.

4   Business Process-Based Determination of the Resource Importance

Based on any given business process structure, we developed a method to determine the importance of a resource in the given organizational context. The importance indicates the organizational impact if the considered resource is not longer able to conduct its designated tasks (we focus on the availability aspect). The unit which is used to express the resource importance depends on the unit used to describe the importance of the overall business process. Monetary (e.g., Euros per hour) or qualitative (e.g., high, medium, and low) ratings are amongst others an option to express the importance of the business processes and the required resources. Assigning a value for the overall business process importance is usually done by the process owner in collaboration with the management. While various factors, such as business process profit, reputation or service level agreements, can influence the decision, the final figures depend on the organization's focus. Likewise, a decision for quantitative or qualitative ratings is based on the focus and available information. This high level of flexibility allows organizations to target their individual requirements. Despite this flexibility, once an organization has made a decision, it is necessary to follow a consistent rating process throughout the organization over all processes to guarantee consistent results. Our approach expects consistent process ratings, and calculates resource importance values, dependent on the resources' business process involvement and the business process structure.

4.1   Assumptions

Before going into the details of the proposed calculation model, we have to state some requirements: the considered business process has to (i) indicate which resources are required by the included activities, (ii) be correctly modeled so that it can be mapped to a valid Petri net, and (iii) provide an importance value for the considered organizational context. Each resource has (i) a business process-wide, local importance value $I_L(R_i)$, and (ii) an organization-wide, global importance value $I_G(R_i)$. The calculation model for these variables is described in the following subsections.

4.2   Determining the Resource's Local Importance

Let A_i be Activity i, P_i Place i, R_i Resource i, $E_{P_iA_j}$ the Edge which connects Place i and Activity j, and $E_{A_iP_j}$ the Edge which connects Activity i and Place j. The local resource importance $I_L(R_i)$ refers to the resource's importance in the context of the analyzed business process. While $I_L(R_i)$ is expressed in either quantitative or qualitative values, the local importance of an activity $I_L(A_i)$ is always expressed by a value between 0 and 1. $I_L(A_i)$ is calculated by summing up the local importance values of its ingoing edges E_{PA_i} and dividing it by the amount of ingoing edges $|E_{PA_i}|$:

$$I_L(A_i) = \frac{\sum_{j=1}^{|E_{PA_i}|} I_L(E_{P_jA_i})}{|E_{PA_i}|} \tag{1}$$

Similar to $I_L(A_i)$, the local importance I_L of place P_i is determined by summing up the local importance values of its ingoing edges (how we calculate the local importance values of edges is described in Equations 3, 4, and 5). Set E_{AP_i} includes the ingoing edges E of place P_i. If $|E_{AP_i}|$ is empty, $I_L(P_i)$ is set to one (this would be the first place in the Petri net).

$$I_L(P_i) = \begin{cases} 1 & , E_{AP_i} = \emptyset \\ \sum_{j=1}^{|E_{AP_i}|} I_L(E_{A_jP_i}) & , E_{AP_i} \neq \emptyset \end{cases} \tag{2}$$

According to the previous equations, we need the local importance of all ingoing edges E of place P_i and activity A_i to calculate their local importance value $I_L(P_i)$ and $I_L(A_i)$. If edge E connects an activity and a place (potential AND-split) the local importance $I_L(E_{A_iP_j})$ equals the importance of the edge origin element A_i:

$$I_L(E_{A_iP_j}) = I_L(A_i) \tag{3}$$

If edge E connects a place and an activity (potential OR-split) the local importance $I_L(E_{P_iA_j})$ is calculated by dividing the importance of the edge origin element P_i by the amount of outgoing edges $|E_{P_iA}|$:

$$I_L(E_{P_iA_j}) = \frac{I_L(P_i)}{|E_{P_iA}|} \tag{4}$$

The developed calculation model assigns each activity, place, and edge within the considered business process a local importance value ($I_L(A_i)$, $I_L(P_i)$, $I_L(P_iA_j)$, and $I_L(A_iP_j)$). Basically these values reflect the probability that the process passes through these elements. Currently, the model assumes an uniform distribution regarding potential process execution flows at conditional routing (OR-split) elements. Example: if there is an OR-split element with two outgoing edges, each edge has a 50% chance of being used (compare Place *P3* in Figure 1).

To improve our business process-based resource importance results regarding their fit to the real world, we introduce two additional edge parameters at each OR-split: (i) pass probability for each outgoing edge $PP(E_{P_i A_j})$, and (ii) value-adding potential of each outgoing edge $VAP(E_{P_i A_j})$. $I_L(E_{P_i A_j})$ is determined by calculating the average of the pass probability $PP(E_{P_i A_j})$ and the value adding potential $VAP(E_{P_i A_j})$ of edge $E_{P_i A_j}$.

$$I_L(E_{P_i A_j}) = \frac{PP(E_{P_i A_j}) + VAP(E_{P_i A_j})}{2} \tag{5}$$

PP and VAP are expressed by a value between 0 and 1. The pass probability of all outgoing edges has to sum up to 1. The value-adding potential of all outgoing edges has to sum up to 1. By combining both values in $I_L(E_{P_i A_j})$ we are able to express besides the pass probability the value-adding potential of potential process execution flows. Each outgoing OR-split edge has to be assessed by the business process owner in a manual manner to determine (i) its pass probability based on historical process execution data, and (ii) its value-adding potential based on available relevant data and/or the business process owner's experience.

After determining the importance of each activity which is included in the considered business process we can calculate the importance of the involved resources. We assume that data about activities' resource usage is available in set M_{R_i} and that for each activity an ordered list L, containing all previous edges originating from an OR-split place, exists. For example: in the context of the business process shown in Figure 1, activity $A15$ would be associated with the list $L_{A_{15}} = \{E_{P_3 A_3}, E_{P_3 A_4}, E_{P_{16} A_{15}}\}$. For any activity combination A_x and A_y in M_{R_i} we check if L_{A_x} is included in L_{A_y} or if L_{A_y} is included in L_{A_x}. If L_{A_x} is a subset of L_{A_y} or L_{A_y} is a subset of L_{A_x} we further inspect the last element (edge) of the subset and keep the place P that it is connecting. If the superset contains exactly one edge that connects place P we can infer that the importance value of the superset is already included in the subset. Therefore, we exclude the activity that corresponds to the superset from M_{R_i}. In the next step we sum up in G_{R_i} the local importance values of those activities which share a common starting pattern and contain exactly one edge starting from the same place but differ in the targeted activity. Importance values of those activities which do not comply with the above rule (share a common starting pattern and contain exactly one edge starting from the same place but differ in the targeted activity), are also added to G_{R_i}. The local importance I_L of Resource R_i in context of Process p equals the highest importance value e included in G_{R_i}, times the overall importance I of the considered business process p.

$$I_{L_p}(R_i) = max\{e \in G_{R_i}\} * I(p) \tag{6}$$

Consider the following example in the context of the business process shown in Figure 1: $M_{R_i} = \{A3, A14, A15\}$, $L_{A_3} = \{E_{P_3 A_3}\}$, $L_{A_{14}} = \{E_{P_3 A_3}, E_{P_3 A_4}, E_{P_{16} A_{14}}\}$, $L_{A_{15}} = \{E_{P_3 A_3}, E_{P_3 A_4}, E_{P_{16} A_{15}}\}$. According to the definition above, we search for subsets but cannot find any in M_{R_i}. In the second step we build new groups starting with L_{A_3}; L_{A_3} does not share an edge only differing in its target activity and thus we create a new element in G_{R_i} with L_{A_3}'s importance value. Continuing with $L_{A_{14}}$, we find in $L_{A_{15}}$ an identical starting pattern $(E_{P_3A_3}, E_{P_3A_4})$ and the same place with a differing target activity $(E_{P_{16}A_{14}}$ and $E_{P_{16}A_{15}})$, thus we add a new element to G_{R_i} summarizing the local importance values of $L_{A_{14}}$ and $L_{A_{15}}$. As described in Section 4.1 the analyzed business process has to provide an importance value for the considered organizational context. This importance value can be quantitative (e.g. Euro per hour) or qualitative (e.g. high, medium, or low) and determines the way how the importance of the involved resources is represented. The business process owner and the management define the importance of the considered business process. Again, the importance indicates the organizational impact if the considered business process is not longer able to deliver the expected output (we focus again on the availability aspect).

4.3   Determining the Resource's Global Importance

Let P be the total number of business processes and $I_{L_p}(R_i)$ the local importance I_L of Resource R_i in context of Process p. The global importance I_G of resource R_i is calculated by summing up its local importance values $I_{L_p}(R_i)$ in the given organizational context:

$$I_G(R_i) = \sum_{p \leq P} I_{L_p}(R_i) \tag{7}$$

Finally, $I_G(R_i)$ provides a comprehensible figure on the resource's importance. The following section demonstrates the developed approach by applying it to three real-world business processes.

5   Proof of Concept

We use BOC's ADONIS tool to model the business processes for the proof of concept. ADONIS allows for attaching resource elements to business process activities and provides an export functionality which is capable of exporting the entire business process representation as an easily accessible XML file. After parsing the ADONIS business process representation into a valid Petri net, we were able to start the developed resource importance calculation. The business processes, overall importance values, and involved resources which are used in the course of the proof of concept are shown in Table 1. As an example Figure 1 shows a Petri net representation of the *Register Damage* business process. At each OR-split (Place *P3* and *P16*) we used equally distributed values for pass probability PP and value-adding potential VAP. Therefore the local importance of each outgoing edge at the places *P3* and *P16* is 0.5 (e.g., $I_L(P_3A_3) = \frac{PP(E_{P_3A_3})+VAP(E_{P_3A_3})}{2} = \frac{0.5+0.5}{2} = 0.5$). The following activity/resource combinations exist in the *Register Damage* business process: M_{PCC} ={A6, A7, A8, A9, A11, A12, A16, A17, A18}, M_{NS} ={A16, A17}, M_{CD} ={A6, A7, A18}, M_{PD} ={A8}, M_{HD} ={A9}, and M_{ED} ={A16, A17}.

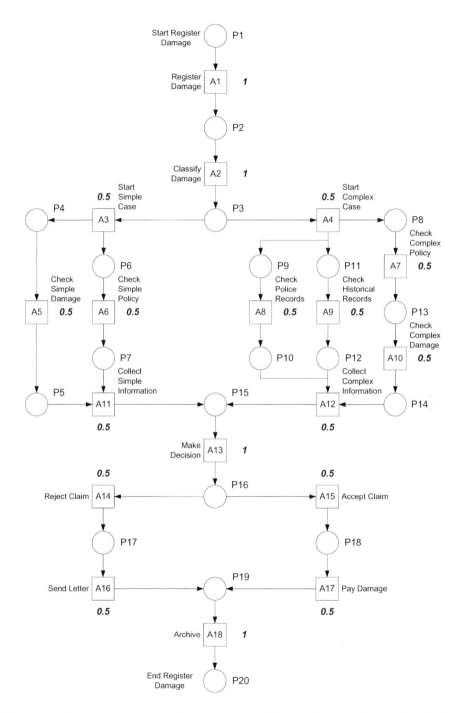

Fig. 1. Petri net representation of business process 'Register Damage' (taken from [21])

Table 1. Business processes, their organization-wide importance and involved resources

Business Process	Importance	Resources
Register Damage	300 €/h	PC-Consultant (PCC) Notification-Server (NS) Client-Data (CD), Police-Data (PD), Historical-Data (HD), Employee-Data (ED)
Consultant Assignment	100 €/h	PC-Reception (PCR) Notification-Server (NS) Client-Data (CD), Employee-Data (ED), Appointment-Data (AD), Historical-Data (HD)
Conclusion of Contract	200 €/h	PC-Consultant (PCC) Notification-Server (NS) Client-Data (CD)

The XML-representation of each business process has been used as input data for the prototype. According to the developed calculation schemes the local importance of each resource is calculated. As an example we will show how the importance of the Police-Data (PD) resource in the context of the Register Damage business process is calculated.

1. **Context Determination:** According to Table 1 and M_{PD}, the Police-Data resource is used in activity A8 of the Register Damage process.
2. **Local activity importance determination:** The local importance of A8 is determined by its incoming edge $I_L(E_{P_9 A_8}) = \frac{I_L(P_9)}{|E_{P_9 A}|} = \frac{0.5}{1} = 0.5$. The importance of place P9 has been calculated based on the importance of activity A4, which has been calculated based on the importance of place P3 and so on. As described in the previous paragraphs the outgoing edge importance of P3 has been calculated based on equally distributed values for pass probability and value-adding potential.
3. **Local resource importance determination:** After creating G_{PD} on basis of M_{PD} (cf. Section 4.2) the local, i.e. business process wide, importance I_L of the Police-Data resource PD in context of the Register Damage process equals the highest importance value included in G_{PD}, times the overall importance I of the Register Damage process p: $I_{L_{RD}}(PD) = max\{e \in G_{PD}\} * I(p) = 0.5 * 300€/h = 150€/h$.
4. **Global resource importance determination:** The global importance of the Police-Data resource is calculated by summing up its local importance values $I_{L_p}(PD)$ in the given organizational context. Since the Police-Data resource is only used in the Register Damage process the global importance equals its local importance in the context of the Register Damage process: $I_G(PD) = \sum_{p \leq P} I_{L_p}(PD) = 150€/h$.

Table 2. Local Resource Importance Results

Business Process	Local Resource Importance
Register Damage	PC-Consultant (1)
	Notification-Server (1)
	Client-Data (1), Police-Data (0.5), Historical-Data (0.5), Employee-Data (1)
Consultant Assignment	PC-Reception (1)
	Notification-Server (1)
	Client-Data (1), Employee-Data (1), Appointment-Data (0.5), Historical-Data (0.25)
Conclusion of Contract	PC-Consultant (1)
	Notification-Server (1)
	Client-Data (1)

Table 3. Global Resource Importance Results

Resource	Global Resource Importance
PC-Consultant	300€/h + 200€/h = **500€/h**
PC-Reception	100€/h = **100€/h**
Notification-Server	300€/h + 100€/h + 200€/h = **600€/h**
Client-Data	300€/h + 100€/h + 200€/h = **600€/h**
Police-Data	**150€/h**
Historical-Data	150€/h + 25€/h = **175€/h**
Employee-Data	300€/h + 100€/h = **400€/h**
Appointment-Data	**50€/h**

Table 2 shows the local resource importance value results in the context of the given business processes. Each value (potential range: 0 - 1) is derived, as shown in the previous example, from the business process activity involving the considered resource and having the maximum activity local importance value.

The local importance values of each resource are used to aggregate them to an organization-wide global resource importance value. Table 3 shows the calculation results. Based on the structure and importance of the considered business processes the results show the organization-wide impact if one of the involved resources is not longer available to the organization.

According to the results, the notification server, client data, and consultant PC are the most valuable resources in the organization (600€/h and 500€/h). The unavailability of appointment data would cause the least impact on the organization (50€/h).

6   Evaluation

To evaluate the developed concepts we compare the results of the corresponding prototypical implementation to the results gained in the course of a

Table 4. Global Resource Importance Evaluation Results

Resource	Participant 1	Participant 2	Participant 3
PC-Consultant	500€/h (83%)	500€/h (83%)	114.55€/h (86%)
PC-Reception	100€/h (17%)	100€/h (17%)	19.3€/h (14%)
Notification-Server	600€/h (100%)	600€/h (100%)	133.85€/h (100%)
Client-Data	600€/h (100%)	600€/h (100%)	133.85€/h (100%)
Police-Data	150€/h (25%)	150€/h (25%)	22.5€/h (17%)
Historical-Data	175€/h (29%)	175€/h (29%)	23.6€/h (18%)
Employee-Data	400€/h (67%)	400€/h (67%)	75.55€/h (56%)
Appointment-Data	50€/h (8%)	50€/h (8%)	7.5€/h (6%)

traditional workshop-based assessment. Three business processes including their organization-wide importance and required resources (see Section 5) have been provided to the workshop participants. The following steps have been performed at the workshop-based assessment: (i) introduction and definition of the workshop goal → business process-based determination of resource importance values, (ii) definition of the *importance* term in the context of the workshop, (iii) manual process analysis by workshop participants → each participant is required to determine the importance of the resources involved in each business process, and (iv) determination of organization-wide resource importance values → the participants are required to aggregate the results of the previous step to organization-wide resource importance values.

Table 4 shows the global resource importance results of each workshop participant. Participant 1 and 2 intuitively use an approach similar to our proposed solution. Since Participant 3 used another calculation model to determine the global resource importance we related each global resource importance result to the most important one. Although Participant 3 used a different calculation model, the relative results differ only slightly from ours. It took every participant about 9 minutes to calculate the importance values of each resource in the local and global context. The subsequent discussion has been dominated by the limitations of our proposed calculation model: (i) the model does not incorporate down-time costs of activities; it ignores the fact that resource down-times of later activities are normally associated with less costs than resource down-times of early activities, (ii) the model does not incorporate the duration of activities; similar to Limitation (i) the model ignores that resources required by long activities are more crucial than resources required by short activities, and (iii) it is not guaranteed that the calculation results reflect the real world importance of the considered resources. Although, Limitation (i) and (ii) could be easily incorporated into the existing calculation model, we decided to accept these limitations at this stage of research since we want to keep the necessary input data at a minimum. Limitation (iii) reflects the fundamental problem of modeling the reality by business processes. As organizations and their work flows are dynamic, business processes have to be continuously adapted to match reality. Using business processes for the resource importance determination in the

ISRM context requires up-to-date and realistically modeled business processes to calculate realistic importance values for the involved resources.

7   Conclusions

ISRM heavily depends on realistic impact values representing the resources' importance in the overall organizational context. Business processes are widely used as a structured flow of organizational activities, which support business goals and are enabled by resources (cf. [22]). Therefore, the central research question of this paper was: How can business processes be used to determine resources' importance in the overall organizational context? Our paper makes a first step towards a business process-based resource importance determination. Based on the organization's business processes, their overall organizational importance, and the resources required by their activities, the proposed method automatically determines the organization-wide importance of the involved resources.

The *advantages* of the developed solution are: (i) the necessary input data is restricted to machine-interpretable business process representations including required resources and the importance of the business process, and (ii) assuming that the required input data is already available our approach provides ISRM with fast results regarding resource importance, which are based on the business processes' structure and resource involvement. The conducted evaluation reveals the following *limitations* of our contribution: (i) activity down-time costs are not incorporated, (ii) activity duration is not considered, and (iii) it is not guaranteed that the calculation results reflect the real-world resource importance, due to the fundamental problem of business process modeling: reflecting the dynamic reality by a model. Although, our model could be easily extended to address Limitation (i) and (ii) we decided to accept the limitations at the current stage of research to keep the necessary input data to a minimum.

Further research will empirically test our proposed solution by conducting case studies in the Austrian social security insurance sector. The gathered research results will be used to refine our approach for determining resource importance values in the ISRM context. Second, we will address the identified limitations and extend this approach to integrate the time factor (e.g., down-time costs and activity duration). Third, we will research on how to express the importance of business processes and resources. Although we used quantitative units in this paper, we do not want to exclude qualitative rating schemes. Fourth, we will extend our research from the availability to the confidentiality perspective. One of the next research questions will be: How can business processes be used to determine resources' confidentiality in the overall organizational context?

Acknowledgment

The authors would like to thank Sigrun Goluch, Gernot Goluch, Stefan Jakoubi, and Simon Tjoa. This work was supported by grants of the Austrian Government's FIT-IT Research Initiative on Trust in IT Systems under the contract

813701 and was performed at the research center Secure Business Austria funded by the Federal Ministry of Economy, Family and Youth of the Republic of Austria and the City of Vienna.

References

1. Gerber, M., von Solms, R.: Management of risk in the information age. Computers & Security 24, 16–30 (2004)
2. Commission of the European Communities: Communication from the Commission to the Council, The European Parliament, The European Economic and Social Committee and the Committee of the Regions 'A strategy for a Secure Information Society - Dialogue, partnership and empowerment". COM (2006) 251 final (2006)
3. Cavusoglu, H., Mishra, B., Raghunathan, S.: The effect of internet security breach announcements on market value: Capital market reactions for breached firms and internet security developers. International Journal of Electronic Commerce 9(1), 69–104 (2004)
4. Stoneburner, G., Goguen, A., Feringa, A.: Risk management guide for information technology systems. NIST Special Publication 800-30, National Institute of Standards and Technology (NIST), Gaithersburg, MD 20899-8930 (2002)
5. Voorhoeve, M., Van der Aalst, W.: Ad-hoc workflow: problems and solutions. In: Proceedings of the Eigth International Workshop on Database and Expert Systems Applications, pp. 36–40. IEEE Computer Society, Los Alamitos (1997)
6. van der Aalst, W.: Generic workflow models: How to handle dynamic change and capture management information? In: Conference on Cooperative Information Systems, pp. 115–126 (1999)
7. Mills, S.: The future of business - aligning business and it to create an enduring impact on industry. Technical report, IBM (2007)
8. Sackmann, S.: A reference model for process-oriented it risk management. In: 16th European Conference on Information Systems, ECIS 2008 (2008)
9. Al-Mashari, M.: Business process management - major challenges. Business Process Management Journal 8, 411–412 (2002)
10. Farquhar, B.: One approach to risk assessment. Computers and Security 10(10), 21–23 (1991)
11. Fredriksen, R., Kristiansen, M., Gran, B.A., Stølen, K., Opperud, T.A., Dimitrakos, T.: The CORAS framework for a model-based risk management process. In: Anderson, S., Bologna, S., Felici, M. (eds.) SAFECOMP 2002. LNCS, vol. 2434, pp. 94–105. Springer, Heidelberg (2002)
12. Alberts, C., Dorofee, A., Stevens, J., Woody, C.: Introduction to the OCTAVE approach. Technical report, Carnegie Mellon - Software Engineering Institute, Pittsburgh, PA 15213-3890 (2003)
13. DCSSI: Expression des Besoins et Identification des Objectifs de Sécurité (EBIOS) - Section 2 - Approach. General Secretariat of National Defence Central Information Systems Security Division, DCSSI (2004)
14. ISO/IEC: ISO/IEC 27005:2007, Information technology - Security techniques - Information security risk management (2007)
15. Sackmann, S.: Assessing the effects of it changes on it risk - a business process-oriented view. In: Multikonferenz Wirtschaftsinformatik (MKWI 2008), pp. 1137–1148. GITO-Verlag, Berlin (2008)

16. Asnar, Y., Giorgini, P.: Analyzing business continuity through a multi-layers model. In: Dumas, M., Reichert, M., Shan, M.-C. (eds.) BPM 2008. LNCS, vol. 5240, pp. 212–227. Springer, Heidelberg (2008)
17. Reijers, H.A., Limam, S., van der Aalst, W.M.P.: Product-based workflow design. J. Manage. Inf. Syst. 20(1), 229–262 (2003)
18. Eom, J.-H., Park, S.-H., Han, Y.-J., Chung, T.-M.: Risk assessment method based on business process-oriented asset evaluation for information system security. In: Shi, Y., van Albada, G.D., Dongarra, J., Sloot, P.M.A. (eds.) ICCS 2007. LNCS, vol. 4489, pp. 1024–1031. Springer, Heidelberg (2007)
19. van der Aalst, W., van Hee, K.: Business process redesign: a petri-net-based approach. Computers in Industry 29, 15–26 (1996)
20. van der Aalst, W.: The application of Petri nets to workflow management. The Journal of Circuits, Systems and Computers 8(1), 21–66 (1998)
21. van der Aalst, W.: Process-oriented architectures for electronic commerce and interorganizational workflow. Information Systems 24(8), 639–671 (1999)
22. zur Muehlen, M., Rosemann, M.: Integrating risks in business process models. In: ACIS 2005 Proceedings (2005)

Case Study and Maturity Model for
Business Process Management Implementation

Michael Rohloff

University of Potsdam, August-Bebel-Str. 89,
14482 Potsdam, Germany
michael.rohloff@wi.uni-potsdam.de

Abstract. This paper presents the implementation of Business Process Management in a large international company. The business case illustrates the main objectives and approach taken with the BPM initiative. It introduces a process management maturity assessment which was developed to assess the implementation of Business Process Management and the achievements. The maturity model is based on nine categories which comprehensively cover all aspects which impact the success of Business Process Management. Some findings of the first assessment cycle are pinpointed to illustrate the benefits and best practice exchange as a result of the assessment.

Keywords: Business Process Management Implementation, Maturity Model.

1 Introduction

Business Process Management is a management practice which encompasses all activities of identification, definition, analysis, design, execution, monitoring & measurement, and continuous improvement of business processes. Consequently Business Process Management encompasses not only the analysis and modeling of business processes but also the organizational implementation, leadership and performance controlling [1, 2, p. 7f.]. Although it is a well-known and largely used practice there is an ongoing discussion on how to best implement Business Process Management. Due to the comprehensive nature of BPM a variety of different approaches exist (e.g. Business Process Reengineering (BPR); Continuous Process Improvement, Workflow Management, reference modeling and implementation of standard enterprise applications).

Facing the importance and vital role of Business Process Management for the transformation and organizational change of enterprises the question arises how different organizations perform in their development of Business Process Management. The notion of maturity has been proposed in other approaches to assess an organizations state in terms of implementing a specific program or the quality of a process.

A prominent and widely used model is the Capability Maturity Model developed by the Software Engineering Institute at Carnegie Mellon University [3]. This model was originally developed to assess the maturity of software development processes. Over the years it was extended to other domains. Today the Capability Maturity

U. Dayal et al. (Eds.): BPM 2009, LNCS 5701, pp. 128–142, 2009.
© Springer-Verlag Berlin Heidelberg 2009

Model Integration is an approach for the assessment and improvement of product development processes in general. A number of additional maturity models were developed which cover other areas like the CMMI Acquisition Model (CMMI-AM) or the People Capability Maturity Model (P-CMM) for personal management and development to name a few. Today, CMMI is widely used in practice to evaluate and to improve (software) development processes [4, 5, 6, 7, 8].

CMMI uses standardized question catalogues and evaluation criteria to assess an organizations product development process and to work out the strengths and weaknesses. It helps to define improvement measures and to plan the implementation in an organization. The CMMI introduces the concept of five maturity levels defined by special requirements that are cumulative.

In recent years a number of maturity models for Business Process Management have been proposed [9, 10, 11, 12, 13, 14, 15, 16, 17]. Most of the models focus on only one dimension for measuring BPM maturity and very few applied studies are known. Exceptions are the Business Process Management Maturity Model (BPMM) of the OMG [9], the Process Audit of Hammer [11], and the maturity model of Rosemann et al. [13, 14, 15, 16].

This paper presents the implementation of Business Process Management in a large international company, undertaken as a corporate, company wide project within Siemens AG.

The next section outlines the objectives and the overall approach for implementing business process management. It introduces the process framework including the reference process house and the overall structure and content of the BPM implementation process.

Section 3 gives an overview of the process management maturity assessment model which was developed in order to assess and to derive improvement measures for the Business Process Management in the company. The assessment process and some results of the assessments are presented to illustrate some benefits of the approach.

2   Implementation of Business Process Management

2.1   The Business Process Management Initiative at Siemens AG

The Siemens AG is engaged in different business sectors with a very broad and diverse product and service spectrum. It is a global company with regional representations in more than 190 countries (for a short overview see [18]). Over the years the process and IT landscape has developed differently in the business units and regions. With the Business Process Management activities a redesign, alignment and optimization of business processes and a better process standardization and utilization of synergies is intended.

Central element of the Business Process Management Initiative was the development of a Siemens Process Framework [19] which consists of a reference process house (RPH) and common methods for process management across the company. These activities, with the development of a reference process house (RPH) in its core, are part of a comprehensive process management initiative [18, 2, pp. 241-252]. The initial company wide activities for process standardization started in

2000 with the E-Business initiative "Generic Business Processes". The primary focus was on the definition of the Supply Chain Management processes based on the Supply Chain Operational Model (SCOR). In the following years the process activities where extended to the Customer Relationship Management and the Product Lifecycle Management. Finally, the activities were taken up and consolidated under the leadership of corporate CIO and the development of a comprehensive reference process house covering all business processes was accomplished [19]. The primary objective was to leverage synergies and cost potentials with a common organization and process coordination, and the definition of reference processes.

Reference models are increasingly used in industrial practice and leave the area of research ([20, 21], see the overview in [22, pp. 393f.], for reference modeling projects see [23]). In practice reference models for processes have particular relevance, e.g. [24, 25, 26, 27]. For the development of the Siemens Reference Process House the Supply Chain Operational Model [24] was a fundamental basis.

The Siemens Process Framework (SPF, figure 1), with its binding set of principles and definitions for the overarching management of processes, provides the basis for a uniform implementation of process management within Siemens. The core component of the SPF is the Reference Process House (RPH). It contains the definitions of all processes and is structured into the following process categories:

- Management Processes
- Customer Relationship Management (CRM) Processes
- Supply Chain Management (SCM) Processes
- Product Life Cycle Management (PLM) Processes
- Support Processes

These reference process definitions are fundamental for process standardization and provide a stable basis for process management. They are subject to a cascaded rollout and refinement in the business groups and regions. Incorporating process definitions, guidelines for documentation and modeling of processes, and a binding decision structure for process standardization, the framework is the basis for:

- Configuration and design of specific business processes (e.g. CRM, PLM, SCM) and end-to-end business process chains
- Redesign of processes based on commonly defined standards for to-be processes
- Common language and common understanding of processes
- Realization of the saving potentials identified through - faster implementation of standard processes - alignment of applications - standardization and cost reduction across matrix organization (synergy effects)
- Comprehensive benchmarking and best practice sharing.

The process management methods of the Siemens Process Framework represent a comprehensive set of tools (including ARIS [27]), concepts, conventions, procedures, and guidelines which are needed for any implementation and operation of process management in the Siemens organization. With the description of all roles and responsibilities required for effective process management on strategic and operational levels the SPF provides a blueprint for process management organization in the groups and regions. It ensures clear communication, decision, and escalation processes.

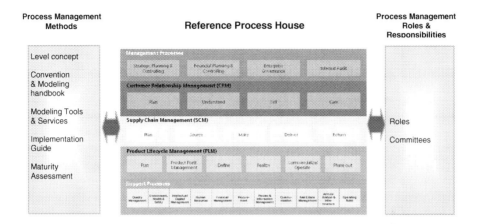

Fig. 1. Siemens Process Framework (SPF)

The main objective of the introduction of Business Process Management is to increase the effectiveness and efficiency of all business processes of the organization. From an operational point of view, process management is about having defined processes, measuring their performance, and improving them incrementally as part of daily business. It is also about defining performance goals for processes "top-down", based on benchmarking results or strategic goals derived from corporate initiatives, and performing major re-engineering activities on processes to close existing performance or cost gaps. Process standards and a common process framework are a fundamental basis for a systematic design and optimization of results, processes, and resources.

Most efficiency and effectiveness problems in an organization have their origin in non-mastered processes. A proper implementation leads to the mastery of processes with regard to lower non-conformance, as well as to high reliability and safety, and results in reduction of process costs, process cycle times, and improvement of quality.

Process standardization affects the strategic levers operational excellence and active management of synergies and supports the vertical and horizontal strategies of Siemens. This is achieved by the cascaded definition and rollout approach of the Process Initiative based on the Reference Process House. The implementation of Business Process Management based on the Siemens Process Framework results in a number of benefits which where pursued with the Process Initiative.

- Establish a Process Management Community within the business units and regions to coordinate and optimize local, regional, and headquarter process improvement initiatives.
- Provide a common reference framework for supporting and coordinating all process related projects in the business units and regions created by different initiatives.
- Present a uniform appearance to customers and business partners through Siemens wide standardized process implementation.
- Provide standard service levels to the global customers.

- Enable best practice sharing across all business units and regions.
- Provide opportunity for shared services and an improved lean IT landscape through process standardization.

2.2 BPM Process and Implementation Topics

Experience shows that business transformations are often a consequence of good process management. Thus, the implementation of process management itself has to be organized as a business transformation program covering all relevant aspects of an organization's development. These aspects have to be addressed by implementation topics which are dependent on each other with regard to their contents. All theses issues are addressed by Business Process Management implementation guidelines (see Process Management Implementation Guide [28]). The following gives a short overview on the different implementation topics.

- *Process Management Organization:* Establish process management roles & bodies according to the Siemens Process Framework and assign the responsible persons.
- *Process Documentation & Standardization:* Develop consistent and organization-wide valid process definitions at least for the portfolio processes. Drive the standardization and alignment of business processes. Establish a process house based on the Reference Process House and where necessary more detailed process definitions addressing at least the portfolio processes. Initiate process improvement initiatives for relevant processes of the process portfolio covering: visualization of as-is processes as required, derivation of improvement potentials & measures, design & implementation of to-be processes.
- *Process Portfolio & Optimization:* Select, assess, and prioritize the processes which have to be standardized and optimized.
- *Target Setting & Incentives:* Check and amend target setting and incentive systems. Define process harmonization/ standardization and process performance goals. Implement process target agreements, define related incentives.
- *Methods & Tools:* Provide standard methods and tools required for the operation of process management and according to the Siemens Process Framework guidelines (e.g. a RPH database and ARIS tools).
- *Qualification & Training:* Derive competency development measures for the persons involved in process management. Define and conduct target group specific qualification programs. Verify the success.
- *Communication:* Provide target group specific information about objectives, content, roles & responsibilities, and progress of process management to create awareness and support the implementation.
- *Process Performance Controlling:* Define key performance indicators (KPI) and metrics for the portfolio processes derived from business goals and strategies. Introduce a continuous KPI-based performance measurement and assessment for the processes.
- *Process Management Maturity Assessment:* Conduct process management maturity assessments of the organization. Derive & implement improvement measures. Repeat process management maturity the objectives assessments periodically.

Only if each of these topics are planned and implemented to a certain degree and in a coordinated way, the effects necessary for overall success are achieved. The overall maturity degree of a process management implementation is therefore directly linked to the maturity degree of each of the implementation topics (see next section). Of course, the business situation, the cultural environment, and the readiness of an organization are additional boundary conditions which have to be considered in the setup of the contents and the timeframe of the implementation program.

3 A Maturity Model for Business Process Management

3.1 Process Management Maturity Assessment

The assessment of the maturity of all activities related to Business Process Management is an essential element of the BPM implementation process. The so-named "Process Management Maturity Assessment" (PMMA) has its focus on the assessment of the organizational implementation of all Business Process Management activities. In contrast most maturity models solely focus on the performance assessment of a specific business process. The process performance of a business process is addressed as a separate category in the implementation process. In this respect the business process performance measurement is one category among others to be addressed in a BPM maturity assessment.

The Process Management Maturity Assessment [29] provides a methodology for a structured analysis and objective assessment of the achieved implementation status of process management (process management maturity) and the compliance with the Siemens Process Framework (SPF [19]) standards [18, p. 107 f., 2, p. 337 f.]. The major objective of the PMMA is the identification of need for action and derivation of measures for process management improvement, as well as the identification of requirements for further support. It serves as a driver for the process initiative. The following objectives are pursued with the PMMA approach:

- to assess the maturity of Business Process Management and the processes
- to monitor the advancement of the process initiative and to derive further fields of actions
- to reveal the potential for best practice sharing
- to motivate and increase the awareness for process management among the involved parties like management, process drivers, and users.

At the time of implementing the Process Initiative no holistic process management maturity model existed which would cover all relevant BPM implementation issues outlined in section 2.2. The BPMM model of the OGM, the maturity model of Rosemann et al. and the Process Audit of Hammer evolved in parallel to the own development of the Process Management Maturity Assessment.

The PMMA follows the principle structure of the Capability Maturity Model Integration Method of the Software Engineering Institute at Carnegie Mellon University (CMMI) but provides a holistic assessment of all areas relevant for BPM based on a comprehensive set of criteria. As an indicator for process maturity, a five step model is applied in the same fashion as the CMMI model.

The PMMA consists of nine categories with one to three sub-categories each. The PMMA categories and sub-categories correspond to the implementation topics of the Process Management Implementation Guide:

- Process Portfolio & Target Setting
- Process Documentation
- Process Performance Controlling
- Process Optimization
- Methods & Tools
- Process Management Organization
- Program Management, Qualification, Communication
- Data Management
- IT-Architecture

For every sub-category, each maturity level 1-5 is clearly defined in a to-be status by a set of criteria. These descriptions, as well as examples for questions and possible deliverables, are combined in worksheets. A tool based on MS-Office products was developed to support the assessment process.

Figure 2 outlines the five overall PMMA maturity levels which consolidate the detailed maturity levels of the categories.

For a sub-category, all defined criteria of a maturity level must be met to achieve the respective level. The overall result of a PMMA will be stated in a maturity level grade (e.g. 3,2). The pre-decimal position states that 100% of all sub-categories fulfill the criteria of level 3 (bottleneck is the lowest value for a sub-category). The decimal place states the percentage of fulfilled sub-categories of the successive level (e.g. 20% of level 4). The achievement of higher levels in sub-categories (e.g. 5) is not reflected in the overall grade.

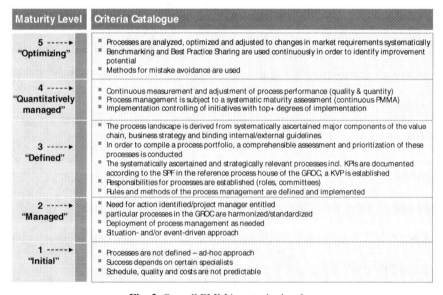

Maturity Level	Criteria Catalogue
5 ----► "Optimizing"	※ Processes are analyzed, optimized and adjusted to changes in market requirements systematically ※ Benchmarking and Best Practice Sharing are used continuously in order to identify improvement potential ※ Methods for mistake avoidance are used
4 ----► "Quantitatively managed"	※ Continuous measurement and adjustment of process performance (quality & quantity) ※ Process management is subject to a systematic maturity assessment (continuous PMMA) ※ Implementation controlling of initiatives with top+ degrees of implementation
3 ----► "Defined"	※ The process landscape is derived from systematically ascertained major components of the value chain, business strategy and binding internal/external guidelines. ※ In order to compile a process portfolio, a comprehensible assessment and prioritization of these processes is conducted ※ The systematically ascertained and strategically relevant processes incl. KPIs are documented according to the SPF in the reference process house of the GROC, a KVP is established ※ Responsibilities for processes are established (roles, committees) ※ Rules and methods of the process management are defined and implemented
2 ----► "Managed"	※ Need for action identified/project manager entitled ※ particular processes in the GROC are harmonized/standardized ※ Deployment of process management as needed ※ Situation- and/or event-driven approach
1 ----► "Initial"	※ Processes are not defined – ad-hoc approach ※ Success depends on certain specialists ※ Schedule, quality and costs are not predictable

Fig. 2. Overall PMMA maturity levels

While the maturity levels of figure 2 document the overall assessment and consolidate the maturity assessment of the different categories, a more detailed look on each of the categories is provided by radar screens (see figure 4). Detailed criteria and a set of questions exist to assess the maturity level for each of the categories. Table 1 summarizes on what needs to be accomplished for a maturity level 3 in each category.

In general, most CMMI based maturity models define five maturity levels and associate a higher level with a higher maturity and a better performing organization. Crawford [30] argues that this can be a misleading interpretation. An organization should aim for a particular maturity level in relation to its organizational strategies and objectives. A detailed view on the implications of the current maturity level based on the identified shortcomings and weaknesses is proposed in order to derive strategies for improvement.

Table 1. PMMA categories and maturity level 3 achievements

PMMA Scope	PMMA Content of Maturity Level 3
Process Portfolio & Target Setting	In order to compile a process portfolio, a comprehensible assessment and prioritization of these processes is conducted
Process Documentation	The systematically ascertained and strategically relevant processes incl. KPIs are documented according to the SPF in the reference process house.
Process Performance Controlling	A systematic procedure to identify KPIs out of the numerous metrics is defined.
Process Optimization	Benchmarks are defined and improvement levers identified.
Methods & Tools	The process landscape is derived from systematically ascertained major components of the value chain, business strategy and binding guidelines.
Process Management Organization	Responsibilities for processes and process management are established
Program Management, Qualification, Communication	The activities for introduction and further development of process management are coordinated systematically by a program and project management.
Data Management	Harmonization/ standardization of data content and formats, clearly defined responsibilities for data definition, content and consistency.
IT Architecture	Requirements from process management are definitive for IT target architecture. The migration requirement for the IT architecture is derived from deviations between as-is and target architecture.

3.2 Maturity Assessment: Initial Study and Findings

In addition to the workout of the PMMA, a qualification and training program was set up to build a pool of certified assessors who can conduct the PMMA. A roadmap was defined when to assess each organizational unit, eventually covering the entire organization. It is planned to repeat the PMMA once a year to track and drive the improvement. Between two and three days are required to prepare, conduct, and evaluate the process management assessment for a particular unit under review. The PMMA will be conducted based on interviews with the management of the units, the Process Owners/ Process Executives for the Business-, Management & Support Processes, and the Process Framework Executive.

The initial assessment analyzed 22 organizational units in the business groups and 29 in the regions in 2006. The PMMA result can be documented in a radar chart showing the level achievement for each category. Moreover, with the help of PMMA highlights and lowlights for each category and suitable actions can be derived and initiated to improve the implementation status of process management (process management maturity).

The results for the analyzed units of the business groups in figure 4 show an overall maturity level ranging below maturity level 3. Although all units participated in the Process Initiative and have implemented Business Process Management it shows that it is quite some effort in terms of time, resources, and people involved to achieve organizational performance. Also, one has to keep in mind that due to the method of measurement the overall maturity level cannot be higher than the lowest maturity level in any category.

A more detailed view is provided by a radar chart showing the level of achievement for each category. Figure 3 shows the assessment for two selected units providing insights in strengths and shortcomings; e.g. one organizational unit is quite strong in Process Portfolio & Target Setting (level 4) and in Process Management Organization (level 5) and the other in Process Documentation (level 5).

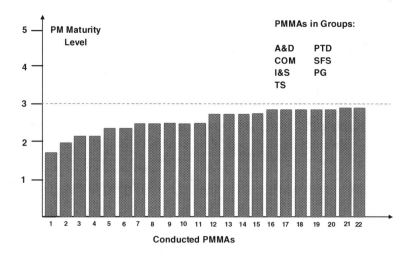

Fig. 3. PMMA assessment for analyzed units (consolidated excerpt)

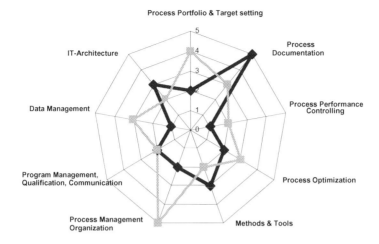

Fig. 4. Detailed PMMA of different categories (example for two units)

The assessment provides a detailed analysis which helps to identify strength and weaknesses and allows comparing the performance of organizations. Thus, it provides a sound basis for best practice sharing.

Table 2 summarizes some strengths and weaknesses for the different categories revealed across the assessed organizational units.

Experiences with the first assessment cycle are promising in terms of acceptance, ease of use, and coverage of BPM impact factors. The Process Management Maturity Assessment is regarded by management and employees as an important part of the overall BPM implementation process in the company. It underlines the importance of coherent Business Process Management activities for company performance.

The assessment results gained with limited effort provided a reasonable transparency on the BPM activities and performance of the assessed organizational units. In general the assessment helps organizations to learn from one another in terms of good and poor performance by understanding the performance of an organization and the underlying reasons. In the case of Siemens it helped to identify best practices in BPM within the company which could be adopted by other organizational units in order to improve performance.

3.3 Comparison of Maturity Models

The proposed Process Management Maturity Assessment advances most of the maturity models which are based on a limited set of criteria, Only the Business Process Maturity Model of the OMG, the Process Audit of Hammer, and the maturity model of Rosemann et al. cover also a broader range of BPM factors. All three models were in progress of development at the time of PMMA development.

Table 2. Strengths and weaknesses in the BPM categories

Category	Strength	Weakness
Process Portfolio & Target Setting System	Specific tools, e.g. scorecards, as basis for deployment from business strategy	No systematic deployment of process portfolio Individual Training necessary Objectives are often monetary
Process Documentation	Process description contains all relevant information (e.g. Input/Output, Interfaces)	Sometimes lacking parts (milestones, metrics or interfaces)
Process Performance Controlling	Milestones and metrics are defined and used for controlling of most processes	No integrated measurement system; focusing on process cost drivers to be enhanced
Process Optimization	CMMI Assessments in PLM Process Benchmarking with internal and external partners	Organizational obstacles for end-to-end process optimization (interfaces!)
Methods & Tools	ARIS often in use Several process management methods are used (e.g. six sigma)	Process description not based on RPH or at least level 4 processes not linked to RPH or documented in ARIS. Level concept/ conventions not used
Process Management Organization	Process Management Roles are defined; organization is process oriented	Process responsibility not clearly defined; no systematic job rotation between roles
Program Management, Qualification, Communication	Process Management reports directly to BU Head; communication plan regarding process management	Roadmap for migration to SPF is missing; no qualification plan available No internal communication
Data Management	Responsibility for data content and format defined Necessary measures are set up	No mechanism to check data quality or integrity No alignment with process landscape Too few resources
IT-Architecture	Requirements of process management are fully covered Migration measures derived	IT architecture not defined, nor communicated – process to derive the to-be it-architecture not defined

End of 2007 the Object Management Group (OMG) released the Business Process Management Maturity Model [9]. It is a model to assess the maturity of business process management. The model is structured into five process area threats:

- Organizational Process Management: foundation and development of process management
- Organizational Business Management: planning, steering and resource allocation at enterprise level
- Domain Work Management: management of product & service deployment and delivery
- Domain Work Performance: operational level of product & service delivery and support
- Organizational Support: all supporting activities for controlling the core activities

BPMM defines objectives for each process area thread. This is supplemented by practices how to reach these objectives. Overall the BPMM offers a variety of recommendations for a Business Process Management implementation. On the other hand it leaves some deficiencies in areas like process organization and process accountability. The important role of IT support is not covered in the BPMM model.

The other two models cover a similar range comparable to the PMMA but with a different clustering of the impact factors. Rosemann et al. identified five factors which are perceived as covering and characterizing BPM [13, 14, 15, 16]. In the progress of defining the model these factors have been restructured and renamed by Rosemann et al. and are finally defined as

- Strategic Alignment: Alignment of process management to strategic objectives
- Governance: Organizational implementation of BPM and responsibilities for assigned tasks
- Methods: Methods for all BPM relevant tasks
- Technology: Technologies e.g. I&C which supports and enables BPM
- People: Competencies of people involved in BPM
- Culture: Common values towards BPM and process change

Hammers Process Audit is based on the Process and Enterprise Maturity Model (PEMM) which he developed in cooperation with a number of companies [11]. Hammer has identified two distinct groups of characteristics that are needed for a good performance of business processes to perform exceptionally well over a long period of time. Process enablers affect individual processes and determine how well a process is able to function. The process enablers include:

- Design: how the process is to be executed,
- Performers: the knowledge and skills of the people involved
- Owner: the senior executive responsible for the process,
- Infrastructure: the systems that support the process
- Metrics: the measurements used to track the performance of the process

In addition a company must also possess or establish organizational capabilities that allow the business to offer a supportive environment:

- Leadership: Senior executives who support the process
- Culture: Emphasis on a customer focus, teamwork, and willingness to change
- Expertise: Skills and methodology needed for process redesign
- Governance: Mechanisms for managing complex projects and change initiatives.

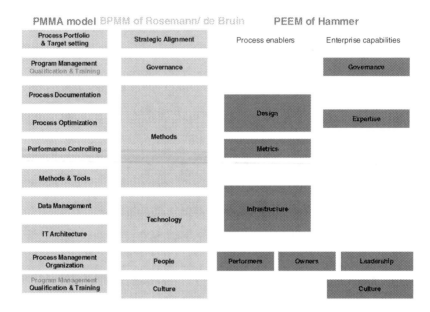

Fig. 5. PMMA Mapping to BPMM and PEMM Maturity Model

Figure 5 maps the nine categories of the Process Management Maturity Assessment with the BPM Maturity Model of Rosemann/ de Bruin and the Process Audit of Hammer. All five factors of the Rosemann et al. model can be mapped to the nine categories of the PMMA. Rosemann/ de Bruin and Hammer explicitly address culture as an impact factor which in the PMMA model is addressed in parts in terms of qualification & training. Hammer emphasizes the process management organization and people issue by addressing performers, owner and leadership as separate factors. At least on the high level clustering of enablers and capabilities Hammer does not identify the strategic alignment of processes to strategy and business as an issue. In all, the comparison gives evidence that all three models cover the essential impact factors for Business Process Management Success.

The mapping can be only a rough indication of the range of factors covered by the models on a high level. A detailed analysis of the underlying criteria and questions for assessment provided they are made public available would show the common ground, possible differences, and additions.

4 Summary and Outlook

Business Process Management is an important management practice for business transformation and organizational change. This paper outlined the implementation of Business Process Management in a large international company, undertaken as a corporate, company wide project within Siemens AG.

The paper introduced a Process Management Maturity Assessment (PMMA) which was developed to assess the implementation of Business Process Management and the performance of organizations in this respect. The maturity model is based on the assessment of nine categories which comprehensively and entirely cover all aspects which impact the success of Business Process Management.

Since the PMMA is based on the principal structure of CMMI using defined maturity levels, structured questionnaires and work sheets, it is easy to use and an assessment can be undertaken in a timeframe of 3 days. A limitation of the CMMI approach is the consolidation of criteria to a single maturity level which may result in misleading interpretations. It is recommended using a detailed view on the assessment and maturity level of each of the nine categories in order to derive a more differentiated picture for improvement measures and best practice exchange, like it was outlined in the example from the business case.

The PMMA was developed to suit the BPM implementation approach which in parts, like the Siemens Process Framework, is company specific. However, the PMMA approach proved to cover all relevant factors for Business Process Management and can be adapted with little effort to a maturity model for general use. This could go in hand with a detailed cross check with the criteria and questions of the maturity model of Rosemann et al., the Business Process Maturity Model of the OMG, and the Process Audit of Hammer.

Overall experiences using PMMA for the assessments are promising in terms of acceptance, ease of use, and coverage of BPM impact factors. The PMMA fits into the overall BPM implementation process in the company and provides an important link to Business Process Management success.

References

1. Becker, J., Kugeler, M., Rosemann, M. (eds.): Process Management: A Guide for the Design of Business Processes, Berlin et al. (2003)
2. Schmelzer, H., Sesselmann, W.: Geschäftsprozessmanagement in der Praxis: Kunden zufrieden stellen – Produktivität steigern – Wert erhöhen, Munich (2008)
3. Paulk, M., Weber, C., Curtis, B., Crissis, M.: Capability Maturity Model for Software, Version 1.1. Software Engineering Institute, Carnegie Mellon, Pittsburgh (1993), http://www.sei.cmu.edu (called 2009-01-31)
4. CMMI: Capability Maturity Model Integration (CMMI) of Carnegie Mellon University, http://www.sei.cmu.edu/cmmi/ (called 2009-01-31)
5. Ahern, D., Clouse, A., Turner, R.: CMMI distilled: A practical introduction to integrated process improvement. Addison-Wesley, Boston (2004)
6. Chrissis, M., Konrad, M., Shrum, S.: CMMI. Guidelines for Process Integration and Product Improvement. Addison-Wesley, Boston (2006)
7. Foegen, M., Solbach, M., Raak, C.: Der Weg zur professionellen IT. Springer, Berlin (2007)
8. Hofmann, H., Yedlin, D., Mishler, J., Kushner, S.: CMMI for Outsourcing: Guidelines for Software, Systems, and IT Acquisition. Addison-Wesley, Boston (2007)
9. BPMM, Business Process Management Maturity Model (BPMM) of OMG, http://www.omg.org/docs/formal/08-06-01.pdf (called 2009-01-31)
10. Fisher, D.M.: The Business Process Maturity Model. A Practical Approach for Identifying Opportunities for Optimization (2004), Business Process Trends: http://www.bptrends.com/resources_publications.cfm (called 2009-01-31)

11. Hammer, M.: The Process Audit. Harvard Business Review, 111–123 (April 2007)
12. Lee, J.-H., Lee, D.H., Kang, S.: An overview of the business process maturity model (BPMM). In: Chang, K.C.-C., Wang, W., Chen, L., Ellis, C.A., Hsu, C.-H., Tsoi, A.C., Wang, H. (eds.) APWeb/WAIM 2007. LNCS, vol. 4537, pp. 384–395. Springer, Heidelberg (2007)
13. Rosemann, M., de Bruin, T., Power, B.: A Model to Measure BPM Maturity and Improve Performance. In: Jeston, J., Nelis, J. (eds.) Business Process Management, Butterworth-Heinemann (2006)
14. Rosemann, M., de Bruin, T., Hueffner, T.: A Model for Business Process Management Maturity. In: ACIS 2004 Proceedings of the Australasian Conference on Information Systems (2004)
15. Rosemann, M., de Bruin, T.: Towards a Business Process Management Maturity Model. In: Proceedings of the 13th European Conference on Information Systems (ECIS 2005), Regensburg (2005)
16. Hüffner, T.: The BPM Maturity Model- Towards A Framework for assessing the Business Process Management Maturity of Organisations, master thesis University of Karlsruhe. GRIN Publishing (2007)
17. Smith, H., Fingar, P.: Process Management Maturity Models (2004), Business Process Trends: http://www.bptrends.com/resources_publications.cfm
18. Feldmayer, J., Seidenschwarz, W.: Marktorientiertes Prozessmanagement: Wie Process Mass Customization Kundenorientierung und Prozessstandardisierung integriert, Munich (2005)
19. Siemens Process Framework, Siemens AG CIO internal documentation, Munich (2005)
20. Becker, J., Delfmann, P. (eds.): Reference Modeling - Efficient Information Systems Design Through Reuse of Information Models, Berlin et al. (2007)
21. Fettke, P., Loos, P. (eds.): Reference Modeling for Business System Analysis. Idea Group (2007)
22. von Brocke, J.: Internetbasierte Referenzmodellierung: State-of-the-Art und Entwicklungsperspektiven. Wirtschaftsinformatik 46(5), 390–404 (2004)
23. RefMod. CC reference modeling, http://www.ercis.de/ERCIS/research/competencecenter/refmod/index.html (called 2009-01-31)
24. SCOR. Supply Chain Operations Reference Model, Version 9, http://www.supply-chain.org/cs/root/home (called 2009-01-31)
25. Fettke, P., Loos, P., Zwicker, J.: Business process reference models: Survey and classification. In: Bussler, C.J., Haller, A. (eds.) BPM 2005. LNCS, vol. 3812, pp. 469–483. Springer, Heidelberg (2006)
26. Scheer, A.-W.: Business Process Engineering: Reference Models for Industrial Enterprises, Berlin et al. (1994)
27. Scheer, A.-W.: ARIS – Business Process Modeling, Berlin et al. (2000)
28. Process Management Implementation Guide, V. 1.0., Siemens AG CIO internal documentation, Munich (2005)
29. Process Management Maturity Assessment, V. 3.1., Siemens AG CIO, internal documentation, Munich (2006)
30. Crawford, J.: Project Management Maturity Model, New York (2001)

Discovering Process Models from Unlabelled Event Logs

Diogo R. Ferreira[1] and Daniel Gillblad[2]

[1] IST – Technical University of Lisbon
[2] Swedish Institute of Computer Science (SICS)
diogo.ferreira@ist.utl.pt, dgi@sics.se

Abstract. Existing process mining techniques are able to discover process models from event logs where each event is known to have been produced by a given process instance. In this paper we remove this restriction and address the problem of discovering the process model when the event log is provided as an unlabelled stream of events. Using a probabilistic approach, it is possible to estimate the model by means of an iterative Expectaction–Maximization procedure. The same procedure can be used to find the *case id* in unlabelled event logs. A series of experiments show how the proposed technique performs under varying conditions and in the presence of certain workflow patterns. Results are presented for a running example based on a technical support process.

1 Introduction

One of the fundamental principles of workflow and BPM systems is the ability to execute multiple instances of a business process where the behaviour of those instances is governed by a predefined process model [1,2]. The goal of *process mining* [3] is to rediscover the process model from the run-time behaviour of process instances, assuming that it is possible, namely: to record events as tasks are performed, and to identify the process instance that produced each event.

These requirements are usually met when the run-time behaviour is recorded in an event log containing a sequence of entries in the form $<case\ id,\ task\ id>$ where *case id* identifies the process instance and *task id* specifies the task that has been performed [3]. Such event logs can be obtained from workflow and case-handling systems, but in applications where there is limited support from process-aware systems it may become difficult to retrieve log data in that form.

In general, it may be possible to record a vast array of events without being able to correlate them to specific process instances. In this scenario, the *case id* attribute is absent and the event log becomes an unlabelled stream of events. Within this stream of events it becomes uncertain whether two events are related or not, as consecutive events may come from different process instances. Also, the number of process instances is unknown.

Our goal is to investigate whether it is possible to discover the process models from such unlabelled event logs. Clearly, the problem of finding the process model in these circumstances is under-defined. However, business processes have

U. Dayal et al. (Eds.): BPM 2009, LNCS 5701, pp. 143–158, 2009.

distinctive sequential patterns [4] and process instances essentially repeat these patterns over and over again. Based on these premises, it is possible to estimate a probabilistic model that is likely to explain the observed behaviour. In this paper we develop a probabilistic framework for that purpose (section 3).

Similar approaches are not common. In [4] the authors describe two experiments where they had to deal with events without an associated *case id*. In both experiments they resorted to application-specific techniques such as context and data attributes to establish the connection between events. In [5] the authors propose an *iterative workflow mining* approach that could be regarded as being related to the expectation–maximization approach we describe here, but it is used for a different purpose, which is to associate low-level events with high-level tasks. In this work we focus on the specific problem of finding the *case id* in unlabelled event logs (section 4). Section 5 discusses the working assumptions and section 6 concludes the paper.

2   Running Example

Fig. 1 illustrates the technical support process for a software product, adapted from a real case study [6]. Basically, the customer calls the vendor to report a problem, the call-center checks if there is an existing contract and records the complaint to be analyzed by the technical support team. The support engineer that is handling the case may either provide a solution or request the development team to fix some bugs. Should the latter become necessary, the development team will have to schedule the release of the bug fix in one of the forthcoming product versions.

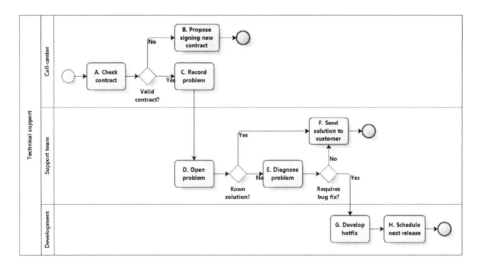

Fig. 1. Technical support process

Whenever the call-center receives a call, a new instance of this process is created. For easier reference, the activities have been labelled with symbol letters from A to H. Let us assume that these activities are recorded in such a way that whenever an activity is completed, the corresponding symbol is recorded in an event log. According to the process model depicted in fig. 1, there are four kinds of possible behaviour: AB, $ACDF$, $ACDEF$ and $ACDEGH$. Each process instance will generate one of these sequences, and the sequences from several process instances may become interleaved since, at any point in time, there may be a number of cases running concurrently.

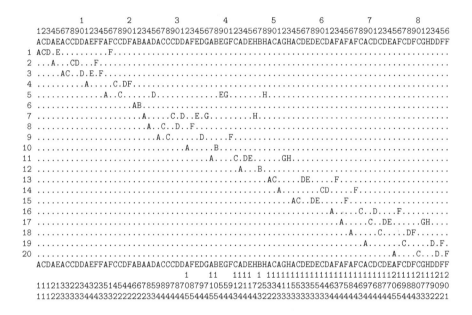

```
         1         2         3         4         5         6         7         8
1234567890123456789012345678901234567890123456789012345678901234567890123456
ACDAEACCDDAEFFAFCCDFABAADACCCDDAFEDGABEGFCADEHBHACAGHACDEDECDAFAFAFCACDCDEAFCDFCGHDDFF
 1 ACD.E......F...................................................................
 2 ...A...CD...F..................................................................
 3 .....AC..D.E.F.................................................................
 4 ..........A.....C.DF...........................................................
 5 ..............A..C......D...........EG.......H.................................
 6 .................AB...........................................................
 7 ....................A.....C.D..E.G.........H..................................
 8 ....................A..C..D..F................................................
 9 ......................A.C......D.....F........................................
10 ..........................A....B..............................................
11 .............................A...C.DE......GH.................................
12 .................................A...B........................................
13 ....................................AC.....DE......F..........................
14 .........................................A........CD....F.....................
15 ...........................................AC..DE.....F.......................
16 ...................................................A.....C..D....F............
17 .....................................................A.....C..DE......GH......
18 .......................................................A.....C.....DF.........
19 ...........................................................A.......C.....D.F.
20 ..............................................................A....C...D.F
ACDAEACCDDAEFFAFCCDFABAADACCCDDAFEDGABEGFCADEHBHACAGHACDEDECDAFAFAFCACDCDEAFCDFCGHDDFF
          1        11     1111 1 11111111111111111111111111211112111212
11121332234323514544667859897870879710559121172533411553355446375846976877069880779090
111223333344433322222222334444444554445544434444322233333333334444443444444554443332221
```

Fig. 2. Events recorded for 20 instances of the technical support process

Fig. 2 shows the events recorded during the execution of 20 process instances. A total of 86 symbols have been recorded, as shown in the first two lines. The complete sequence – called the *symbol sequence* – is shown in the third line, and again below the 20 separate instances. The last three lines display two distinct features: the two lines before the last contain the instance number for each recorded event – this is called the *source sequence* – and it refers to the same set of numbers as printed in the leftmost column of the figure. The very last line displays a count of the total number of instances running concurrently at the time when each event was recorded. In this example it can be seen that there were at most 5 instances running concurrently at different points in time.

If both the symbol sequence and the source sequence are known then we have the equivalent of an event log with *task id* and *case id*, respectively, and it is possible to discover the process model using existing process mining techniques.

What we want to investigate is whether it is possible to discover the process model when only the symbol sequence is known.

The source sequence may be unknown for a number of reasons, including the fact that the business process may lack an appropriate support system – this is especially true for organizations with a fragmented IT infrastructure comprising several disparate tools and applications. The source sequence may also have to be removed from the event log for privacy reasons, for example to avoid identifying customers, citizens or medical patients. Finally, it could be the case that the event log is captured by systems that forward tasks without being aware of the process logic. In these scenarios it becomes useful to have a technique that is able to discover the hidden logic behind an unlabelled stream of events.

3   Probabilistic Approach

Let K be the number of sources[1] that produce symbols according to the same underlying process model. The output of all sources is recorded in a common event log, where symbols produced by any source may become interleaved with symbols produced by other sources. Let $x = \{x_1, x_2, \ldots, x_N\}$ be the symbol sequence of length N where each symbol x_n comes from one of the available K sources. Let $s = \{s_1, s_2, \ldots, s_N\}$ be the unknown source sequence where each element s_n says which source produced the symbol x_n.

For the purpose of estimating the process model from the symbol sequence x alone, we will use a probabilistic approach based on a first-order Markov chain augmented with special start ($\circ$) and stop ($\bullet$) states. Fig. 3 shows one possible representation for the technical support process shown earlier. The transition matrix M specifies the conditional probabilities for the transition between any two symbols; for example, the probability of producing symbol E after symbol D is given by $p(E|D) = M(D, E) = 0.47$. The conditional probabilities in each row add up to 1.0 except in the last row that represents the stop state.

$M =$	$\circ$	A	B	C	D	E	F	G	H	$\bullet$
$\circ$	–	1.0	–	–	–	–	–	–	–	–
A	–	–	0.15	0.85	–	–	–	–	–	–
B	–	–	–	–	–	–	–	–	–	1.0
C	–	–	–	–	1.0	–	–	–	–	–
D	–	–	–	–	–	0.47	0.53	–	–	–
E	–	–	–	–	–	–	0.5	0.5	–	–
F	–	–	–	–	–	–	–	–	–	1.0
G	–	–	–	–	–	–	–	–	1.0	–
H	–	–	–	–	–	–	–	–	–	1.0
$\bullet$	–	–	–	–	–	–	–	–	–	–

Fig. 3. Transition matrix for the technical support process

---

[1] From this point onwards, we will refer to *process instances* as *sources*.

The special start and stop states do not produce symbols; instead, they are used solely for the purpose of representing the probability of the process beginning or ending with certain symbols. Note that $p(\circ|...) = p(...|\bullet) = p(\bullet|\circ) \triangleq 0$, i.e., there are no transitions to the start state, no transitions from the stop state, and no direct transitions from the start to the stop state, respectively.

The following sections describe how to estimate the transition matrix M and the source sequence s from a given symbol sequence x. Since the estimation involves several steps, we proceed incrementally by first explaining how to compute M from both x and s (section 3.1), then how to estimate s from x and M (section 3.2) and finally how to use the two previous steps to iteratively estimate both M and s from x alone (section 3.3). Note that when only x is given, there must be some way to initialize M in order to get the procedure running. This leads to the concept of M^+ in section 3.3.

3.1   Estimating M Given x and s

If both the symbol sequence x and the source sequence s are given, it becomes straightforward to estimate the transition matrix shown in Fig. 3. With both x and s it is possible to separate x into the symbol sequences produced by each source. We define $y^{(k)} = \{y_1^{(k)}, y_2^{(k)}, \ldots, y_{m_k}^{(k)}\}$ of length m_k as the symbol sequence produced by source k alone, where $y_1^{(k)} = \circ$ and $y_{m_k}^{(k)} = \bullet$. Each sequence $y^{(k)}$ can be easily compiled by picking up the symbols x_n for which $s_n = k$ and adding the special start and stop states.

In its simplest form, the joint probability of x and s can be expressed as:

$$p(x; s) = \prod_{k=1}^{K} \prod_{i=1}^{m_k-1} M(y_i^{(k)}, y_{i+1}^{(k)}) \tag{1}$$

For any given pair of symbols a and b, it can be shown that the estimator $\hat{M}(a, b)$ that maximizes $p(x; s)$ is given by[2]:

$$\hat{M}(a, b) = \frac{\sum_k \eta_{(a,b)}(y^{(k)})}{\sum_k \sum_{b'} \eta_{(a,b')}(y^{(k)})} \tag{2}$$

where a and b are symbols and $\eta_{(a,b)}(y^{(k)})$ is the number of times that the transition from a to b occurs in sequence $y^{(k)}$.

From the event log in Fig. 2 we have:

$$y^{(1)} = y^{(3)} = y^{(13)} = y^{(15)} = \circ ACDEF \bullet$$
$$y^{(2)} = y^{(4)} = y^{(8)} = y^{(9)} = y^{(14)} = y^{(16)} = y^{(18)} = y^{(19)} = y^{(20)} = \circ ACDF \bullet$$

---

[2] For this purpose it is convenient to use the log-likelihood $L(M) \triangleq \log p(x; s \mid M) = \sum_k \sum_i \log M(y_i^{(k)}, y_{i+1}^{(k)})$. Maximizing this expression in terms of $M(a, b)$ requires the use of a Lagrange multiplier to find the solution of $\partial L / \partial M(a, b) = 0$ subject to the constraint $\sum_{b'} M(a, b') = 1$. The solution is the maximum likelihood estimator (MLE) shown in equation (2).

$$\boldsymbol{y}^{(5)} = \boldsymbol{y}^{(7)} = \boldsymbol{y}^{(11)} = \boldsymbol{y}^{(17)} = \circ ACDEGH \bullet$$
$$\boldsymbol{y}^{(6)} = \boldsymbol{y}^{(10)} = \boldsymbol{y}^{(12)} = \circ AB\bullet$$

and therefore $\hat{\boldsymbol{M}}(D, E) = \frac{4 \times 1 + 9 \times 0 + 4 \times 1 + 3 \times 0}{4 \times 1 + 9 \times 1 + 4 \times 1 + 3 \times 0} \simeq 0.47$ as before.

3.2   Estimating s Given x and M

If $\boldsymbol{M}$ would be known, then it would be possible to estimate the source sequence $\boldsymbol{s}$ for a given symbol sequence $\boldsymbol{x}$. In principle we would be interested in finding the optimal source sequence $\hat{\boldsymbol{s}} = \arg\max_{\boldsymbol{s}}\{p(\boldsymbol{x}; \boldsymbol{s})\}$ that maximizes the joint probability of $\boldsymbol{x}$ and $\boldsymbol{s}$. Unfortunately, finding $\hat{\boldsymbol{s}}$ is a combinatorial optimization problem where one would have to test all possible source sequences in order to find the set of sequences $\boldsymbol{y}^{(k)}$ that maximize $p(\boldsymbol{x}; \boldsymbol{s})$ according to equation (1).

In practice, it is possible to obtain an approximation of $\hat{\boldsymbol{s}}$, denoted by $\tilde{\boldsymbol{s}}$, by following a greedy procedure to pick the most likely source for each symbol in $\boldsymbol{x}$. This procedure is based on the idea that if we know the previous symbol ε_k for every source k, then symbol x_n should be assigned to the source k that is able to produce x_n with the highest transition probability. That is, we choose to make $\tilde{s}_n \leftarrow \arg\max_k\{\boldsymbol{M}(\varepsilon_k, x_n)\}$.

After assigning symbol x_n to source k, the previous symbol ε_k for source k is updated to x_n (i.e. $\varepsilon_k \leftarrow x_n$) and we move on to the next symbol x_{n+1}. We find source $\tilde{s}_{n+1}$ by the same procedure, i.e. $\tilde{s}_{n+1} \leftarrow \arg\max_k\{\boldsymbol{M}(\varepsilon_k, x_{n+1})\}$, and so on, until all symbols in $\boldsymbol{x}$ have been assigned to some source.

Whenever symbol x_n is such that $\boldsymbol{M}(\circ, x_n)$ is higher than $\boldsymbol{M}(\varepsilon_k, x_n)$ for every source k, then a new source is *activated* and x_n is assigned to that newly created source. On the other hand, whenever symbol x_n is such that the transition probability to the stop state $\boldsymbol{M}(x_n, \bullet)$ is higher than the transition probability to any other symbol, then source s_n is *deactivated* and removed from the set of active sources.

The following examples are based on the event log shown in Fig. 2 and the transition matrix in Fig. 3:

- At position 40 there are four active sources whose previous symbols are $\varepsilon_5 = E$, $\varepsilon_7 = G$, $\varepsilon_9 = D$ and $\varepsilon_{11} = A$. The present symbol is $x_{40} = G$ and the probabilities of each active source producing this symbol are $\boldsymbol{M}(E, G) = 0.5$, $\boldsymbol{M}(G, G) = 0$, $\boldsymbol{M}(D, G) = 0$ and $\boldsymbol{M}(A, G) = 0$, respectively. Hence, symbol x_{40} gets assigned to source 5, which sets $\tilde{s}_{40} \leftarrow 5$ and $\varepsilon_5 \leftarrow G$. Note that $\boldsymbol{M}(\circ, G) = 0$, so activating a new source is not an option at this point.
- At position 41 the present symbol is $x_{41} = F$ and the best candidate is source 9 with $\boldsymbol{M}(D, F) = 0.53$, which sets $\tilde{s}_{41} \leftarrow 9$ and $\varepsilon_9 \leftarrow F$. After this, source 9 gets deactivated because $\boldsymbol{M}(F, \bullet) = 1$ and therefore it cannot produce any additional symbols.
- At position 42 there are only 3 active sources with previous symbols $\varepsilon_5 = G$, $\varepsilon_7 = G$ and $\varepsilon_{11} = A$. Symbol $x_{42} = C$ gets assigned to source 11 which has the highest transition probability $\boldsymbol{M}(A, C) = 0.85$. We have $\tilde{s}_{42} \leftarrow 11$ and $\varepsilon_{11} \leftarrow C$.

– At position 43 the present symbol is $x_{43} = A$ and the transition probability from the previous symbol to symbol A is zero for all active sources: sources 5 and 7 have $\varepsilon_5 = \varepsilon_7 = G$ and $M(G, A) = 0$; source 11 has $\varepsilon_{11} = C$ and $M(C, A) = 0$. However, $M(\circ, A) = 1$ and therefore a new source with number 12 is created, setting $\tilde{s}_{43} \leftarrow 12$ and $\varepsilon_{12} \leftarrow A$.

More formally, this procedure can be described as shown in Algorithm 1. Note that at line 4 the set of candidate sources becomes the set of all active sources except those that have previously produced a symbol equal to x_n. In other words, we are assuming that each source does not produce the same symbol more than once (this assumption will be discussed ahead in section 5).

Algorithm 1. Greedy algorithm to compute $\tilde{s} = \{\tilde{s}_1, \tilde{s}_2, \ldots, \tilde{s}_N\}$

Input: symbol sequence x and transition matrix M
Let Ψ be the set of currently active sources
Let ψ be the set of candidate sources ($\psi \subseteq \Psi$)
Let K be the total number of sources used
Let Ω be the set of distinct symbols in x
1:   $\Psi \leftarrow \emptyset$
2:   $K \leftarrow 0$
3:   **for** $n = 1$ to N **do**
4:     $\psi \leftarrow \Psi \setminus \{k : x_n \in y^{(k)}\}$
5:     **if** $(\psi = \emptyset) \vee (\forall_{k \in \psi} : M(\circ, x_n) > M(\varepsilon_k, x_n))$ **then**
6:       $K \leftarrow K + 1$
7:       $\Psi \leftarrow \Psi \cup \{K\}$ // activate a source
8:       $y^{(K)} \leftarrow \{\circ\}$
9:       $\tilde{s}_n \leftarrow K$
10:    **else**
11:       $\tilde{s}_n \leftarrow \arg\max_{k \in \psi}\{M(\varepsilon_k, x_n)\}$
12:    **end if**
13:    $\varepsilon_{\tilde{s}_n} \leftarrow x_n$
14:    $y^{(\tilde{s}_n)} \leftarrow y^{(\tilde{s}_n)} \cup \{x_n\}$
15:    **if** $(\forall_{b \in \Omega} : M(x_n, \bullet) > M(x_n, b))$ **then**
16:       $\Psi \leftarrow \Psi \setminus \{\tilde{s}_n\}$ // deactivate a source
17:       $y^{(\tilde{s}_n)} \leftarrow y^{(\tilde{s}_n)} \cup \{\bullet\}$
18:    **end if**
19: **end for**
Output: source sequence $\tilde{s}$ and separate source sequences $y^{(1\ldots K)}$

3.3   Estimating M and s from x Alone

Equipped with equation (2) and Algorithm 1 it is possible to devise an iterative procedure to estimate both M and s when only the symbol sequence x is known. Provided with an initial estimate for M we use Algorithm 1 to obtain $\tilde{s}$; then we use $\tilde{s}$ to separate x into $y^{(k)}$ and by equation (2) we compute $\hat{M}$; these two steps complete one iteration. By repeating these steps, we continuously improve

$\hat{M}$ and $\tilde{s}$ until finally none of them changes anymore; at this point a solution has been found. This procedure is described in Algorithm 2.

Algorithm 2. Expectation–Maximization procedure to estimate $\hat{M}$ and $\tilde{s}$

Input: symbol sequence x
1: initialize $\hat{M} \leftarrow M^+$
2: **repeat**
3:    (*E-step*) use $\hat{M}$ in Algorithm 1 to obtain $\tilde{s}$ and $y^{(1...K)}$
4:    (*M-step*) use $y^{(1...K)}$ in equation (2) to update $\hat{M}$
5: **until** ($\hat{M}$ does not change)
Output: transition matrix $\hat{M}$ and source sequence $\tilde{s}$

Algorithm 2 is essentially an Expectaction–Maximization technique [7] to estimate the model parameters M from the incomplete data x, where s is the missing data. The question now is how to initialize $\hat{M}$ in order to get the procedure running. The simplest way to do this is to randomize $\hat{M}$ subject to the stochastic constraints $\sum_{b'} M(a, b') = 1$. However, this random initialization will, in general, lead to a sub-optimal solution as there are many local maxima of the likelihood function where Algorithm 2 will converge. Instead, we need a better way to initialize $\hat{M}$ in order to have a starting point that is actually closer to an optimal solution.

Let

$$M^+(a, b) \triangleq \frac{\eta_{(a,b)}(x)}{\sum_{b'} \eta_{(a,b')}(x)} \qquad (3)$$

be the *global model* where $\eta_{(a,b)}(x)$ is the number of times that transition a to b occurs in sequence x (i.e. where a and b are consecutive symbols). This global model captures the transition probabilities as if the symbol sequence x had been produced by a single source. Even if x is the result of interleaving a number of sources, their underlying behaviour will be present in M^+ since consistent behaviour will stand out with stronger transition probabilities than the spurious effects of random interleaving. Therefore, M^+ is a good initial guess for the estimation of M.

3.4   Example

From the process shown in Fig. 1, an event log of 300 sources was generated, having at most 5 overlapping sources. The event log was generated via simulation, using the same ratios as in Fig. 2, i.e. about $9/20 = 45\%$ of $ACDF$, $4/20 = 20\%$ of $ACDEF$, $4/20 = 20\%$ of $ACDEGH$, and $3/20 = 15\%$ of AB. After running Algorithm 2 on the symbol sequence, the transition matrix in Fig. 4 was obtained[3].

---

[3] Source code for the algorithms and instructions for running examples similar to this one can be found at: http://web.tagus.ist.utl.pt/~diogo.ferreira/mimcode/

$$\hat{M} =$$

	○	A	B	C	D	E	F	G	H	●
○	–	0.97	–	–	–	0.03	–	–	–	–
A	–	–	0.16	0.84	–	–	–	–	–	–
B	–	–	–	–	–	–	–	–	–	1.0
C	–	–	–	–	1.0	–	–	–	–	–
D	–	–	–	–	–	0.41	0.59	–	–	–
E	–	–	–	–	–	–	0.38	0.55	–	0.07
F	–	–	–	–	–	–	–	–	–	1.0
G	–	–	–	–	–	–	–	–	1.0	–
H	–	–	–	–	–	–	–	–	–	1.0
●	–	–	–	–	–	–	–	–	–	–

Fig. 4. Estimated transition from an unlabelled a symbol sequence

Also, from $y^{(1...K)}$ the algorithm estimates that 48.1% of sources produce $ACDF$, 19.8% produce $ACDEGH$, 15.9% produce AB, 13.6% produce $ACDEF$, and there is a small fraction (2.6%) of a single-step sequence E. This last case has some effects on $\hat{M}$, where $\hat{M}(\circ, E) = 0.03$ and $\hat{M}(E, \bullet) = 0.07$; also, this is the reason why the number of sources $K = 308$ was found to be slightly higher than 300.

The transition matrix $\hat{M}$ is shown in graphical form on Fig. 5, where the width of each arc is made proportional to the transition probability. Except for the arcs labeled 0.03 and 0.07 that involve symbol E, the graph depicts the same behaviour as the process model in Fig. 1.

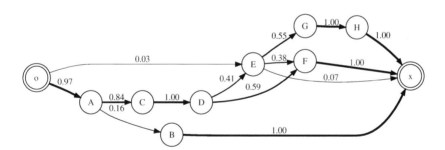

Fig. 5. Estimated model for the technical support process

4   Finding the *case id* in Unlabelled Event Logs

Up to this point we have focused on M as the main outcome of Algorithm 2. However, it is clear that the separate source sequences $y^{(1...K)}$ represent, on their own, a complete event log with both *case id* and *task id*. This suggests that Algorithm 2 can be used as a means to find the *case id* for the activities recorded in an unlabelled event log.

Let $\mathbb{Z}$ be the set of *distinct sequences* in $\boldsymbol{y}^{(1...K)}$. We associate the probability $q(\boldsymbol{z})$ of a sequence $\boldsymbol{z} \in \mathbb{Z}$ with the number of times $\boldsymbol{z}$ occurs in $\boldsymbol{y}^{(1...K)}$. Basically, $q(\boldsymbol{z})$ is the percentage of sequences equal to $\boldsymbol{z}$ in $\boldsymbol{y}^{(1...K)}$. For example, from the results in the example of section 3.4 we have: $q(ACDF) = 0.481$, $q(ACDEGH) = 0.198$, $q(AB) = 0.159$, $q(ACDEF) = 0.136$ and $q(E) = 0.026$.

Let $p(\boldsymbol{z})$ denote the actual distribution one would get if both $\boldsymbol{x}$ and the true source sequence $\boldsymbol{s}$ were known. Then the following metric based on the geometric mean of both distributions can be used to determine how good Algorithm 2 is as a labelling mechanism:

$$G(p \parallel q) \triangleq \sum_{\boldsymbol{z} \in \mathbb{Z}} \sqrt{p(\boldsymbol{z}) \cdot q(\boldsymbol{z})} \tag{4}$$

From the example in section 3.4 we have: $G(p \parallel q) = \sqrt{0.45 \times 0.481} + \sqrt{0.2 \times 0.198} + \sqrt{0.15 \times 0.159} + \sqrt{0.2 \times 0.136} + \sqrt{0 \times 0.026} \cong 0.98$, i.e. in this example the algorithm was able to achieve about 98% accuracy in labelling the symbol sequence $\boldsymbol{x}$ with the estimated source sequence $\tilde{\boldsymbol{s}}$.

Once the log is labelled, then it becomes possible to apply process mining techniques such as the α-algorithm [8], the heuristics miner [9], the genetic miner [10], the fuzzy miner [11], or other techniques available in the ProM framework [12]. In general, all these techniques require a labelled event log. If only an unlabelled log is available, then Algorithm 2 can be used as a first pre-processing stage. Also, Algorithm 2 is able to produce a model $\hat{\boldsymbol{M}}$ in the form of a transition matrix but once the log is labelled other process mining techniques can be used to extract other kinds of models such as Petri nets, heuristic nets, etc.

4.1   Accuracy and Performance

The metric $G(p \parallel q)$ provides a scoring measure which evaluates the degree of similarity between a complete event log, where both $\boldsymbol{x}$ and $\boldsymbol{s}$ are known, and an incomplete event log $\boldsymbol{x}$ that has been labelled by the estimated source sequence $\tilde{\boldsymbol{s}}$. We will call this metric the G-score; it is a measure of the accuracy of Algorithm 2 as a labelling mechanism for incomplete event logs. In general, this accuracy will depend on the total number of sources in the event log, and on the number of overlapping sources. In principle, the higher the number of sources, the easier it becomes to discover consistent behaviour in the event log. On the other hand, the higher the number of overlapping sources, the more difficult it is to separate the events belonging to different sources.

Fig. 6 shows the results of running Algorithm 2 over symbol sequences with varying number of sources, all having at most 5 overlapping sources. From Fig. 6(a) it is clear that accuracy tends asymptotically to 1.0 as the number of sources (and hence the length of sequence $\boldsymbol{x}$) increases. Fig. 6(b) suggests that execution time evolves quadratically with sequence length; however, it should be noted that the average time per run is below 1 sec. in all experiments.

Fig. 7 shows the results of running Algorithm 2 over symbol sequences of 300 sources with varying number of overlapping sources. From Fig. 7(a) it is

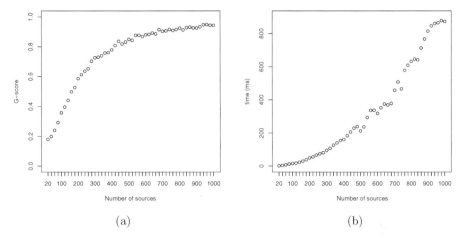

Fig. 6. Average G-score (a) and average runtime (b) for input event logs with varying number of sources and having at most 5 overlapping sources. Each point has been obtained by averaging over 1000 synthetically-generated logs.

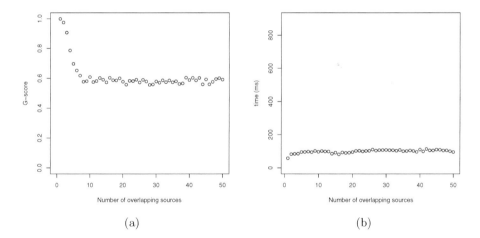

Fig. 7. Average G-score (a) and average runtime (b) for input event logs with 300 sources and varying number of overlapping sources. Each point has been obtained by averaging over 1000 synthetically-generated logs.

apparent that accuracy is exceedingly good when there is little or no overlap at all; it drops dramatically as the number of overlapping sources approaches the length of the sequences produced by each source; but it remains fairly constant and above 0.5 no matter how much the overlap is further increased. Fig. 7(b) suggests that execution time is rather independent of the amount of overlap.

Table 1. Estimation results on different patterns. In all experiments, symbol sequences have been generated using 300 sources and at most 5 overlapping sources.

Pattern	$p(z)$	No. symbol sequences	Average G^*-score	Best G^*-score	Best $q(z)$
Parallelism	ABCEDF : 0.5 ABECDF : 0.3 ABCDEF : 0.2	1000	0.716	0.854	ABCEDF : 0.398 ABCDEF : 0.180 ABECDF : 0.158 ABCDF : 0.062 ABCDE : 0.037 ABEDF : 0.034 ECDF : 0.031 ABCE : 0.028 ABCEF : 0.025 EDF : 0.019 ABEF : 0.009 CDF : 0.006 EF : 0.003 CEDF : 0.003 E : 0.003 CDEF : 0.003
Loop-3	ABCDE : 0.5 ABCDBCDE : 0.25 ABCDBCDBCDE : 0.125 ABCDBCDBCDBCDE : 0.125	1000	0.503	0.539	BCDEA : 0.581 BCD : 0.400 A : 0.010 BCDE : 0.010
Loop-2	ABCDE : 0.5 ABCDCDE : 0.25 ABCDCDCDE : 0.125 ABCDCDCDCDE : 0.125	1000	0.500	0.538	CDEAB : 0.578 CD : 0.402 CDE : 0.010 CDAB : 0.006 AB : 0.004
Loop-1	ABCE : 0.5 ABCCDE : 0.25 ABCCCDE : 0.125 ABCCCCDE : 0.125	1000	0.498	0.537	CDEAB : 0.578 C : 0.401 CDE : 0.010 CAB : 0.006 AB : 0.002 EAB : 0.002 CDAB : 0.002
Non-local dependency	ABCDE : 0.6 AFCGE : 0.4	1000	0.840	0.909	ABCDE : 0.507 AFCGE : 0.320 AFCDE : 0.087 ABCGE : 0.087

4.2 Parallelism, Loops and Non-local Dependencies

There are a number of workflow patterns [13] that business processes often contain and that may be difficult to capture using process mining techniques. In [14] the authors address the problem of discovering parallel behaviour; in [15] the authors address the problem of mining *short loops* of length one and two; and in [10] the authors present the *drivers license* example where there are non-local dependencies between log events, i.e. where the current symbol depends on a past symbol that has been produced before the immediately previous one.

These and other workflow patterns can become quite challenging to discover since first-order Markov models capture behaviour in terms of the previous state alone. However, experiments suggest that Algorithm 2 can still provide useful insight into the behaviour of processes that contain such patterns. Table 1 presents the results on simple experiments with these patterns.

In models with parallelism it is possible to capture the behaviour by a set of independent sequences. As shown in the last column of Table 1, the top three sequences match the original behaviour in $p(z)$. There is, however, some amount of mislabelling in the remaining sequences. In particular, the algorithm finds it difficult to establish a relationship between symbol E and the remaining symbols. This explains why $ABCDF$ becomes the fourth strongest sequence and why there are so many sequence variations involving symbol E.

Models with loops pose special problems, as they involve a repetition of symbols. Since Algorithm 1 does not assign repeating symbols to the same source, the solution provided by Algorithm 2 tends to isolate loop behaviour into separate sources, i.e. each loop iteration is assigned to a different source. This is apparent in the loop experiments reported in Table 1, where the second strongest source corresponds to the loop body: BCD for the loop of length 3, CD for the loop of length 2, and C for the loop of length 1.

As for the strongest source, this corresponds to the linear sequence without looping. There is, however, a mismatch between this sequence and the first sequence in the original model: the sequence seems to have been shifted-left with respect to the original behaviour. This can be explained by the fact that the looping behaviour increases the probability of the start symbol being the first symbol in the loop, hence all sequences tend to be shifted to that symbol.

By looking at $p(z)$ and $q(z)$ in the loop experiments, it becomes apparent that $G(p \parallel q)$ is zero, since there are no common sequences between both distributions. To be able to determine the best solution in these experiments, we relax $G(p \parallel q)$ in order to include the shifting of sequences in $q(z)$. For example, in order to match the sequences of both distributions in the loop-3 experiment, we consider a new entity $q^*(z)$ where $q^*(BCDEA) = q^*(ABCDE) = q^*(EABCD) = q^*(DEABC) = q^*(CDEAB) = 0.581$. This leads to the definition of the G^*-score as $\sum_z \sqrt{p(z) \cdot q^*(z)}$ whose results are reported in Table 1. For the parallelism and non-local dependency experiments, the G^*-score results are equal to the G-score values.

For non-local dependencies, the algorithm is able to capture the most recurring sequences with relative ease, with only a small percentage of incorrect sequences.

5 Working Assumptions

While choosing the source $\tilde{s}_n$ for each symbol x_n Algorithm 1 considers all active sources except those that have already produced symbol x_n earlier on. This means that no source is allowed to produce the same symbol more than once, and therefore the solutions found by Algorithm 2 will have this same characteristic. This restriction is intended to reduce the search space for the source sequence $\tilde{s}$. Without this assumption, every active source remains as a possible candidate for any given symbol, so it becomes more difficult to assign the correct source to each symbol. Also, without this restriction there would be much more local maxima of the likelihood function, making it extremely difficult for Algorithm 2 to find to the optimal solution.

Table 2. Estimation results in an experiment involving log L3 of [16]. Symbol sequences have been generated using 1000 sources and at most 20 overlapping sources.

Pattern	$p(z)$	No. symbol sequences	Average G-score	Best G-score	Best $q(z)$
Duplicate tasks	BDE : 24 / 61 $\simeq$ 0.393 AABHF : 7 / 61 $\simeq$ 0.115 CHF : 15 / 61 $\simeq$ 0.246 ADBE : 6 / 61 $\simeq$ 0.098 ACBGDFAA : 1 / 61 $\simeq$ 0.016 ABEDA : 8 / 61 $\simeq$ 0.131	1000	0.196	0.591	BDE : 0.381 A : 0.355 CHF : 0.169 BHF : 0.056 BD : 0.010 B : 0.009 F : 0.009 G : 0.009 DE : 0.002 BE : 0.001

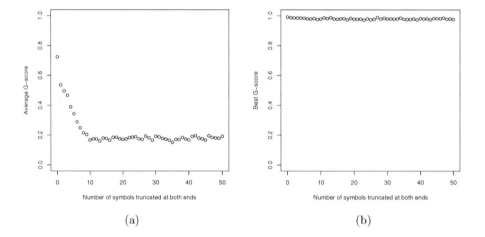

(a)　　　　　　　　　　　　　　　　　(b)

Fig. 8. Average G-score (a) and best G-score (b) for input event logs generated from the technical support process with 300 sources, 5 overlapping sources, and varying number of symbols truncated at both ends

Nevertheless, the algorithm can still capture behaviour in the presence of repeating symbols, although it will forcefully disconnect a sequence when a repeated symbol occurs. Examples appear in the loop experiments in Table 1, where loop iterations are captured as a separate source, contributing to the overall probability of the second strongest sequence in all three experiments.

The same behaviour is due to happen in the presence of *duplicate tasks* [16], i.e. when two different activities are represented by the same symbol. Table 2 shows the results of an experiment using an event log (L3) taken from [16]. Some sequences in the original log have duplicate tasks. Algorithm 2 is able to capture some recurring patterns such as BDE, CHF, and BHF but the remaining sequences are broken due to the presence of repeating symbol A.

A second working assumption is that the symbol sequence x contains the complete sequences, from the first to the last symbol, produced by each source.

In practice this assumption may not hold, since the symbol sequence x may be an excerpt of recorded behaviour during a period of time. It could be that at the beginning of sequence x some sources were already active, so the first symbols from these sources are missing; at the end of sequence x, the last symbols from some of the active sources could also be missing.

To account for this possibility, we consider that x may be truncated at both ends by a certain amount of symbols. Fig. 8 shows the results of running Algorithm 2 on truncated symbol sequences. As the first symbols are truncated, the average G-score drops sharply and then stabilizes around 0.18 when the transient behaviour has been removed and x is left with steady-state behaviour. On the other hand, Fig. 8(b) shows that the best G-score attained remains fairly constant no matter how many symbols are truncated. This means that truncating x makes it more difficult, on average, to find the source for each event, but it does not diminish the ability of the algorithm to find the optimal solution.

6   Conclusion

In this paper we described an Expectation–Maximization approach to estimate the transition matrix M that represents the process model extracted from an unlabelled event log where the *case id* is missing. The probabilistic framework used for this purpose comprises a set of sources that are instances of M and that produce events which become randomly interleaved in the output symbol sequence x. Finding the source for each event is a prerequisite for estimating M, so the proposed approach can also be used as a labelling mechanism to find the *case id* in unlabelled event logs.

Since M is a first-order Markov model it may be unable to represent certain workflow patterns, but once the log is labelled it is possible to leverage the use of existing process mining techniques to obtain other kinds of models. The experiments reported in this paper show that the proposed approach is capable of labelling log events even in the presence of workflow patterns that M is unable to explicitly represent. This means that the proposed technique can become a valuable aid in the discovery of process models when log data is available as an unlabelled stream of events.

References

1. Hollingsworth, D.: The workflow reference model. Document Number TC00-1003, Workflow Management Coalition (1995)
2. van der Aalst, W.M.P., ter Hofstede, A.H.M., Weske, M.: Business process management: A survey. In: van der Aalst, W.M.P., ter Hofstede, A.H.M., Weske, M. (eds.) BPM 2003. LNCS, vol. 2678, pp. 1–12. Springer, Heidelberg (2003)
3. van der Aalst, W.M.P., van Dongen, B.F., Herbst, J., Maruster, L., Schimm, G., Weijters, A.J.M.M.: Workflow mining: A survey of issues and approaches. Data and Knowledge Engineering 47(2), 237–267 (2003)

4. Ferreira, D., Zacarias, M., Malheiros, M., Ferreira, P.: Approaching process mining with sequence clustering: Experiments and findings. In: Alonso, G., Dadam, P., Rosemann, M. (eds.) BPM 2007. LNCS, vol. 4714, pp. 360–374. Springer, Heidelberg (2007)
5. Buffett, S., Geng, L.: Bayesian classification of events for task labeling using workflow models. In: Proceedings of the 4th Workshop on Business Process Intelligence, BPI 2008 (2008)
6. Ferreira, D., Mira da Silva, M.: Using process mining for ITIL assessment: a case study with incident management. In: Proceedings of the 13th Annual UKAIS Conference, Bournemouth University (April 2008)
7. Dempster, A.P., Laird, N.M., Rubin, D.B.: Maximum likelihood from incomplete data via the EM algorithm. Journal of the Royal Statistical Society 39(1), 1–38 (1977)
8. van der Aalst, W.M.P., Weijters, A.J.M.M., Maruster, L.: Workflow mining: Discovering process models from event logs. IEEE Transactions on Knowledge and Data Engineering 16(9), 1128–1142 (2004)
9. Weijters, A.J.M.M., van der Aalst, W.M.P., Alves de Medeiros, A.K.: Process mining with the heuristics miner algorithm. BETA Working Paper Series WP 166, Eindhoven University of Technology (2006)
10. Alves de Medeiros, A.K., Weijters, A.J.M.M., van der Aalst, W.M.P.: Genetic process mining: An experimental evaluation. Data Mining and Knowledge Discovery 14(2), 245–304 (2007)
11. Günther, C.W., van der Aalst, W.M.P.: Fuzzy mining – adaptive process simplification based on multi-perspective metrics. In: Alonso, G., Dadam, P., Rosemann, M. (eds.) BPM 2007. LNCS, vol. 4714, pp. 328–343. Springer, Heidelberg (2007)
12. van Dongen, B.F., de Medeiros, A.K.A., Verbeek, H.M.W(E.), Weijters, A.J.M.M.T., van der Aalst, W.M.P.: The proM framework: A new era in process mining tool support. In: Ciardo, G., Darondeau, P. (eds.) ICATPN 2005. LNCS, vol. 3536, pp. 444–454. Springer, Heidelberg (2005)
13. van der Aalst, W.M.P., ter Hofstede, A.H.M., Kiepuszewski, B., Barros, A.P.: Workflow patterns. Distributed and Parallel Databases 14(3), 5–51 (2003)
14. Cook, J.E., Du, Z., Liu, C., Wolf, A.L.: Discovering models of behavior for concurrent workflows. Computers in Industry 53, 297–319 (2004)
15. Alves de Medeiros, A.K., van Dongen, B.F., van der Aalst, W.M.P., Weijters, A.J.M.M.: Process mining: Extending the α-algorithm to mine short loops. BETA Working Paper Series WP 113, Eindhoven University of Technology (2004)
16. Rozinat, A., van der Aalst, W.M.P.: Conformance checking of processes based on monitoring real behavior. Information Systems 33(1), 64–95 (2008)

Abstractions in Process Mining: A Taxonomy of Patterns

R.P. Jagadeesh Chandra Bose[1,2] and Wil M.P. van der Aalst[1]

[1] Department of Mathematics and Computer Science, University of Technology,
Eindhoven, The Netherlands
[2] Philips Healthcare, Veenpluis 5-6, Best, The Netherlands

Abstract. Process mining refers to the extraction of process models from event logs. Real-life processes tend to be less structured and more flexible. Traditional process mining algorithms have problems dealing with such unstructured processes and generate spaghetti-like process models that are hard to comprehend. One reason for such a result can be attributed to constructing process models from raw traces without due pre-processing. In an event log, there can be instances where the system is subjected to similar execution patterns/behavior. *Discovery of common patterns* of invocation of activities in traces (beyond the immediate succession relation) can help in improving the discovery of process models and can assist in defining the conceptual relationship between the tasks/activities.

In this paper, we characterize and explore the manifestation of commonly used process model constructs in the event log and adopt pattern definitions that capture these manifestations, and propose a means to form abstractions over these patterns. We also propose an iterative method of transformation of traces which can be applied as a *pre-processing step for most of today's process mining techniques*. The proposed approaches are shown to identify promising patterns and conceptually-valid abstractions on a real-life log. The patterns discussed in this paper have multiple applications such as trace clustering, fault diagnosis/anomaly detection besides being an enabler for hierarchical process discovery.

1 Introduction

Process mining refers to the extraction of process models from event logs [1]. An event log corresponds to a bag of process instances of a business process. A process instance is manifested as a trace (a trace is defined as an ordered list of activities invoked by a process instance from the beginning of its execution to the end). Process mining techniques can deliver valuable, factual insights into how processes are being executed in real life.

Real-life processes tend to be less structured than expected. Traditional process mining algorithms have problems dealing with such unstructured processes and generate spaghetti-like process models that are hard to comprehend. One reason for such a result can be attributed to constructing process models from raw traces without due pre-processing. A majority of process mining techniques

U. Dayal et al. (Eds.): BPM 2009, LNCS 5701, pp. 159–175, 2009.

in the literature are purely syntactic in nature. From the viewpoint of existing process mining techniques all of the activities are different and unrelated. The activity names are treated simply as strings that typically do not have any semantics attached to them. However, in reality subsets of activities are related in that their usage is confined to certain contexts, and cater to some functionality. Recent efforts in process mining on Semantic MXML try to address this problem by incorporating semantics in the log specification [2]. It requires the domain expert to come up with the ontologies describing the domain concepts and relationships between them. Ontologies can assist in defining hierarchies of concepts over activities and there by provide abstractions. Asking a domain expert to build this from scratch would at times be too much to ask for, especially in real-life domains such as healthcare and finance where the complexity of the system/domain is too high.

The discovery of process models is based on the dependency relations that can be inferred among the activities in the log. More specifically, the dependency that is often explored is the succession relation (activity B succeeds activity A); process models are generated by assigning the control-flow links between tasks, i.e., activities based on the succession relation. Considering activities in isolation contributes to the "spaghettiness" of the discovered process models to a certain extent. Moreover, the context in which the activity is executed is not considered fully. In an event log, there can be instances where the system is subjected to similar execution patterns/behavior (where the pattern can manifest as a larger subsequence of tasks/activities), and instances where unrelated cases are executed. Discovery of common patterns of invocation of activities in traces (beyond the immediate succession relation) can help in improving the discovery of process models and can assist in defining the conceptual relationship between the tasks/activities.

In this paper, we first characterize and explore the manifestation of commonly used constructs (of building a process model) in the event log and propose pattern definitions that capture these manifestations. Some of these pattern definitions have been in existence in the string-processing and bioinformatics literature. We adopt these pattern definitions to the process mining domain and propose a means to form abstractions over these patterns. We propose a novel iterative method of transformation of traces which can be applied as a preprocessing step for most of the process mining analysis. The proposed approach first identifies the looping constructs in traces and replaces the repeated occurrences of the manifestation of the loop by an abstracted entity (activity) that encodes the notion of a loop. The second step involves the identification of sub-processes or common functionality in the traces and replacing the sub-processes/common functionality with abstract entities. We also present means to deal with complex process model constructs involving combination of choice, parallelism and loops. Fig 1 depicts two process models: one is obtained by mining the original log (Fig 1(a)), and the other is based on the log with abstractions (Fig 1(b)). We have used the heuristics miner plugin (with default settings) in

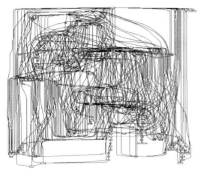 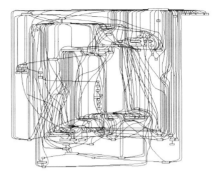

(a) Process model mined from original log

(b) Process Model mined from log with abstractions over loop construct patterns

Fig. 1. Process models obtained by applying the heuristic miner to an event log of Philips Healthcare

ProM[1] tool to mine the models. The model mined on the orignal log had 141 activities and 2901 arcs and had the fitness measures of 0.295 and −0.693 for the continuous semantics (*cs*) and improved continuous semantics (*ics*) metrics respectively. On the other hand, the model mined on the abstracted log had 99 activities and 537 arcs with fitness measures of 0.344 and 0.443 for the *cs* and *ics* metrics respectively. It is evident that the abstracted log is less spaghetti-like, more expressive, and more comprehensible.

We evaluate the goodness of the patterns proposed in this paper on a real-life log of Philips Healthcare. Philips Healthcare collates logs from their medical systems across the globe. These logs contain information about user actions, system events etc. The number of such log-recording systems in conjunction with the fine grained nature of logging makes the dataset available extremely large i.e., in the order of a few thousand logs per day. The patterns and abstractions presented in this paper are shown to be quite effective in that they are able to group activities pertaining to common functionality and also identify patterns of abnormal usage.

The rest of the paper is organized as follows. In Section 2, we introduce the notations used in the paper. Section 3 defines a few pattern definitions and correlates these signatures with the process model constructs. In Section 4, we propose one approach to form abstractions based on the patterns. Pattern definitions catering to the manifestation of complex process model constructs such as choice/parallelism within loops and sub-processes are discussed in Section 5. Approaches to discover these patterns from the event log are presented in Section 6. In Section 7, we propose an iterative approach of transforming traces which can be used as a pre-processing step for process mining analysis. In Section 8,

---

[1] ProM is an extensible framework that provides a comprehensive set of tools/plugins for the discovery and analysis of process models from event logs. See http://www.processmining.org for more information and to download ProM.

we present and discuss the patterns uncovered in a case study of a real-life log of Philips Healthcare. We discuss related work in Section 9. Finally, we conclude in Section 10.

2 Notations

Let $\mathcal{A}$ denote the set of activities. $|\mathcal{A}|$ is the number of activities. $\mathcal{A}^+$ is the set of all non-empty finite sequences of activities from $\mathcal{A}$. A trace, T is an element of $\mathcal{A}^+$. For $i \leq j$, $T(i,j)$ denotes the subsequence from the ith position to the jth position in the trace T. An event log, $\mathcal{L}$, corresponds to a multi-set (or bag) of traces from $\mathcal{A}^+$.

As an example, let $\mathcal{A} = \{a, b, c\}$ be the set of activities; $|\mathcal{A}| = 3$. $T =$abcabb is a trace of length 6. $T(2,5) =$bcab is a subsequence of T from positions 2 to 5. $\mathcal{L} = \{$aba, aba, abba, baca, acc, cac$\}$ represents an event log.

3 Taxonomy of Patterns

In this section, we will introduce various definitions of patterns and correlate them to the manifestation of process model constructs. Discovering such model constructs would help in answering questions such as *Are there loops within loops in my process model?*, *What are the most commonly used functionalities in my model?*, and also can assist in mining models bottom-up from primitive model constructs.

3.1 Loops as Tandem Arrays

Simple loops manifest as the repeated occurrence of an activity or subsequence of activities in the traces. In other words, an activity or a sequence of activities constituting a loop manifest themselves in a tandem fashion in a trace.

- *Tandem Array:* A tandem array in a trace T is a sub-sequence $T(i,j)$ of the form α^k with $k \geq 2$ where α is a sequence that is repeated k times. The *subsequence* α is called a tandem repeat type. We denote a tandem array by the triple (i, α, k) where the first element of the triple corresponds to the *starting position* of the tandem array, the second element corresponds to the *tandem repeat type*, and the third element corresponds to the *number of repetitions*.
- *Maximal Tandem Array:* A tandem array $T(i,j)$ of the form $\alpha^k (k \geq 2)$, is called a maximal tandem array if there are no additional copies of α before or after $T(i,j)$.
- *Primitive Tandem Repeat Type:* A tandem repeat type α is called a primitive tandem repeat type if and only if α is not a tandem array. i.e., $\alpha = \beta^p$, for some non-empty sequence β only if $p = 1$.
- *Primitive Tandem Array:* A tandem array $T(i,j)$ of the form α^k $(k \geq 2)$, is a primitive tandem array iff α is a primitive tandem repeat type.

For example, consider the trace $T=$gdabcabcabcabcabcafica. $(3, \text{abc}, 4)$, $(3, \text{abcabc}, 2)$, $(4, \text{bca}, 4)$, $(4, \text{bcabca}, 2)$, $(5, \text{cab}, 3)$ are the tandem arrays in T. The corresponding tandem repeat types are abc, abcabc, bca, bcabca, cab respectively. The primitive tandem repeat types are abc, bca, cab.

3.2   Sub-processes as Conserved Regions

Finding similar regions (sequence of activities) common within a trace and/or across a set of traces in an event log signifies some set of common functionality accessed by the process. In other words, a region of high similarity shared within a process instance or by two or more process instances might be evidence of common functionality (often abstracted as a sub-process). In order to find these commonalities across the traces in the entire event log, we first construct a single sequence which is obtained by the concatenation of traces in the event log with a distinct delimiter between the traces. Let us denote this concatenated sequence by S. Multiple invocations of a sub-process within a trace can be detected by finding similar regions manifested within a trace.

- *Maximal Pair:* A maximal pair in a sequence, S is a pair of identical subsequences α and β such that the symbol to the immediate left (right) of α is different from the symbol to the immediate left (right) of β. In other words, extending α and β on either side would destroy the equality of the two strings. A maximal pair is denoted by the triple (i, j, α) where i and j correspond to the starting positions of α and β in S with $i \neq j$.
- *Maximal Repeat:* A maximal repeat in a sequence, S is defined as a subsequence α that occurs in a maximal pair in S.
- *Super Maximal Repeat:* A super maximal repeat in a sequence is defined as a maximal repeat that never occurs as a substring of any other maximal repeat.
- *Near Super Maximal Repeat:* A maximal repeat α is said to be a near super maximal repeat if and only if there exist at least one instance of α at some location in the sequence where it is not contained in another maximal repeat.

Consider the event log, $\mathcal{L} = \{$aabcdbbcda, dabcdabcbb, bbbcdbbbccaa, aaadabbccc, aaacdcdcbedbccbadbdebdc$\}$ over the alphabet $\mathcal{A} = \{$a, b, c, d, e$\}$. Table 1 depicts the maximal, super maximal and near super maximal repeats present in each trace of the event log. For trace T_1, the set of maximal repeats $= \{$a, b, bcd$\}$. Since maximal repeat b, is subsumed in maximal repeat bcd, b does not qualify to be a super maximal repeat. The occurrence of maximal repeat b at position 6 in T_1 does not overlap with any other maximal repeat. Hence b qualifies to be a near super maximal repeat. Similarly for trace T_3, all occurrences of maximal repeats b and bb coincide with the maximal repeat bbbc. Hence neither qualify for near super maximal repeat. The occurrence of maximal repeat c at position 10 in T_3 does not coincide with any other maximal repeat. Hence, c qualifies to be a near super maximal repeat.

Table 2 depicts the maximal/super maximal/near super maximal repeats present in the entire event log, $\mathcal{L}$. These are the repeats in the sequence obtained

Table 1. Maximal, Super Maximal and Near Super Maximal Repeats in each trace of the event log $\mathcal{L}$

Id	Trace	Maximal Repeat Set	Super Maximal Repeat Set	Near Super Maximal Repeat Set
T_1	aabcdbbcda	{a, b, bcd}	{a, bcd}	{a, b, bcd}
T_2	dabcdabcbb	{b, dabc}	{dabc}	{b, dabc}
T_3	bbbcdbbbccaa	{a, b, c, bb, bbbc}	{a, bbbc}	{a, c, bbbc}
T_4	aaadabbccc	{a, b, c, aa, cc}	{b, aa, cc}	{a, b, aa, cc}
T_5	aaacdcdcbedbcc-badbdebdc	{a, b, c,d, e, aa, bd, cb, db, dc, cdc}	{e, aa, bd, cb, db, cdc}	{a, c, e, aa, bd, cb, db, dc, cdc}

Table 2. Maximal, Super Maximal and Near Super Maximal Repeats in the Event Log $\mathcal{L}$

Maximal Repeat Set	{a, b, c, d, e, aa, ab, ad, bb, bc, bd, cb, cc, cd, da, db, dc, aaa, abc, bbc, bcc, bcd, cdc, dab, abcd, bbbc, bbcc, bbcd, bcda, dabc, bcdbb}
Super Maximal Repeat Set	{e, ad, bd, cb, aaa, cdc, abcd, bbbc, bbcc, bbcd, bcda, dabc, bcdbb}
Near Super Maximal Repeat Set	{e, aa, ad, bb, bd, cb, cc, db, dc, aaa, bcc, cdc, dab, abcd, bbbc, bbcc, bbcd, bcda, dabc, bcdbb}

by concatenation of all traces in the event log. Near super maximal repeats are a hybrid between maximal repeats and super maximal repeats in that it contains all super maximal repeats and those maximal repeats that can occur in isolation in the sequence without being part of any other maximal repeat. Near super maximal repeats can assist in identifying *choice constructs* in the process model. Let us denote the set of maximal repeats, super maximal repeats and near super maximal repeats by M, SM and NSM respectively. The following relation holds between the three.

$$SM \subseteq NSM \subseteq M$$

The set $NSM \setminus SM$ (the set difference) depicts all maximal repeats that occur both in isolation and are also subsumed in some other maximal repeat. For any repeat $r \in NSM \setminus SM$, a super maximal repeat r^s which contains (subsumes) r can be either of the form αr or $r\beta$ or $\alpha r\beta$ (where α and β are subsequences of activities). This indicates that r can be a common functionality which might occur in conjunction with α and/or β. In other words, it indicates that α and β can potentially be optional (sequence of) activities in the context of r.

3.3 Mapping Primitive Tandem Repeats and Conserved Regions into Equivalence Classes

We consider both a *primitive tandem repeat type* and all variants of *maximal repeat* as a *repeat* in this section. For a repeat, r, let repeat alphabet $\Gamma(r)$, denote the set of symbols/activities that appear in the repeat. For example, for the repeats abba, abdgh, and adgbh, the repeat alphabets correspond to $\{a, b\}$, $\{a, b, d, g, h\}$, and $\{a, b, d, g, h\}$ respectively. Different repeats can share a common repeat alphabet. In the above example, the repeats abdgh and adgbh share the same repeat alphabet $\{a, b, d, g, h\}$. We can define equivalence classes on repeat alphabet.

$$[X] = \{r \mid r \text{ is a repeat and } \Gamma(r) = X\}$$

For the above example, $[\{a, b, d, g, h\}] = \{abdgh, adgbh\}$. Furthermore, the equivalence class under repeat alphabet will capture any variations in the manifestation of a process execution due to *parallelism*.

Reducing the number of features. Large data sets and data sets with large alphabet might contain abundant repeats. But not all of them might be characteristically significant. For example, there might be repeats which occur only in a small fraction of traces. One way to tackle this is to filter the repeats. One can retain only those repeats that are contained in a large fraction of traces in the event log, i.e., repeats that have a high support in the event log. Other means of feature reduction can also be thought of.

4   Abstractions of Patterns

Subprocess abstractions can be discovered by considering a *partial ordering* on the repeat alphabet. Subsumption is used as the *cover relation*. A repeat alphabet ra_1 is set to cover another repeat alphabet ra_2 if $ra_2 \subset ra_1$. For example, consider the repeat types abcd and abd. It is most likely for activity c to represent a functionality similar to that of a, b, and d, since c occurs within the context of a, b and d. By defining a partial order on the repeat alphabets and generating a Hasse diagram on the partial ordering, one can form abstractions by considering the *maximal* elements in the poset. Fig 2 depicts the partial ordering on the repeat alphabets as a Hasse diagram. $\{a,b,c\}$ and $\{a,c,d\}$ are the maximal elements of the partial ordering. Maximal elements can be considered as abstractions of processes. Let us denote these two maximal elements with abstract activities A and B respectively. Repeat alphabets under a maximal element can all be represented with the abstraction of the maximal element. Repeat alphabets that contribute to more than one maximal element can either be put in one of the maximal elements or can define an abstraction in itself. It can be considered as a (sub-)functionality that is used in a larger functionality. In our example, let us assume that the repeat alphabet $\{a,c\}$ is assigned to the maximal element $\{a,b,c\}$. With this abstraction, all repeats with repeat alphabets $\{a,b\}$, $\{a,c\}$, $\{b,c\}$, and $\{a,b,c\}$ are represented by the abstract activity A while the repeats with repeat alphabets $\{a,d\}$ and $\{a,c,d\}$ are represented by the abstract activity B in all the traces. There can be instances where two maximals of the parital ordering on the repeat alphabets share a lot in common. In order to reduce the total number of abstract activities introduced, one can define extended joins on the maximal elements. Fig 3 depicts the scenario where

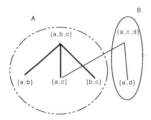

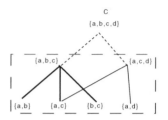

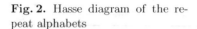

Fig. 2. Hasse diagram of the repeat alphabets

Fig. 3. Hasse diagram of the repeat alphabets with extended joins

an extended join has been introduced on the maximal elements of Fig 2. The two maximal elements of Fig 2 viz., {a,b,c} and {a,c,d} are extended to join at {a,b,c,d}. The extended join now covers all repeat alphabets. Let us denote the extended join with abstract activity C. All repeats with repeat alphabets {a,b}, {a,c}, {b,c}, {a,d}, {a,b,c} and {a,c,d} can now be represented by a single abstraction viz., C in all the traces.

Let ra_1^m and ra_2^m be repeat alphabets corresponding to two *maximal* elements in the Hasse diagram. Different criteria for extending the maximal elements can be defined. For example, one can choose to extend two maximal elements provided they share a set of common elements above a particular threshold and also when the differences between them is less. In other words, extend the maximal elements only if $|ra_1^m \cap ra_2^m| \geq \delta_c$ and $|(ra_1^m \setminus ra_2^m) \cup (ra_2^m \setminus ra_1^m)| \leq \delta_d$. δ_c corresponds to the threshold on the number of common elements which can either be a fixed constant or a fraction of the cardinality of the participating maximal elements such as $0.6 \times \min(|ra_1^m|, |ra_2^m|)$. δ_d corresponds to the threshold on the number of differences between the two maximal elements.

5   Patterns in the Manifestation of Complex Process Model Constructs

The pattern definitions defined above (both tandem arrays, maximal repeats and its variants) capture some important manifestations of the process model constructs, but they are not sufficient enough to cater to complex model constructs where there is a parallelism or choice within other constructs. We call the above pattern definitions to be *exact*. In order to deal with complex constructs, the pattern definitions need to be more flexible and robust. In this section, we address some of these pattern definitions and call these *approximate*.

5.1   Approximate Tandem Arrays

In a trace T, an *approximate tandem array* is a concatenation of sequences $\alpha = s_1 s_2 s_3 \ldots s_k$ for which there exists a sequence s_c such that each s_i ($1 \leq i \leq k$) is approximately similar to s_c. The notion of similarity can be defined in multiple ways (such as Hamming distance, string edit distance). For example, two sequences with string edit distance [3] less than δ (for some threshold, δ) can be considered to be similar. Here, s_c can be different from each and every s_i; alternatively, we may constrain that s_c be equal to at least one s_i ($1 \leq i \leq k$). s_c is called as the *primitive approximate tandem repeat type*, and the approximate tandem array α is represented by the triple (j, s_c, k) where j signifies the starting position of α in T. Edit distance is defined as the minimum number of operations required to transform one sequence into the other (where the operations correspond to substitution, deletion or insertion of activities). Generic edit distance uses a cost function where different costs can be associated to the edit operations. Levenshtein distance (LD) is a specific case of generic edit distance where all the symbols are treated equally and the cost of each edit operation is 1. Levenshtein distance might not be a good metric in most scenarios as it does not consider the context for edit operations. We have discussed some of the pitfalls of Levenshtein edit distance and proposed an automated approach to derive the costs of edit operations in [4]. One can use the generic edit distance framework [3] to define robust notions of similarity for approximate patterns.

- *Choice within Loops:* Approximate tandem arrays can be used to detect choices within loops. For example, consider the process model construct depicted in Figure 4(iii). In this example, we have a choice construct over the activities b and c inside the loop. $S = \mathtt{abdacdacdabd}$ is one manifestation of the process model that constitutes an approximate tandem array with abd or acd as an approximate primitive tandem repeat type. The similarity criteria is Levenshtein distance, $LD(s_c, s_i) \leq 1$.

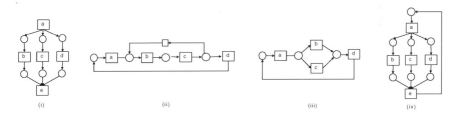

Fig. 4. Few Examples of Complex Process Model Constructs

- *Parallelism within Loops:* Parallelism within loops can also be handled in a similar fashion by approximate tandem arrays. However, defining an appropriate notion of similarity is crucial for the success of this approach. A too lenient notion might generate too many false positives while a stringent notion will miss certain constructs. This problem is compounded by the number of activities involved in the parallelism construct. A more practical way is to process the traces iteratively as would be discussed in Section 7.

5.2 Approximate Conserved Regions

Just like the approximate tandem arrays, we can define notions of approximation for the non-tandem repeats (maximal repeats, super-maximal and near-super maximal repeats). Approximate repetitions are specified by authorizing some number of errors between repeated copies. The set of allowed errors can be defined under the hamming distance and edit distance framework. If only replacements are allowed, this yields the classic Hamming distance, defined as the number of mismatches between the two sequences; if both replacements and insertions/deletions are permitted, then we are operating in the edit distance framework.

A repeat pair (α, α') is a $k-$approximate repeat if and only if the distance between them $d(\alpha, \alpha') \leq k$. Consider Fig 4(i) which contains a parallelism construct. $T_1 = $ abcde, $T_2 = $ acdbe, $T_3 = $ adcbe are some of the traces of the construct. The pairs (abcde, acdbe), (acdbe, adcbe) and (abcde, adcbe) are all $2-$approximate under the Levenshtein edit distance.

6   Approaches for Discovering the Patterns

Maximal, super maximal and near super maximal repeats can be efficiently discovered in linear time using suffix trees for strings [5], [6]. Repeats that exist across the traces in the event log can be determined by applying the repeat identification algorithms on the sequence obtained by concatenating the traces in the event log with a delimiter not present in the alphabet $\mathcal{A}$. Such a concatenation of traces might yield a very long sequence. One can adopt efficient suffix-tree construction techniques such as [7] to handle very long sequences. Approximate repetitions can be found by first identifying exact repetitions and searching for all sub-sequences within a distance of k with the exact repetitions.

Gusfield and Stoye [8] proposed a linear time algorithm based on suffix trees to detect tandem repeats. Discovering tandem arrays takes $\mathcal{O}(n+z)$ time, where n is the length of the trace and z is the number of primitive tandem repeat types in the trace. Sokol et al [9] proposed an approach for a variant of approximate tandem arrays under the edit distance in $\mathcal{O}(nk \log k \log(n/k))$ time and $\mathcal{O}(n+k^2)$ space (where k is the threshold on the edit distance for similarity). The generic problem of approximate tandem arrays is still an open research problem. We have adopted Ukkonen's algorithm [10] for the construction of suffix-trees in linear-time.

7   Pre-processing Traces and Resolving Complex Constructs

7.1   Pre-processing Traces with Abstractions

The discovery of process models is based on the dependency relations that can be inferred among the activities in the log. More specifically, the dependency that is often explored is the succession relation (activity B succeeds activity A); process models are generated by assigning the control-flow links between tasks/activities based on the succession relation. Invocations of an activity in different contexts are treated alike. The fan-in/fan-out of the control-flow links on an activity increases by such a treatment thereby making the final model look spaghetti-like. Spaghettiness of process models can be reduced by first mining common functionalities/constructs, abstracting them and then discovering process models on the abstracted log. By doing so, *multiple invocations of an activity can be distinguished based on the context of its occurrence.* Algorithm 1 presents a single-phase of the pre-processing. The *basic idea is to first process for any loop constructs* (find patterns pertaining to loops viz., tandem arrays and approximate tandem arrays) *and replace the manifestation of loops with abstract entities.* Subsequence patterns that are conserved within a trace and/or across the event log (signifying common functionality) are then discovered and abstracted. This can be iterated over any number of times with the event log for iteration $i + 1$ being the output event log of iteration i.

Algorithm 1. Single-phase preprocessing of traces

```
 1: Given an event log L = {T_1, T_2, T_3, ..., T_m}.
 2: Remove duplicate traces from L. Let the set of unique traces be
    L' ⊆ L = {T'_1, T'_2, ..., T'_n}; Each T'_i ∈ L.
 3: Let L'' = φ
 4: {Identify loop manifestations}
 5: for all T'_i ∈ L' do
 6:    Identify all primitive tandem arrays, approximate tandem repeats in T'_i. Let PTR_i
       denote the set of all primitive tandem repeat types in trace T'_i.
 7: end for
 8: Let PTR = ∪ⁿ_{i=1} PTR_i.
 9: Find abstractions over the set of repeat alphabets of PTR. Let A be the set of such
    abstractions. For each abstraction a_i ∈ A, there exist a set of repeats that constitute
    the abstraction. Let f : PTR → A be the function defining the abstraction for each
    repeat.
10: {Process the traces and replace the loop manifestation with abstract entities}
11: for all T'_i ∈ L' do
12:    Let T''_i be an empty trace
13:    for j = 1 to |T'_i| do
14:       if there exits a maximal primitive tandem array (j, α, k), α ∈ PTR, k ≥ 1 then
15:          {check whether there exist any larger tandem array overlapping with this one}
16:          if there exist another primitive tandem array (j', β, k') such that |β| > |α| and
             j' ≤ j + k * |α| then
17:             Set k = ⌊(j' − j)/|α|⌋.
18:          end if
19:          Append f(α) to T''_i
20:          Set j = j + k * |alpha|
21:       else
22:          Append T'_i(j) to T''_i
23:       end if
24:    end for
25:    L'' = L'' ∪ {T''_i}
26: end for
27: Find conserved regions across all traces in L''
28: Let CR be the set of conserved regions
29: Find abstractions over the set of repeat alphabets of CR. Reuse abstractions already
    defined over PTR for repeats that are common to both PTR and CR. Let A' be the set
    of complete set of abstractions. For each abstraction a_i ∈ A', there exist a set of
    repeats that constitute the abstraction. Let g : CR → A' be the function defining the
    abstraction for each repeat.
30: Process the traces and replace the conserved regions with abstract entities
```

The algorithm is straightforward but the steps $14 - 20$ pertaining to the treatment of overlapping loop manifestations deserve attention. Fig 5 depicts a scenario. $(1, \mathsf{ab}, 5)$ and $(9, \mathsf{abcd}, 3)$ are two tandem arrays in the trace. The prefix of the second loop manifestation overlaps with the suffix of the first loop. In this case, we shorten the first tandem array to $(1, \mathsf{ab}, 4)$ and give preference to longer patterns.

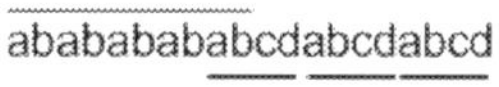

abababababcdabcdabcd

Fig. 5. Overlap of primitive tandem arrays

7.2   Iterative Approach to Resolve Complex Constructs

Consider the process model depicted in Figure 4(iv) which consists of a parallelism construct within a loop. Consider two traces $T_1 = \mathsf{abcdeacbdeabcde}$ and $T_2 = \mathsf{acbdeabcde}$. $r_1 = \mathsf{abcde}$ and $r_2 = \mathsf{acbde}$ constitute the maximal repeats present in the two traces (obtained in the concatenated sequence of T_1 and T_2). Now, repeats r_1 and r_2 are equivalent under the repeat alphabet $\{\mathsf{a, b, c, d, e}\}$. Let us represent this equivalence class with an abstract entity, say A. Processing traces T_1 and T_2 and replacing all occurrences of repeats within this equivalence class with the abstract entity, we get the transformed traces $T_1' = \mathsf{AAA}$ and $T_2' = \mathsf{AA}$. In the second iteration of preprocessing, the loop construct can be discovered in the transformed traces.

Loops within loops can also be discovered using a multi-phase approach. Consider Fig 4(ii). $T_1 = \mathsf{abcdabcbcbcd}$, $T_2 = \mathsf{abcbcd}$ are two of the traces pertaining to the construct. $(6, \mathsf{bc}, 3)$ and $(2, \mathsf{bc}, 2)$ would be identified as tandem arrays in traces T_1 and T_2 respectively. Let us assume that the tandem repeat type bc is represented with an abstract entity A. Replacing all occurrences of tandem arrays with tandem repeat type bc in the log with the abstract entity, we get the transformed traces $T_1 = \mathsf{aAdaAd}$ and $T_2 = \mathsf{aAd}$. Now in the second iteration, aAd would be identified as a tandem array. Thus loops within loops can be uncovered. Though this example relates to simple loops within loops, iterative approach of identifying loops and conserved regions (both *exact* and *approximate*) alternatively and performing consistent abstractions over them would help in realizing more complex constructs.

8   Experimental Results and Discussion

We have analyzed the significance of the patterns described in this paper over a large set of event traces (of real systems) with varying alphabet sizes. We present one such study here where we have considered a set of 1372 event traces of a health care system. The traces correspond to the commands of clinical usage logged by the system. There were a total of 213 distinct commands (activities/event classes) in the event log (alphabet size, $|\mathcal{A}| = 213$) and the entire event log had $215, 623$ events.

In this study, we have done analysis only on the exact repetitions. The analysis of approximate repetitions (both for the manifestation of loop constructs and conserved regions) is underway. Table 3 depicts a few examples of primitive tandem repeat types identified in the log. It can be seen that the commands involved in the loop manifestation all belong to a common functionality. The primitive tandem repeat types 1 and 2 correspond to some *image processing* functionality. The difference between 1 and 2 being that in the former, an *image reverse* operation is performed where as in the latter an *image forward* operation is invoked. The primitive tandem repeat type 3 corresponds to a functionality of *beam/detector* movement while that of 4 corresponds to a *wedge* movement functionality. Primitive tandem repeat type 5 corresponds to *geometry* functionality. There were a total of 826 primitive tandem repeat types in the event log. The shortest primitive tandem repeat type is of length 1 (signifying a loop over a single activity/command) while the longest spans over 13 activities. Under the equivalence class of repeat alphabets, the number of distinct classes were 363.

Table 3. A few examples of primitive tandem repeat types

S.No	Primitive Tandem Repeat Type	Frequency
1	(SetReplayScope, SetReplayType, SetSpeedAndDirection, StartReplay, StartStepImgRev, StopStepImgRev, StartStepRunFwd, StopStepRunFwd)	206
2	(SetReplayScope, SetReplayType, SetSpeedAndDirection, StartReplay, StartStepRunFwd, StopStepRunFwd, StartStepImgFwd, StopStepImgFwd)	319
3	(MoveDetectorLateral.Move, AngulateBeamLateral.Move, RotateBeamLateral.Move, AngulateBeamLateral.Move)	43
4	(BLWedge2RotateClockwise, BLWedge2TranslateIn, BLWedge2TranslateStop, BLWedge2RotateStop)	51
5	(ResetGeo.Start, ResetGeo.Stop)	85

There were a total of 170 maximal elements in the Hasse diagram on the repeat alphabet over the primitive tandem repeat types. A lot of these maximal elements were found to be similar. For example, the three maximal element repeat alphabets, {StartStepImgRev, StopStepImgRev, AngulateBeamFrontal.Move}, {StartStepImgRev, StopStepImgRev, MoveDetectorFrontal.Move} and {Start StepImgRev, StopStepImgRev, RotateBeamFrontal.Move} are all similar in that there exist some beam limitation related movements in conjunction with some image analysis. Using extended joins, these three maximal elements can be combined to a extended repeat alphabet {StartStepImgRev, StopStepImgRev, AngulateBeamFrontal.Move, RotateBeamFrontal.Move, MoveDetectorFrontal .Move}. The number of abstractions can be reduced thus.

Table 4 depicts some of the examples of near super maximal repeats (*nsm*) over the data set. It can be clearly seen that all image processing related commands used as a functionality are captured in *nsm* 1. The *nsm* 2 pertains to

Table 4. Few examples of near super maximal repeats over the dataset

S.No	Near Super Maximal Repeats
1	(StartStepRunFwd, StopStepRunFwd, SetSpeedAndDirection, StartReplay, StartStepRunFwd, StopStepRunFwd, StartStepImgFwd, StopStepImgFwd, StartStepRunRev, StopStepRunRev, SetSpeedAndDirection, StartReplay, StartStepImgFwd, StopStepImgFwd, StartStepRunFwd, StopStepRunFwd, SetSpeedAndDirection, StartReplay, StartStepRunFwd, StopStepRunFwd)
2	(BLWedge1TranslateIn, BLWedge1RotateClockwise, BLWedge1RotateStop, BLWedge1TranslateStop, BLWedge1TranslateIn, BLWedge1RotateCounterClockwise, BLWedge1RotateStop, BLWedge1TranslateStop, BLWedge1RotateCounterClockwise)
3	(StartStepImgRev, StopStepImgRev, StartStepImgFwd, StopStepImgFwd, SetZoomFactor, SelectView, SetZoomCentre)

wedge related movements while *nsm* 3 depicts the zoom functionality used in succession/conjunction with the image processing functionality which can be easily imagined from an application point of view. In the data set there were a total of 26,000 near super maximal repeats (5391 repeat alphabets) after the first iteration. Though the numbers sound large, a lot of these are similar which can be seen from the fact that there were only 1129 maximal elements in the Hasse diagram on these repeat alphabets. Using (δ_c and δ_d) as parameters, these can further be reduced using extended joins. Using one such parameter setting, we were able to reduce the number of abstractions to 40.

Fig 6 depicts a process model mined using the heuristics miner in the ProM tool on the abstracted log of Philips Healthcare. The original log was first filtered to remove highly infrequent activities (frequency of occurrence less than 0.005%). This resulted in a log with 141 distinct event classes and 215,399 total number of events. The abstractions are defined over the exact tandem array patterns capturing the manifestation of loop constructs. For the abstracted log we conducted only one iteration of pre-processing. The process model from the original log (without the abstraction) is shown in Fig 1(a). It is evident that the process model mined from the abstracted log is more comprehensible (less spaghetti-like). Further, the abstractions were formed over activities that are related by a functionality. For example, all the shutter movement operations were grouped to an abstract entity. Similarly, wedge related movements, commands pertaining to image processing functionality have been grouped as an abstract entity. In other words, we were to able to identify *conceptually-valid* abstractions.

The approximate notions of patterns induces the flexibility and thereby the variety over the class of patterns. Another notion of flexibility is introduced in the approach for abstraction over the repeat pattern alphabet. Recall that we have introduced the notion of parameterized extended joins (over the fraction of common/different elements between two repeat alphabets). By choosing different thresholds for the approximation (similarity) between patterns, one can form a multi-level abstraction. The assumption that we make over the definition of

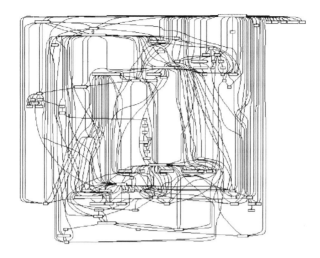

Fig. 6. Process model mined using heuristics miner on log abstracted with loop construct manifestations

these patterns is that each functionality (process model construct, sub-process etc) gets manifested at least twice in the event log (either within the same trace or across traces), which is a reasonable assumption to make. While the notion of approximate patterns induces the flexibility, it also acts as a weakness, as choosing a right notion of approximation is non trivial. For the edit-distance based approximation, choosing a right cost function for edit operations is critical. However, it can be mitigated with approaches for automated derivation of costs such as in [4].

9 Related Work

Greco et al [11], [12] proposed an approach to mine hierarchies of process models that collectively represent the process at different levels of granularity and abstraction. The basic idea of their approach is to cluster the event log into different partitions based on the homogeneity of traces and mine process models for each of the clusters. Clustering induces a hierarchy in the form of a tree with the root node depicting the entire log and the leaf nodes corresponding to traces pertaining to concrete usage scenarios. Two kinds of abstractions over activities viz., *is-a* and *part-of* is then done by traversing the tree bottom-up and considering every pair of activities and checking whether they can be *merged* without adding too many spurious control flow paths among the remaining activities. This approach tries to analyze the mined process models (post-processing) for identifying activities that can be abstracted. However, for large complex logs, the mined process models (even after clustering) can be quite spaghetti-like thereby increasing the complexity of such analysis. In contrast, the approach proposed in this paper analyzes the raw traces and defines abstraction (pre-processing) and thereby reduces the spaghettiness of the mined process model. Our approach can be used complementarily as a precursor to [11], [12]. It is conjectured that such a hybrid approach will yield better results. Polyvyanyy et al [13] have proposed a slider approach for enabling flexible control over various process model abstraction criteria (such as activity effort, mean occurrence of an activity, probability of a transition etc.). The slider is employed for distinguishing significant process model elements from insignificant ones. Taking cartography as a metaphor, Günther and Aalst [14] have proposed a process mining approach to deal with the "spaghettiness" of less structured processes. The basic idea here is to assign significance and correlation values to activities and transitions, and depicting only those edges/activities whose significance/correlation is above a certain threshold. Less significant activities/edges are either removed or clustered together in the model. Günther and Aalst [14] too have used a slider based approach to specify the threshold and thereby alter the levels of abstraction. Approaches such as [13], [14] looks at abstraction from the point of retaining highly significant information and discarding less significant ones in the process model. In contrast, the approach proposed in this paper looks at abstraction from a functionality point of view. The approach proposed in this paper can be used as a preprocessing step for the logs and can be seamlessly integrated with other approaches for abstraction [11], [12], [14] as well as with approaches for process discovery.

10   Conclusions and Future Work

In this paper, we have presented a few pattern definitions and correlated them to the manifestation of process model constructs. We have also presented an approach to form abstractions of activities in the log based on the patterns. Further, a multi-phase approach for pre-processing the traces with the patterns and abstractions was presented. We have applied the proposed techniques on a real-life log and the results are promising. The pattern definitions proposed in this paper have multi-faceted applications such as enabling of hierarchical process mining (thereby reducing the spaghettiness of mined models), trace clustering and fault diagnosis.

Acknowledgments. The authors are grateful to Philips Healthcare for funding the research in Process Mining.

References

1. van der Aalst, W., Weijters, A., Maruster, L.: Workflow Mining: Discovering Process Models from Event Logs. IEEE Trans. Knowl. Data Eng. 16(9), 1128–1142 (2004)
2. de Medeiros, A.K.A., van der Aalst, W., Pedrinaci, C.: Semantic Process Mining Tools: Core Building Blocks. In: 16th European Conference on Information Systems, pp. 1953–1964 (2008)
3. Ristad, E.S., Yianilos, P.N.: Learning String-Edit Distance. IEEE Trans. PAMI 20(5), 522–532 (1998)

4. Bose, R.P.J.C., van der Aalst, W.: Context Aware Trace Clustering: Towards Improving Process Mining Results. In: SIAM International Conference on Data Mining, pp. 401–412 (2009)
5. Gusfield, D.: Algorithms on Strings, Trees, and Sequences: Computer Science and Computational Biology. Cambridge University Press, Cambridge (1997)
6. Kolpakov, K.: Finding Maximal Repetitions in a Word in Linear Time. In: IEEE Symposium on Foundations of Computer Science (FOCS), pp. 596–604 (1999)
7. Cheung, C.F., Yu, J.X., Lu, H.: Constructing Suffix Tree for Gigabyte Sequences with Megabyte Memory. IEEE Trans. Knowl. Data Eng. 17(1), 90–105 (2005)
8. Gusfield, D., Stoye, J.: Linear Time Algorithms for Finding and Representing all the Tandem Repeats in a String. Journal of Computer and System Sciences 69, 525–546 (2004)
9. Sokol, D., Benson, G., Tojeira, J.: Tandem Repeats Over the Edit Distance. Bioinformatics 23(2), e30–e36 (2007)
10. Ukkonen, E.: On-Line Construction of Suffix Trees. Algorithmica 14(3), 249–260 (1995)
11. Greco, G., Guzzo, A., Pontieri, L.: Mining Hierarchies of Models: From Abstract Views to Concrete Specifications. In: Business Process Management, pp. 32–47 (2005)
12. Greco, G., Guzzo, A., Pontieri, L.: Mining Taxonomies of Process Models. Data Knowl. Eng. 67(1), 74–102 (2008)
13. Polyvyanyy, A., Smirnov, S., Weske, M.: Process Model Abstraction: A Slider Approach. In: Enterprise Distributed Object Computing, pp. 325–331 (2008)
14. Günther, C.W., van der Aalst, W.M.P.: Fuzzy Mining - Adaptive Process Simplification Based on Multi-perspective Metrics. In: Business Process Management, pp. 328–343 (2007)

Aggregating Hierarchical Service Level Agreements in Business Value Networks

Irfan ul Haq, Altaf Huqqani, and Erich Schikuta

Department of Knowledge and Business Engineering
University of Vienna, Austria
{irfan.ul.haq,huqqana3,erich.schikuta}@univie.ac.at

Abstract. Business scenarios such as Business Value Networks and Extended Enterprises pose new challenges for service choreographies across
heterogeneous Virtual Organizations. In such scenarios, services compose
together hierarchically in a producer-consumer manner to form service
supply-chains of added value. Service Level Agreements (SLAs) are defined at various levels in this hierarchy to ensure the expected quality
of service for different stakeholders. Automation of service composition
directly implies the aggregation of their corresponding SLAs. But so far,
the aggregation of SLAs has been treated only as a single layer process
which is insufficient to complement the hierarchical aggregation of services. In this paper we elaborate on the requirement of a hierarchical
aggregation of SLAs corresponding to service choreographies in Business
Value Networks. During the hierarchical aggregation of SLAs, certain
SLA information pertaining to different stakeholders is meant to be restricted and can be only partially revealed to a subset of their business
partners. We introduce the concept of SLA-Views to protect such privacy
concerns. We, then formalize the notion of SLA Choreography and define
an aggregation model based on SLA-Views to enable the automation of
hierarchical aggregation of Service Level Agreements. The aggregation
model has been designed to comply with the WS-Agreement standard.

Keywords: Service Level Agreements, Business Value Networks, Value
Chains, SLA Management.

1 Introduction

Novel concepts such as Cloud Computing, Autonomic Computing, and Business
Grids pursue the same industrial goal: to enable consumers to access the shared
resources on demand. In the notion of commodity computing, services are the
basic building blocks of complex software systems. A Service Level Agreement
(SLA) is a formally negotiated contract between service provider and service
consumer that ensures the expected level of service for the service consumer.
The service consumer can be a client or another service.

In a service-enriched environment such as the Grid or the Cloud Computing
infrastructures, services scattered across various Virtual Organisations (VOs)
under multiple administration domains, can compose together in form of service

U. Dayal et al. (Eds.): BPM 2009, LNCS 5701, pp. 176–192, 2009.

choreographies. During such service choreographies, Service Level Agreements (SLA) are made among different partners on various points of the choreography. These partners may include the client, the Virtual Organizations (VO), or other services. Service composition directly implies the need of composition of their corresponding SLAs. So far, SLA composition has been considered [1] as a single layer process. This single layer SLA composition model is insufficient to describe supply-chain business networks. In a supply-chain, a service provider may have sub-contractors and some of those sub-contractors may have further sub-contractors making a hierarchical structure. This suppy-chain network spanned across various Virtual Organisations may emerge as a Business Value Network. Business Value Networks [2] are ways in which organizations interact with each other forming complex chains including multiple providers/administrative domains in order to drive increased business value. NESSI (Networked European Software and Services Initiative) , which is a consortium of over 300 ICT industrial partners has highlighted the importance of Business Value Networks [2] as a viable business model in the emerging service oriented ICT infrastructures.

In addition to the notion of Business Value Networks, NESSI has pointed out various other possibilities for similar inter-organizational business models; Hierarchical Enterprises, Extended Enterprises, Dynamic Outsourcing, and Mergers to name a few. The process of SLA aggregation in such enterprizes is a hierarchical process. There is no SLA aggregation model till this date, which can describe this type of hierarchical aggregation. To enable these supply-chain networks as Service Oriented Infrastructures (SOI), the case of the Service Level Agreements needs to be elaborated and its issues resolved. SLA@SOI [3]is a European project that focusses on SLA issues in SOI. On its agenda is the provision of such Service aggregators, that offer composed services, manageable according to higher-level customer needs. In SLA@SOI's vision, service customers are empowered to precisely specify and negotiate the actual service level according to which they buy a certain service.

It is not sensible to expose the complete information of SLAs spun across the whole chain of services to all the stakeholders. Not only because of the privacy concerns of the business partners, but also for disclosing it could endanger the business processes creating added value. To achieve this balance between trust and security, we introduce the concept of SLA-views. The inspiration for this concept comes from the notion of business process-views [4][5] and workflow Views [6]. We apply the concept of views on SLA-Choreography. Each business partner will have its own view comprising of its local SLA information. The holistic effect of these views will emerge as the overall SLA-Choreography. In this paper we present a formalized approach based on the concept of SLA-Views and adherent to WS-Agreement standard, to automate the aggregation process of hierarchical SLAs in Business Value Networks. The overall contribution of the paper consists of:

- a privacy model based on the concept of SLA-Views,
- a formal description of hierarchical SLA-Choreographies based on SLA-Views in Business Value Networks,

- a formal model for SLA aggregation in hierarchal SLA-Choreographies, and
- the customization of WS-Agreement to support the hierarchical SLA aggregation model

In section 2, we give a survey of the related work. Section 3 introduces the hierarchical choreography of SLAs. Section 4 formalizes the concept of SLA View and SLA Choreography. Section 5 describes the formal model of hierarchical aggregation of SLAs and section 6 highlights some of its business applications. Section 7 presents a motivational example based on this model. Finally, Section 8 concludes the paper with an overview of our achievements and strategy for the future work.

2   Related Work

The related work spans across three dimensions: aggregation models of SLAs, formal description of SLAs and the privacy of stake-holders in business cooperations.

2.1   SLA Aggregation

Service Level Agreement is a contract between a service and its client; the client being a person or yet another service. Service composition in workflow also demands SLA composition. A little research [1] [7] has been done towards dynamic SLA aggregation of workflows. Blake and Cummings [1] have defined three aspects of SLAs which are Compliance, Sustainability and Resiliency. Compliance means suitability i.e the consumer receives what is expected. Sustainability is the ability to maintain the underlying services in timely fasion. Resiliency directly corresponds to the maintenance of services to ensure their performance over an extended period of time. The authors then subdivide these three categories into six aspects of SLA but this makes their approach rather specific because it does not cover the whole range of SLA aspects. They put forth a model to compose SLAs of services mapping to a workflow but they take into account the services existing only at one level. Frankova [7] has also highlighted the importance of this issue but has just described a vision and not any concrete model. Unger et al's work [8]is directly relevant to our focus of research. They focus on aggregation of SLAs in context with Business Process Outsourcing (BPO). They synchronize their work with Business Process Execution Language (BPEL) and WS-Policy. Their model is based on SLO aggregation of SLAs on a single level. One of the limitation of their approach is that they take into account services related to one process in one enterprise because they focus on BPO. Our approach describes corss-VO SLA aggregation and strictly adheres to WS-Agreement.

2.2   Formal Description of SLA

Aiello et al. [9] present a very nice formal description of SLA. Their approach is based on WS-Agreement. They extend the WS-Agreement standard by introducing a new category of terms called Negotiation Terms. They build automata representation of SLA states to describe the negotiation process. Their formal model is too vague and they do not explain how this model will describe the sub-entities in WS-Agreement. Unger et al. [8] present a rigorous formal model for SLA aggregation. They follow BPEL and WS-Policy whereas our formal model adheres to WS-Agreement standard.

2.3 Workflow Views

For privacy concerns we will coin the notion of SLA-Views, which is similar to the concept of workflow views but is not formally based on it. The concept of Workflow Views is used to maintain the balance between trust and security among business partners. Schulz et al. [10] have introduced the concept of view based cross-organizational workflows and they call it as coalition workflows. Chebbi et al. [11] provide a very comprehensive approach that is view based, web services focused and is applicable to dynamic inter-organizational workflow cooperation. This means that the cooperation across organizations is described through views without specifying the internal structure of participating workflows. Their concept of contracts is similar to that of SLA, however, SLAs are more dynamic due to negotiation, renegotiation and fault tolerance features. Their is some very relevant work done by Chiu et al. [12] interms of a contract model based on workflow views. They demonstrate how management of contracts can be facilitated. They start with an example, highlight domains of different participating organizations and then develop a model to identify the corresponding workflow views. They go on further to develop an e-contract model based on plain text format. Service Level Agreements, represented in XML format are more structured and flexible than the e-contracts. Furthermore their approach starts with defining views in an inter-organizational workflow and then describing e-contracts to enforce the obligatory communication links in the views. Our model allows SLAs to maintain their individual identity. Therefore, we define views directly on the SLA aggregation structure rather than on workflows. Moreover, our approach provides a formal description of hierarchical SLAs and their aggregation model.

3   Hierarchical Choreography of SLAs

A service level agreement is a contract that defines mutual understandings and expectations regarding a service between the service provider and the service consumer. WS-Agreement [13], a standard SLA language from OGF (Open Grid Forum) [14], defines the structure of agreement as depicted in figure 1. The contract should bear an official name. Agreement Context contains information about the initiator, the responder and the provider of the agreement; expiration time of the agreement; and its template Id. Service Terms define the functional attributes of the agreement whereas the Guarantee Terms contain the non functional attributes. Guarantee terms further describe the conditions, service level objectives and business value list related to the agreement. Business value list may express the importance of meeting an objective as well as information regarding penalty or reward.

Fig. 1. structure of an agreement in accordance with WS-Agreement specification

Referring to figure 1, we can formally define the Service Terms, and Guarantee Terms as part of the encapsulating section Terms.

Definition 1 (Service Term). A service term denoted by $term_s$ is an element of the set Service Terms denoted by $STerms$. A $term_s \in STerms$ is a tuple such that,

$$term_s =< name, value, type_a >$$

where name and value denote the name and value of a service term and $type_a$ describes its aggregation type.

We have taken the liberty to implant a new mandatory element to the WS-Agreement standard, namely, $type_a$. The $type_a$ element corresponds to the aggregation function that helps us automate the aggregation of SLAs. We postpone its definition to the latter part of the paper where we will discuss the aggregation process.

Definition 2 (Guarantee Term). A guarantee term denoted by $term_g$ is an element of the set Guarantee Terms i.e, $GTerms$. A $term_g \in GTerms$ is a tuple such that:

$$term_g =< SLO, condition_q, BVL >$$

where SLO represents Service Level Objectives, $condition_q$ represents Qualifying Conditions and BVL represents Business Value List. Combining the above two definitions, now we can define the notion Terms in the WS-Agreement.

Definition 3 (Term). A $term \in Terms$ is a pair such that

$$term = (term_s, term_g)$$

where $term_s \in STerms$ and $term_g \in GTerms$

Following the above definitions, SLA can now be formally defined as:

Definition 4 (SLA). A service Level Agreement (SLA) denoted by sla is a tuple

$$sla = < Name, Context, Terms >$$

where $Terms = \cup_{i=1}^{n} term_i$

and Context is a list of strings. Context defines the names of the SLA provider, the consumer and the initiators. It also contains the duration of the SLA. The parameter *Name* denotes the name of the SLA.

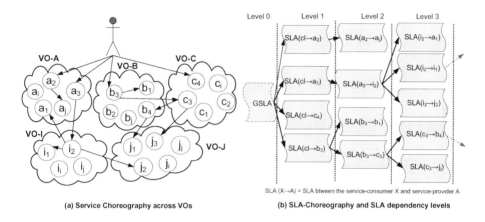

(a) Service Choreography across VOs   (b) SLA-Choreography and SLA dependency levels

Fig. 2. Hierarchical Aggregation of SLAs

A Virtual Organization (VO) in business context, is a temporary or permanent, coalition of geographically dispersed organizations expressing high level mutual trust to collaborate and share their resources and competencies in order to fulfill the customers' requests. Web services scattered across various administrative domains, when composed together, are said to form service choreographies. In these service choreographies, many service-to-service SLAs are formed. The situation becomes even more complex in Business Value Networks, where, services scattered across many such Virtual Organizations (VO) collaborate to enable complex supply chain networks. One way to visulaize this hierarchy is in terms of dependency layers. Deeper a service in this chain is, more dependent its ancestors are. A hierarchy of corresponding SLAs pertains to this chain of services. There is no multi-level SLA model that can describe the hierarchical aggregation of SLAs in such Business Value Network. We will call this hierarchical aggregation of SLAs as SLA-Choreography with relevance to the Service Choreography.

In figure 2, we have presented a simplified picture of a cross-VO choreography. The client (that may be a workflow process) is directly connected to some services, scattered across three VOs: VO-A, VO-B, VO-C. These services are coordinating with other services to carry out their jobs. This coordination results into service chains, distributed across multiple Virtual Organisations. This scenario can be compared with a simple Business Value Network. The partner services play the producer-consumer roles in this service choreography. All of these services establish Service Level Agreements (SLA), thus giving rise to an SLA-Choreography in connection with the underlying service choreography.

Another way to visualize this SLA-Choreography is in terms of hierarchical organization of SLAs. There may be several dependency layers in this SLA-Choreography. The dependency increases along the hierarchy. The aggregated effect of this dependency travels from the very bottom towards the topmost. This SLA aggregation is depicted in Fig. 2. In this hierarchy the SLAs, which are connected to the client process, are said to exist on level 1. This hierarchy indicates a supply chain type of correspondence among the services. These layers also denote the visibility levels of service providers and the client. The client has concern only with the services immediately connected to it and can not see beyond. Similarly a service can see its coordinating services i.e its providers and its consumers with which it is making service level agreements. It has no information about the rest of the service choreography. Despite of its privacy concerns, a service is dependent on its lower services. The effect of SLAs formed among the services at lower levels is bubbled up through the upper layers.

There are many interesting questions that need answers: What trust model will bind together the Business Value Networks? Who will manage this SLA-Choreography? How to monitor and validate this SLA-Choreography? Although these questions are related to our overall research agenda but are beyond the scope of this paper. In this paper we focus on an even more basic problem: To develop a formal model that can describe this SLA-Choreography and construct an aggregation model for hierarchical SLAs while protecting the privacy concerns of the stakeholders at the same time. For this purpose, we introduce the concept of SLA-Views.

4   SLA Views

The concept of *Views* originates from the field of databases and has been successfully adapted in business workflows [11][5]. In workflows, a view can be a subset of that workflow or can be a representation of that workflow in aggregated or abstracted fashion. We have also employed the notion of views to represent a subset of SLA-Choreography. As the matter of fact the notion of SLA-Views is related to that of workflow views in a very general sense. In formal sense, SLA-Views are absolutely different from the workflow views. SLA-Choreography is not a workflow so the rules of workflows are not applicable on it. For instance, in a workflow, rules such as: there should be a single start and single exit or every split should have a join, do not apply on SLA choreography.

A view in an SLA-Choreography represents the visibility of a business partner. Every service provider is limited only to its own view. A partner (for example a service) makes two kinds of SLAs: the SLAs for which it acts as a consumer and the SLAs for which it is a provider. For clarity, we name these two types as the consumer-oriented SLAs and the producer-oriented SLAs respectively.

In figure 3, SLAs are connected to small circles, which we call *aggregation points*, by certain edges called *dependencies*. There are two types of dependencies.

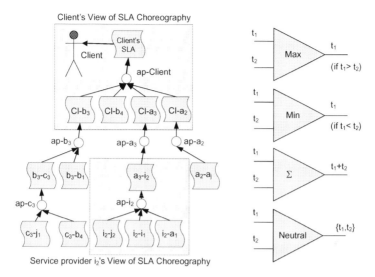

Fig. 3. Different Views in the SLA-Choreography And Some Basic Aggregation Function

Consumer-oriented SLAs are connected to the aggregation points from below by the *sink dependencies* and the producer-oriented SLAs are connected from above by *the source dependencies*. To understand the overall picture of the SLA-Choreography, we need to formalize these concepts.

Definition 5 (Aggregation Point). An Aggregation Point ap is an object such that

$$ap =< aggsla >$$

where $aggsla$ is the aggregated SLA produced by aggregating the consumer-oriented SLAs connected to it. In figure 3, ap-i_2 is an aggregation point. An aggregation point is the point where the consumer-oriented SLAs (of the consumer service) are aggregated and on the basis of their aggregated content, the service is able to decide what it can offer as a provider. The master-slave relationships in Business Value Networks are directly translated to producer-consumer model with one service provider (Enterprise) as a producer and other as a consumer. So both the producer and the consumer enterprises will have their own aggregation points connected together through their mutual SLA. However, for peer-to-peer relationships, both peers act as producer and consumer of services. This issue can be easily resolved by translating peer-to-peer relationships into producer-consumer model. For this purpose, we device the concept of virtual aggregation point (vap) to automate the aggregation process. Virtual aggregation point is discussed in detail in section 6.

Now let us define dependencies which have been shown in figure 3(a) as edges joining the aggregation point with the producer and consumer oriented SLAs. The Aggregation Point ap-i_2 is connected with three consumer-oriented SLAs and one producer-oriented SLA through dependencies.

Definition 6 (Source Dependency). A source dependency dep_{src} is a tuple

$$dep_{src} =< ap, sla >$$

where ap is the aggregation point and sla is the producer-oriented SLA. In figure 3(a), it is represented by the directed edge from the aggregation point ap-i_2 to the producer-oriented SLA, $sla_{a_3-i_2}$.

Each $dep_{src} \in Dep_{src}$, where Dep_{src} is the set of all source dependencies within the SLA-Choreography. Let

$$source : (ap) \rightarrow dep_{src}$$

$source(ap_i)$ is the unique $s \in Dep_{src}$, for which a unique producer-oriented SLA exists with $s = (ap_i, sla_i)$. This means that the function source maps each aggregation point ap_i to a unique SLA through a unique source dependency s.

Definition 7 (Sink Dependency). A sink dependency dep_{sink} is a tuple

$$dep_{sink} =< sla, ap >$$

where ap is the aggregation point and sla is the consumer-oriented SLA. In Figure 3, it is represented by the directed edge from the consumer-oriented SLA i_2-i_1 to the aggregation point ap-i_2. The aggregation point ap-i_2 is connected with three sink dependencies.

Each $dep_{sink} \in Dep_{sink}$, where Dep_{sink} is the set of all sink dependencies within the SLA-Choreography. Let

$$sink : (ap) \rightarrow P(dep_{src})$$

where $P(Dep_{sink})$ is the power set of Dep_{sink}.

$sinks(ap_i)$ is the set $S_{sink} \in P(Dep_{sink})$, i.e. $S_{sink} \subseteq Dep_{sink}$ such that for each $s_i \in S_{sink}$ a unique consumer oriented SLA exists with $s_i = (sla_i, ap_j)$. This means that the function $sinks$ maps a set of consumer-orieted SLAs to a unique aggregation point such that each consumer-oriented SLA sla_i is mapped through a unique sink dependency s_i.

Definition 8 (Dependency). A dependency Dep is a set that is the union of two sets namely Dep_{src} and Dep_{sink} which are pairwise disjoint, i.e.

$$Dep = Dep_{src} \cup Dep_{sink}$$

$$Dep_{src} \cap Dep_{sink} = \phi$$

Based on these definitions, in figure 3, we see that the producer-oriented SLA $(a_3$-$i_2)$ is dependent on the terms of the corresponding consumer-oriented SLAs, aggregated at ap-i_2 . For example the bandwidth and space aggregated at ap-i_2 would be the upper limit of what service i_2 can offer to service a_3. At the same time service i_2 will have to decide about its profit on the basis of the information about total cost in the aggregated SLA. The aggregation point in this sense is also a decision point for a service.

With having all the related concepts formalized, now we are in a position to provide a formal definition of the SLA-View.

Definition 9 (SLA-View). An SLA-View denoted by *slaview* is a tuple such that

$$slaview_i =< sla_p, dep_{sr}, ap_i, SLA_c, Dep_{sn} >$$

where sla_p = producer-oriented SLA, SLA_c= Set of consumer-oriented SLAs, dep_{sr}= source dependency, Dep_{sn}= set of sink dependencies, and ap_i= aggregation point. Each aggregation point ap_i in the SLA-Choreography corresponds to a unique *sla-view$_i$*.

In figure 3, the SLA-Views of the client and a service are highlighted.

Definition 10 (SLA-Choreography). An SLA_{chor} is a tuple such that:

$$SLA_{chor} =< SLA, APoints, Deps >$$

where SLA is set of all *sla* within an SLA-Choreography, $APoints$ is set of aggregation points ap, and $Deps$ is set of dependencies dep. Another way to describe the SLA-Choreography is in terms of SLA-Views, i.e.

$$SLA_{chor} = \cup_{i=1}^n slaview_i$$

This means that the whole SLA-Choreography may be seen as an integration of several SLA-Views. In terms of Business Value Networks, it should be noted that SLA-View defines boundaries of a stakeholder. The aggregation process is performed at every aggregation point. Each aggregation point, which also denotes a dependency level, belongs to one of the service providers. Although each service provider is limited to its own aggregation information, but this information is in fact dependent on the aggregation information at lower levels. The sustainability of this business network requires all the stakeholders to trust each other and their ability to maintain their privacy at the same time. SLA-Views maintain a balance between this privacy and trust.

5   Aggregation Process

In the aggregation process, terms of the consumer-oriented SLAs are aggregated. WS-agreement has no direct support for such an aggregation so we introduced an attribute for aggregation type namely, "$type_a$" in the Definition 1. WS-Agreement gives the liberty to incorporate *any* external schema. Therefore $type_a$ can be made an essential part of the service terms and will describe how the corresponding service will behave during the aggregation process. We can define $type_a$ in a formal way, as follows:

Definition 11 (The function type$_a$). A $type_a \in Types$ is a function that maps a set of tuples to a single tuple which is the aggregation of that set.

$$type_a : tuples(term) \rightarrow term$$

$$type_a(term_1, ...term_n) = term_{agg}$$

We define $type_a$ as an aggregation function that aggregates n terms into one term. Its result is $aggsla$ in the aggregation point (please see Definition 5). Each term in $aggsla$ is computed by applying the type function for that term to the values of the terms for all the dependent (consumer-oriented) SLAs which define that term. In the present context, we define four types of terms namely sumtype, maxtype, mintype and neutral but new types can be added according to the situation, i.e.

$$Types = \{sumtype, maxtype, mintype, neutral\}$$

These functions have been depicted in figure 3(b). The function sumtype can be formally defined as follows.

Definition 11.1 (The function sumtype)

$$sumtype \in Types(\Leftrightarrow sumtype : tuples(term) \rightarrow term$$

$$sumtype(term_1, ...term_n) = \sum_{i=1}^{n} term_i$$

$type_a$ is an aggregation function that aggregates n number of terms into one term. sumtype is of the type of $type_a$ and takes the summation of all terms. Examples include terms for storage space, memory, availability and cost.

Definition 11.2 (The function maxtype)

$$maxtype \in Types(\Leftrightarrow maxtype : tuples(term) \rightarrow term$$

$$maxtype(term_1, ...term_n) = \max_{i=1}^{n} term_i$$

maxtype is an aggregation function that aggregates n number of terms into one term. It does so by picking up the maximum of these terms which represents the aggregation of all the input terms.If several terms addressing the same utility are being aggregated and their type has been declared as maxtype then only the term pertaining to the maximum value will become part of the aggregated SLA. Examples include latency, which may become a bottle neck for the whole process and an activity with highest latency will directly contribute (though a in negative sense) to the throughput of a workflow sequence.

Definition 11.3 (The function mintype)

$$mintype \in Types(\Leftrightarrow mintype : tuples(term) \rightarrow term$$

$$mintype(term_1, ...term_n) = \min_{i=1}^{n} term_i$$

mintype is an aggregation function that aggregates n number of terms into one term. It does so by picking up the minimum of these terms which represents the aggregation of all the input terms. Similar to maxtype, when several terms addressing alike utilities are being aggregated and their type has been declared

as mintype then only the term pertaining to the minimum value will contribute
to the aggregated SLA. Its example can be the bandwidth. In a sequence of
activities the activity pertaining to the minimum bandwidth will become the
bottleneck for the whole sequence making other activities with higher bandwidth
ineffective.

Definition 11.4 (The function neutral)

$$neutral \in Types(\Leftrightarrow neutral : (term) \rightarrow term$$

$$neutral(term_i) = term_i$$

neutral is an aggregation function that includes all the input terms separately
without any processing. This function is applied on those terms which can not
be mixed with other terms and need to preserved in the aggregation process
as separate terms.The terms declared as neutral are unaffected through the aggregation process and are just copied in the aggregated SLA. They represent
services which are independent from similar services, for example identity of
some valuable data in a certain organization or discount in a specific service etc.

So far we have defined only four types of terms but it is important to realize
that this enumeration can be extended without affecting the generic definition
of the $type_a$ function. In certain cases, for example calculating the reward and
penalty expressions, logical operations will also be required. On similar lines,
we can define logical functions such as AND, OR, XOR to integrate the service
level objectives or other constituents of Guarantee Terms to form rule-based
aggregation expressions.

6   A Case for Hierarchical Aggregation of SLAs in Business Applications

NESSI, in their Grand Vision and Strategic Research Agenda (SRA) [2] defines
Value Networks as the ways in which organisations interact with each other
to drive increased business value. Figure 4 shows their example Business Value
Network (BVN) where the Enterprises A and D have been shown to collaborate
on the development of a new product. Enterprise A has subcontractors B and C
whereas the enterprise has E and F as subcontractors. The Enterprises A and D
form a peer-to-peer relationship between themselves.

So far, we have discussed the aggregation of SLAs in context with the composition of services in a producer-consumer manner, along service value chains.
This service level SLA aggregation model can be scaled up to enterprise level.
It can conveniently describe both master-slave and peer-to-peer relationships
in Business Value Networks. Master-slave relationship can be simply mapped
on the producer-consumer model where an SLA is formed between the service
provider and the client. However, in peer-to-peer relationships, the participating
enterprises are acting as the service provider and the client at the same time. To
form a WS-Agreement compliant SLA between them, one party can either be

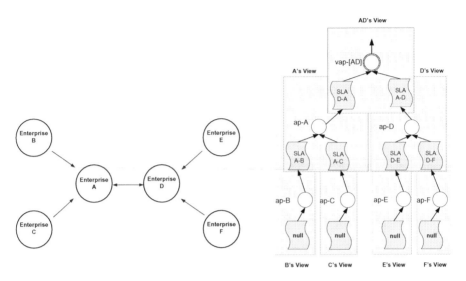

Fig. 4. A Business Value Network and its corresponding SLA Choreography with different Enterprises' Views

treated as a service provider or a service consumer in context with some service. Therefore a peer-to-peer relationship needs to be dissolved into two producer-consumer relationships with a separate SLA associated with each of them. Here we would like to define a Virtual Enterprise Organisation (VEO). According to NESSI's definition [2] VEOs are formed when two or more administrative domains (and hence their Enterprise Grids) overlap and share resources. It further describes that the reality of VEO is that only a subset of the overall Grid within an enterprise is likely to be contributed to this virtual organisation. The underlying relationships among different enterprises within a VEO can be master-slave or peer-to-peer or a combination of both. We will apply the concept of VEO to peer-to-peer relationships in figure 4. If we consider the enterprises A and D to form a Virtual Enterprise Organisations (VEO), their SLAs are aggregated at a virtual aggregation point (vap) that represents this VEO. The virtual aggregation point is important to be represented because it in turn describes the SLA view of the resulting VEO which is different from the SLA views of A and D. The shared functionality of the VEO is described in the aggregated SLA computed within the vap-[AD]. Note that the big brackets have been adopted to highlight the jointly contained capabilities of enterprises A and D. The terms of services are aggregated through aggregation functions described in section 5. The terms marked as neutral are not merged and kept separate in the aggregated SLA. The virtual aggregation point also denotes the decision point of the resulting VO and policies such as distribution of revenue and cost of offered services will also be decided inside it. From a practical perspective, there are numerous issues such as trust, security, heterogeneity related to SLA aggregation among peer-to-peer enterprises. We have provided a conceptual framework including a third party trust manager to address these issues in [15].

Other NESSI models such as Hierarchical Enterprises and Extended Enterprises [2] can be easily described through our model. The concept of intercloud or cloud of clouds [16] is becoming very popular these days, which realizes the virtual collaboration of clouds. Such a virtual collaboration among clouds maps straightforwardly on our SLA aggregation model.

7   Motivational Scenario

In the following section, we will present a motivational scenario of an ad-hoc business value network which is enabled by the aggregation mechanism presented above. Arfa is visiting ULM. She is shooting movies and capturing snapshots with the camera, built in her mobile phone. The mobile device has limited storage space but luckily she knows a web service that can archive, enhance and host her movies online as soon as she completes a recording. She is also very much excited to share her experiences with her family and friends. Therefore she wants to update some blogs with images of the places and their historical description. Her friend told her about an online service that can collect images from her cell phone, print them and send them as postcards. So, she would like to do three tasks: automatically store and host her movies to external storage from where she and her friends can watch anytime using their mobile or static devices; automatically print some selected images as postcards and mail them to her family and friends through regular post; update some blogs with images and their historical descriptions. The SLA-Choreography resulting from this simple workflow is shown in figure 5. There are two services, namely the host-video service and post-photo service. The host video service downloads the video from the mobile device, enhance s it and archives it. Any authenticated user can play the video in a youtube like style. The Post-Photo service makes SLAs with two services: the Print&Post service and E-Post service. E-Post service is able to do its task by contracting two service namely Blog-Service and MMS-Service. The Blog service can automatically update the blogs with the images and automatically generate stories about their historical significance on the basis of their exact address. MMS service sends the selected images to friends on their mobile phones.

The SLA-Choreography resulting from this scenario is depicted in figure 5. We can see the aggregation functions described in figure 3(b) being applied in the scenario shown in figure 5. It is evident that the resolution offered by Host-Video service is the minimum of the three services below it. So at the aggregation point ap-S_1, the aggregation function Min will choose only minimum of the three resolutions as their aggregation types have been declared "min". On the same grounds, the job completion time for E-Post service is the maximum of those of Blog service and MMS service beause it is of "maxtype". The total cost that the client has to pay is the sum of the cost incurred on Host-Video service and the cost spent on Post-Photo service because cost has been declared as "sumtype".

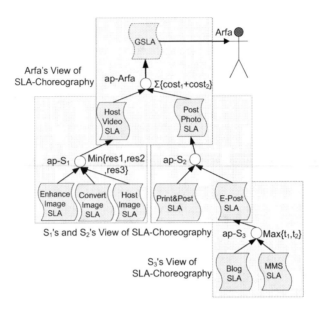

Fig. 5. Different Partners' SLA Views in Motivational Scenario

We take the liberty of importing external schema into WS-Agreement's Service Description Terms' section. The following chunk of Schema allows this.

```
<xs:complexType name="ServiceDescriptionTermType">
   <xs:complexContent>
      <xs:extension base="wsag:ServiceTermType">
         <xs:sequence>
            <xs:any namespace="##other" processContents="strict"/>
         </xs:sequence>
      </xs:extension>
   </xs:complexContent>
</xs:complexType>
```

The above schema enables us to include an XML structure of elements adhering to any external Schema. This makes it possible to incorporate the aggregation type (typea) element inside a Service Description Term. A simple schema to accomplish this can be written as follows.

```
<?xml version="1.0" encoding="utf-16"?> <xs:schema
xmlns:myns="http://schemas.xyz.com" xmlns="http://www.mynamespace.com"
targetNamespace="http://www.mynamespace.com"
xmlns:xs="http://www.w3.org/2001/XMLSchema">
<xs:simpleType name="aggregationType">
 <restriction base="xs:string">
   <enumeration value="Mintype"/>
   <enumeration value="Maxtype"/>
   <enumeration value="sumtype"/>
```

```
  <enumeration value="neutral"/>
  </restriction>
</xs:simpleType> ...
  <xs:element name="Resolution">
    <xs:complexType>
      <xs:sequence>
        <xs:complexType name="ResolutionXY">
          <xs:sequence>
            <xs:element name="ResolutionX" type="xs:integer"/>
            <xs:element name="ResolutionY" type="xs:integer"/>
          </xs:sequence>
        <xs:element name="aggregationType" type="xs:aggregationType"/>
        </xs:complexType">
      </xs:sequence>
    </xs:complexType>
  </xs:element>
... </xs:schema>
```

Then the service Description Term namely "resolution" for the Enhance-Video service may be expressed as follows.

```
<wsag:ServiceDescriptionTerm wsag:Name=Resolution"
wsag:ServiceName="Enhance-Video">
   <myns:ResolutionXY>
      <myns:ResolutionX> 640</myns:ResolutionX>
      <myns:ResolutionY>480</myns:ResolutionY>
   </myns:ResolutionXY>
   <myns:aggregationType> mintype</myns:aggregationType>
</wsag:ServiceDescriptionTerm>
```

The aggregationType (i.e. $type_a$) declares Resolution as a minType term. When it will be aggregated with other minType terms, only the minimum of these terms will become part of the aggregated SLA. Other aggregation types listed in the schema can be expressed and aggregated in a similar fashion.

8   Conclusion

We presented a view based formal model to describe hierarchical Service Level Agreements in supply chain scenarios such as Business Value Networks. SLA-Views help to maintain balance between trust and privacy. Our model identifies basic aggregation constructs that are used in the aggregation of SLAs. The whole aggregation process stays in compliance with the WS-Agreement standard. Due to the limited scope of this paper we could not include various details of our research related to different aspects of Business Value Networks such as flow of value and business models. However, We plan to address these details in context with the Cloud Computing, as a separate research paper. In future, we will continue our work on implementing a secure aggregation and validation framework for SLAs in heterogeneous Virtual Organizations.

Acknowledgements

We are extremely thankful to the reviewers of BPM09 for their valuable guidelines regarding the application areas of our research and thus helped us to produce a much improved Camera Ready Version of our paper. This work was partly supported by the project grant number IP395009, funded by University of Vienna.

References

1. Blake, M.B., Cunnings, D.J.: Workflow composition of service level agreements. In: International Conference on Services Computing, SCC 2007 (2007)
2. NESSI-Grid, http://www.soi-nwg.org/doku.php?id=sra:description (last access: March 12, 2009)
3. Project, S.: (March 12, 2009), http://www.sla-at-soi.org/index.html
4. Liu, D.R., Shen, M.: Workflow modeling for virtual processes: an order-preserving process-view approach. Information Systems 28, 505–532 (2002)
5. Liu, D.R., Shen, M.: Business-to-business workflow interoperation based on process-views. Decision Support Systems 38, 399–419 (2004)
6. Eder, J., Tahamatan, A.: Temporal consistency of view based interorganizational workflows. In: 2nd International United Information Systems Conference, Austria (2008)
7. Frankova, G.: Service level agreements: Web services and security, pp. 556–562. Springer, Heidelberg (2007)
8. Unger, T., Leyman, F., Mauchart, S., Scheibler, T.: Aggregation of service level agreement in the context of business processes. In: Enterprise Distributed Object Computing Conference (EDOC 2008), Munich, Germany (2008)
9. Aiello, M., Frankova, G., Malfatti, D.: What's in an agreement?An analysis and an extension of WS-agreement. In: Benatallah, B., Casati, F., Traverso, P. (eds.) ICSOC 2005. LNCS, vol. 3826, pp. 424–436. Springer, Heidelberg (2005)
10. Schulz, K.A., Orlowska, M.E.: Facilitating cross-organisational workflows with a workflow view approache. Data and Knowledge Engineering 51, 109–147 (2004)
11. Chebbi, I., Dustdar, S., Tata, S.: The view based approach to dynamic inter-organizational workflow cooperation. Data and Knowledge Engineering 56, 139–173 (2006)
12. Chiu, D., Li, K.K.Q., Kafeza, E.: Workflow view based e-contracts in a cross-organisational e-services environment. Distributed and Parallel Databases 12, 193–216 (2002)
13. Ludwig et al: Web service agreement (ws-agreement). gfd.107 proposed recommendation (last access: July 12, 2008)
14. (OGF), O.G.F.: http://www.ogf.org/ (last access: March 12, 2009)
15. ul haq, I., Huqqani, A.A., Schikuta, E.: A conceptual model for aggregation and validation of slas in business value networks. In: The 3rd International Conference on Adaptive Business Information Systems, ABIS 2009 (2009)
16. Jha, S., Merzky, A., Fox, G.: Using clouds to provide grids with higher levels of abstraction and explicit support for usage modes. Concurrency and Computation: Practice and Experience 21(8), 2087–1108 (2009)

Set Algebra for Service Behavior: Applications and Constructions

Kathrin Kaschner and Karsten Wolf

Universität Rostock, Institut für Informatik, 18051 Rostock, Germany
{kathrin.kaschner,karsten.wolf}@uni-rostock.de

Abstract. Compatibility of *behavior*, i.e. the correct ordering of messages, is one of the core aspects for the interaction between services as parts of an inter-organizational business process. In previous work, we proposed formal representations for service behavior (including Petri nets and service automata) and finite representations of sets thereof (*operating guidelines*).

In this article, we show how the basic set operations union, intersection, and complement, as well as membership and emptiness tests, can be implemented on finite representations of (typically infinite) sets of services. We motivate the operations by three examples of applications—service substitution, selection of behavior, and navigation in a behavioral registry.

1 Introduction

Correct interaction between services [1,2,3,4] requires compatibility in several aspects. This includes semantics (compatible interpretation of message contents), non-functional properties (compatible security levels, policies, latencies, etc.), and *behavior* (compatible order of exchanged messages). We contribute to the aspect of behavior. This aspect is particularly important if services implement complex business protocols, for instance as participants in an inter-organizational business process.

The behavior of a service can be formally described in various formalisms including process algebra [5,6], state machines [7,8,9,10], and Petri nets [11,12,13]. There exist strong links from industrial languages like WS-BPEL [14] and BPMN [15] to these formalisms. In previous work [16,17,18], we considered *sets* of service behaviors, most prominently the set of all behaviorally compatible partners to a given service (its *operating guidelines*). We showed that such a (generally infinite) set can actually be finitely represented, using the concept of *annotated automata* [19]. This way, we could provide solutions as well as tool support for several interesting questions including controllability (a sanity check closely related to the soundness of a workflow model) [20], substitutability of services [21,18], test case generation [22], contract-based composition [23], and others.

This article is devoted to the implementation of basic set operations on an extension of annotated automata. This means that, given extended annotated automata representing sets M_1 and M_2 of services, respectively, we are able to

U. Dayal et al. (Eds.): BPM 2009, LNCS 5701, pp. 193–210, 2009.

compute extended annotated automata that represent the intersection $M_1 \cap M_2$, the union $M_1 \cup M_2$, and the complement $\overline{M_1}$ of the given sets. We further study the decision problems membership ($S \in M$?) and emptiness ($M = \emptyset$?).

We motivate the relevance of the proposed operations by sketching three applications. The first application is a novel approach to checking substitutability of services. In contrast to previous techniques [21,18], we obtain a natural counterexample facility in the case of non-substitutability. The second application concerns the joined use of operating guidelines and user-defined requirements in the automated calculation of a partner service to a given service. This yields a more systematic and more general approach than previous work [24]. The third application provides an approach that permits the selection of a service from a service registry according to a behavioral query. So far, behavioral specifications can only be queried in approximations where somehow the control flow for implementing the behavior is "guessed" [25].

While several groups study service behavior, we are not aware of any competing approach that is centered around *sets* of service behaviors and their finite representation.

We start with a presentation of the motivating applications in Sect. 2. In Sect. 3, we introduce the basic formalisms for modeling the behavior of services and sets thereof. Then we study a few preliminary operations in Sect. 4. The actual implementation of set operations is discussed in Sect.5. In Sect. 6, we revisit the motivating applications and discuss their complexity.

2   Motivation

In this section, we sketch a few approaches which all involve the use of set operations on sets of services. All approaches make sense as soon as it is possible to finitely represent operands and results to the basic set operations union, intersection, and complement, and as soon as the problems membership and emptiness are decidable. In subsequent sections, we shall show that these assumptions are indeed valid.

Initial representations of operands may be constructed from scratch, or may stem from the calculation of *operating guidelines* to a given service S [16,17]. Operating guidelines are a finite representation of the set $Partners(S)$ of all correctly interacting partners of the given service. The representation is based on annotated automata which are introduced subsequently.

Substituting a service

If a service S is substituted by another service S' (e.g., a new version as a result of reorganizations or outsourcing), it may be desirable that the substitution is somewhat invisible to the outside world. A valid requirement from the point of view of behavior would be that $Partners(S) \subseteq Partners(S')$. This means that any partner that interacts correctly with S will also interact correctly with S'.

The stated inclusion can be transformed into the emptiness problem as follows: $Partners(S) \subseteq Partners(S') \iff Partners(S) \cap \overline{Partners(S')} = \emptyset$.

As we know from [16,17], the sets $Partners(S)$ and $Partners(S')$ can be finitely represented using annotated automata. Hence, an implementation of the set operations intersection, complement, and the emptiness check would allow us to decide substitutability.

The proposed solution is not the first approach to substitutability. Even operating guidelines have been used before for this task [21,18]. However, these approaches were not able to come up with a counterexample in the case of non-substitutability. A counterexample would be any service in $Partners(S_1)$ but not in $Partners(S_2)$.

In our case, the intermediate expression $Partners(S) \cap \overline{Partners(S')}$ represents the set of all valid counterexamples and our (subsequently presented) implementation of the emptiness check will actually return an element of this set. The returned counterexample may provide useful diagnostic information for non-substitutability.

Querying a set of behaviors

Consider a setting where you want to use a service S without having your own fixed partner service S'. The set $Partners(S)$ of all correctly interacting partner services contains all possible choices and can be finitely represented using the approach of [16,17]. Consequently, it would be desirable to construct S' by simply selecting an element of $Partner(S)$. Of course, S' does only represent control flow and communication events which must then be enhanced with actual code. However, the communication structure is correct by construction and hence an automated construction could indeed be desirable.

As the set $Partners(S)$ is typically infinite, the finite representation is somewhat implicit. The selection of an element from $Partners(S)$ is thus a nontrivial task. A suitable way would be to specify constraints for "desired" behavioral properties and then to compute an element that is in $Partners(S)$ and satisfies the constraints.

Examples for behavioral constraints could be exclusion of certain communication events (e.g., "I want a service where I do not need to send my credit card number"), enforcement of communication events (e.g., "I want a service where I eventually receive a *delivered book* message"), order of events (e.g., "I want to pay only after having received the ordered item") and, many more.

According to a fundamental principle in set theory, any behavioral constraint can be identified with the set of services satisfying it. In [24], we demonstrated that many elementary behavioral constraints, including enforcement, exclusion, and a number of ordering constraints, can in fact be finitely represented using the formalism of annotated automata used subsequently.

At this point, set algebra comes into play. Let P_1 and P_2 be behavioral properties. Let $Sat(P_i)$ be the set of services that satisfy P_i. Then Boolean combinations of P_1 and P_2 obviously correspond to set operations as follows: $Sat(P_1 \wedge P_2) = Sat(P_1) \cap Sat(P_2)$ (conjunction); $Sat(P_1 \vee P_2) = Sat(P_1) \cup Sat(P_2)$ (disjunction); $Sat(\neg P_1) = \overline{Sat(P_1)}$ (negation).

In other words, any language for representing some behavioral requirements can be extended to a language that naturally permits Boolean combinations of these requirements. If the primitives of the language are chosen such that they can be finitely represented then any Boolean combination can be finitely represented as well.

Finally, consider a requirement P. For checking that a service S can actually be used by a partner that satisfies requirement P, we simply need to select an arbitrary element of $Partners(S) \cap Sat(P)$. If this intersection is empty, S cannot be used correctly such that in addition P is satisfied. Otherwise the behaviors in $Partners(S) \cap Sat(P)$ represent exactly the desired behaviors. If this set contains more than one element, we can either select one element arbitrarily or add more requirements. Using this approach, we can interactively choose a correctly interacting partner for a given service S.

Navigating in a behavioral registry

Now, consider the same setting as in the previous subsection, but with a whole service registry containing services $S_1, \ldots, S_n$ instead of a single service S. We assume that, for every S_i, a finite representation of $Partners(S_i)$ is available. We call such a registry a *behavioral registry*.

The considered problem is to find out whether the registry contains a service S_i which can be used by a partner that fits to a specified requirement P. In a naive solution, we need to build the intersections $Sat(P) \cap Partners(S_1)$, ..., $Sat(P) \cap Partners(S_i)$ where S_i is the first service in the registry that yields a nonempty intersection. Hence, we would need to perform up to n intersection operations. Unfortunately, n may get intractably large.

Using set algebra, we show a way to reduce the number of intersection operations to the much smaller number of at most $\log n$. To this end, consider sets $M_{i,j} = \bigcup_{k=i}^{j} Partners(S_i)$. These sets can be computed from the sets $Partners(S_i)$ using the set operation *union*.

Consider now the following sequence of intersections. Start with $Sat(P) \cap M_{1,n}$. If the intersection is empty, we know that no service in the registry can be used such that P is satisfied. Otherwise, we know that there is some service S^* that satisfies P and interacts correctly with one of the services $S_1, \ldots, S_n$. Our implementation of the empiness check will actually return such a service S^*.

Continue checking $S^* \in M_{1,\frac{n}{2}}$. If the intersection is nonempty, we know that a correctly interacting partner for S^* is among the first $\frac{n}{2}$ entries of the registry. Otherwise, such a service must be among the remaining services. Continuing according to this pattern, every membership test divides the search space into halves. After $\log n$ iterations, there is only a single service remaining and we are done.

If we fill up the registry with dummy services such that n becomes a power of 2, we actually only need to compute those $M_{i,j}$ where $k = j + 1 - i$ is a power of 2 as well and j is a multiple of that k. For example, in a registry with 8 elements, we will only compare to $M_{1,8}$ in the first round, to $M_{1,4}$ or $M_{5,8}$ in the second, and $M_{1,2}$, $M_{3,4}$, $M_{5,6}$, or $M_{7,8}$ in the third round. The number of such sets is equal to $n - 1$, so the overhead of storing the $M_{i,j}$ is linear as far as the *number* of represented sets is concerned.

The proposed kind of querying a registry is orthogonal to other approaches which cannot select a service according to its behavior. Semantic approaches typically assume a trivial behavior ("stateless services"). Other approaches like [25] try to approximate behavior with patterns in the WS-BPEL flow of control. There are, however, many different control flow structures that implement the same behavior, so this approach can only partially select a service by a behavioral specification.

3 Behavior of Services

In the remainder of this article, we develop an approach for realizing the basic set operations using finite representations of sets of service behaviors. We use automata in different shapes for the various aspects of our approach. Generally, transition labels correspond to communication activities such as sending or receiving a message from the environment. One of the labels may be the label τ which shall always represent an internal (non-communicating) activity.

Definition 1 (Automaton). $A = [Q, C, \delta, Q_0]$ *is an* automaton *iff Q is a nonempty finite set of* states, C *is a set of* labels, $\delta \subseteq Q \times C \times Q$ *is a* transition relation *such that every state $q \in Q$ is reachable from q_0 via transitive applications of δ, and $\emptyset \subset Q_0 \subseteq Q$ is the nonempty set of* initial states.

We shall also write $q \xrightarrow{x}_\delta q'$ for $[q, x, q'] \in \delta$. We generally use indices to distinguish ingredients of different automata whenever there could be ambiguities.

We describe the behavior of a service as a *service automaton*. Service automata have only one initial state and extend general automata with a notion of final states for modeling completion of a service instance.

Definition 2 (Service automaton). $S = [Q, C, \delta, q_0, \Omega]$ *is a* service automaton *iff $[Q, C, \delta, \{q_0\}]$ is an automaton and $\Omega \subseteq Q$ is a set of* final states.

Figure 1 shows four service automata. Initial states have an incoming arc from nowhere. Final states are depicted bold. The edges are labeled with τ (non-communicating activity) or with messages sent to (preceded with !) or received from (preceded with ?) a partner service. The service automata $S_1, \ldots, S_3$ can be seen as simple online shops with the communication activities !o (for send order), !s (for send special offer), !i (for send invoice), ?a (for receive accept offer), and ?r (for receive reject offer). The service automaton S_4 represents the behavior of a buyer service: It receives an offer or a special offer. While special offers are always accepted, standard orders may be rejected. Upon acceptance of a (special) offer, an invoice is received and the service terminates. Note that receiving events of $S_1, \ldots, S_3$ are sending events of S_4 and vice versa.

For the finite representation of a *set* of service automata, we employ the concept of *annotated automata* [19]. An annotated automaton extends a general

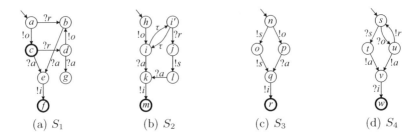

(a) S_1 (b) S_2 (c) S_3 (d) S_4

Fig. 1. Four examples for service automata

automaton with Boolean annotations to states. We use the usual symbols $\wedge, \vee, \neg$ for Boolean operations. $\bot$ denotes the Boolean function that returns *false* to all arguments. $\top$ denotes the Boolean function that returns *true* to all arguments.

Definition 3 (Annotated automaton). *An* annotated automaton $A^\phi = [A, \phi]$ *consists of an automaton* $A = [Q, C, \delta, Q_0]$ *and an annotation function* ϕ, *where, for all* $q \in Q$, $\phi(q)$ *is a Boolean formula with propositions in* $C \cup \{final\}$. A^ϕ *is deterministic if the following condition holds for all* $q_1, q_2 \in Q$ *with* $q_1 \neq q_2$: *If* $\{q_1, q_2\} \in Q_0$ *or there are a state* q *and a label* x *with* $\{[q, x, q_1], [q, x, q_2]\} \subseteq \delta$ *then* $\phi(q_1) \wedge \phi(q_2) \equiv \bot$.

Figure 2 shows an example for a deterministic annotated automaton. Note that an annotated automaton does not have final states. Instead, a proposition *final* in its annotations may constrain final states of represented service automata, as defined below. The notion of determinism generalizes classical automata theory. We may start with, or move to, different states. However, the local annotations of the possible states exclude each other and are subsequently used to resolve the nondeterminism. Using this mechanism, subsequent concepts are well-defined.

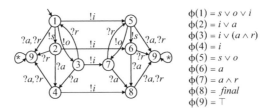

$$\phi(1) = s \vee o \vee i$$
$$\phi(2) = i \vee a$$
$$\phi(3) = i \vee (a \wedge r)$$
$$\phi(4) = i$$
$$\phi(5) = s \vee o$$
$$\phi(6) = a$$
$$\phi(7) = a \wedge r$$
$$\phi(8) = final$$
$$\phi(9) = \top$$

Fig. 2. An annotated automaton A. Annotations are listed at the right hand. "*" means that there is a transition for every element in the alphabet. For reducing the number of edge crossings, we depicted two copies of state 9.

Informally, an annotated automaton A^ϕ represents all those service automata S that can be structurally embedded and respect the annotations. This intuition is formalized in the notion of a matching relation. Structural embedding is formalized in a matching relation which is actually a particular strong simulation relation. Respecting the annotations is checked by evaluating the annotations of A^ϕ in an assignment that corresponds to outgoing edges in S. Determinism of A^ϕ assures that the matching relation is uniquely determined. Throughout the paper, we represent an assignment as a set of propositions with the following meaning: Presence in the set represents the value *true* assigned to this proposition, absence represents the value *false*.

Definition 4 (Matching realation). *Let $S = [Q_S, C, \delta_S, q_{0S}, \Omega_S]$ be a service automaton and $A^\phi = [Q_A, C, \delta_A, Q_{0A}, \phi]$ be a deterministic annotated automaton. We say that S matches with A^ϕ if there is a relation $\varrho_{S,A}$ that satisfies the following requirements:*

(Initial) There is a $q_{0A} \in Q_{0A}$ such that $[q_{0S}, q_{0A}] \in \varrho_{S,A}$.
(Communication) If $[q_S, q_A] \in \varrho_{S,A}$, and $q_S \xrightarrow{x}_{\delta_S} q_S'$, then there is a state q_A' with $q_A \xrightarrow{x}_\delta q_A'$ and $[q_S', q_A'] \in \varrho_{S,A}$.
(Annotation) If $[q_S, q_A] \in \varrho_{S,A}$ then $\phi(q_A)$ is satisfied in the assignment $\beta(q_S)$ with $\beta(q) = \{x \in C \mid there is a state q' with q \xrightarrow{x}_\delta q' or (x = final and q \in \Omega_S)\}$.
If a relation $\varrho_{S,A}$ satisfying the given requirements exists, we call the minimal such relation (w.r.t. set inclusion) the matching relation *between S and A^ϕ. In this case, we further say that S matches with A^ϕ. Let $Match(A^\phi)$ be the set of all service automata that match with A^ϕ.*

If $\varrho_{S,A}$ exists, it is indeed unique. In fact, the requirements can be seen as an inductive definition where *Initial* is the base, *Communication* is the step, and *Annotation* uniquely resolves the ambiguities in the other clauses (the choice of a state in A^ϕ).

Among the services in Fig. 1, S_1 and S_2 match with the annotated automaton in Fig. 2. The corresponding matching relations are $\varrho_{S_1,A} = \{[a, 1], [c, 3], [d, 1], [b, 3], [e, 4], [f, 8], [g, 9]\}$ and $\varrho_{S_2,A} = \{[h, 1], [i, 3], [i', 3], [k, 4], [j, 1], [l, 2], [m, 8]\}$. Examples for assignments are $\beta(i) = \beta(i') = \{?a, ?r\}$, $\beta(j) = \{!s\}$ and $\beta(m) = \{final\}$. In contrast, S_3 does not match. In state o, there is a transition with label $!s$ that is not present in the corresponding state 2 of A. In state p, the assignment $\beta(p) = \{?a\}$ does not satisfy $\phi(3) = i \vee (a \wedge r)$. S_4 does not match as its alphabet does not fit (all receiving events in S_4 are sending events in A and vice versa). Consequently, for the states t and u there is no corresponding state in A which fulfills the communication requirement of Def. 4.

In [16,17], we showed that, for each service automaton S, a deterministic annotated automaton, named *operating guideline OG_S*, can be constructed such that the set of all service automata that match with OG_S is exactly the set of those service automata which interact correctly with S; that is, for the operating guideline OG_S it holds $Match(OG_S) = Partners(S)$. The annotated automaton A in Fig. 2 is actually the operating guideline OG_{S_4} of the service automaton S_4 in Fig. 1. States $1, \ldots, 4$ and 8 in A describe the natural interaction of a partner with S_4. States $5, \ldots, 7$ are there due to our asynchronous model of message passing. This means that the invoice may be sent at any time since it will stay in the mailbox until it is received. State 9 describes useless but harmless behavior. In fact, states corresponding to state 9 in a matching relation are unreachable but harmless. Transitions to state 9 just wait for messages that are simply not sent by S_4 at that time. As long as a partner has other opportunities to proceed (as specified in the annotations), code for receiving such a message does not hurt.

The matching situation with $A = OG(S_4)$ in particular means that S_1 and S_4 as well as S_2 and S_4 compose to a correctly interacting system. The deadlock state g in S_1 is an example of a harmless problem since it is never reached in the composition with S_4. In contrast, S_3 and S_4 do not interact correctly. If S_3 proceeds through states o and q, two messages of type s are sent to S_4. However, only one of them can be consumed by S_4. If S_3 passes to state p instead, then the sequence $s - u - s$ in S_4 leads to a deadlock, i.e. a situation where neither S_3 nor S_4 can proceed.

Unfortunately, set operations, particularly complement, cannot be implemented using annotated automata as such. One of the intuitive reasons is that the annotations represent constraints which hold *whenever* the corresponding state is visited. For violating such a "whenever" constraint, it is sufficient to violate it once. This more existential kind of requirement cannot be expressed with annotated automata.

Consequently, we proceed with an extension of annotated automata that is capable of implementing the set operations. The proposed extension has already been used in [17] for expressing a certain type of behavioral constraints on compatible partners. The main idea of [17] is to add a global Boolean formula with states as propositions.

Definition 5 (Extended annotated automaton). *Let $A^\phi = [Q, C, \delta, Q_0, \phi]$ be a deterministic annotated automaton and χ be a Boolean formula with propositions taken from the set Q. Then, $A^{\phi,\chi} = [A, \phi, \chi]$ is an extended annotated automaton.*

The Boolean formula χ is called the *global constraint* of $A^{\phi,\chi}$ while the $\phi(q_i)$ are called *local constraints*. Propositions of χ evaluate to true if they are "touched" by the matching relation $\varrho_{S,A}$. Fortunately, our matching relation is unique, so this evaluation is well defined.

Definition 6 (Matching with $A^{\phi,\chi}$). *Let S be a service automaton and let $A^{\phi,\chi}$ be an extended (and thus deterministic) annotated automaton. S matches with $A^{\phi,\chi}$ iff S matches with A^ϕ (the annotated automaton without extension) using the matching relation $\varrho_{S,A}$ and χ is evaluated to true by the following assignment γ_S: $\gamma_S = \{q_A \mid$ there is a state $q_S \in Q_S$ such that $[q_S, q_A] \in \varrho_{S,A}\}$. Let $Match(A^{\phi,\chi})$ denote the set of all service automata that match with $A^{\phi,\chi}$.*

Adding the global constraint $\chi = 2 \vee 6$ to the annotated automaton A in Fig. 2, we obtain an extended annotated automaton that matches only with services which *may* send a special offer. Among the services of Fig. 1, only S_2 matches with A and the global constraint χ. It evaluates χ in the assignment $\gamma_{S_2} = \{1, 2, 3, 4, 8\}$ which contains state 2. In contrast, S_1 evaluates χ in the assignment $\gamma_{S_1} = \{1, 3, 4, 8, 9\}$ which contains neither state 2 nor state 6. Indeed, S_1 may not send a special offer.

Every annotated automaton A^ϕ can be canonically transformed into an extended annotated automaton $A^{\phi,\chi}$ by setting $\chi \equiv true$.

4   Preliminary Operations on Annotated Automata

In this section, we study three transformations on annotated automata. The transformations bring an automaton into a normal form that simplifies subsequent constructions. First, we aim at making an annotated automaton *total* and *complete*. An annotated automaton is total if every state has successors for all labels. It is complete if the annotations of the successor states with same label cover all cases.

Definition 7 (total, complete). *An extended annotated automaton $A^{\phi,\chi}$ is total if, for all states q and all labels x, there is a state q' such that $q \xrightarrow{x}_\delta q'$. A^ϕ is complete if the following two conditions hold:*

(1) $(\bigvee_{q \in Q_0} \phi(q)) \equiv \top$
(2) for all $q \in Q$ and all $x \in C$, $(\bigvee_{q':[q,x,q'] \in \delta} \phi(q')) \equiv \top$

In a sequel we shall show that an annotated automaton make total without changing its semantics. In contrast, it is not possible to transform an arbitrary annotated automaton into a complete one. However, we shall demonstrate, that the transformation into a complete automaton is possible for *extended* annotated automata. The advantage of a total and complete automaton is that the existence of the matching relation is no issue. In this case, it is only the formula χ that decides whether or not a service matches.

Corollary 1. *If A^ϕ is total and complete then, for each S operating on the same alphabet as A^ϕ, the matching relation $\varrho_{S,A}$ exists; that is, i.e. all services with the same alphabet as A match with A^ϕ.*

This observation can be verified easily as, in each situation referred to in Def. 4, there is an available successor in A^ϕ to satisfy the requirements.

   Actually, every (extended) annotated automaton $A^{\phi(,\chi)}$ can be transformed into a total one without changing $Match(A^{\phi(,\chi)})$. The following construction implements this transformation. The idea is to insert missing edges with label x but to forbid their use by adding $\neg x$ to the corresponding annotations:

- Invent a fresh "trap" state q_t with annotation $\top$ and successors to itself, i.e. let $q_t \xrightarrow{x}_{\delta_A} q_t$, for all $x \in C$.
- For each state q and each label x such that there is no state q' with $q \xrightarrow{x}_\delta q'$,
  - insert an edge $q \xrightarrow{x}_\delta q_t$;
  - replace $\phi(q)$ with $\phi(q) \wedge \neg x$.

Correctness of this construction can be easily verified by inspecting Def. 6. The costs for the transformation are linear in the number of states of A^ϕ. In this and all subsequent cost considerations, we treat the size of the alphabet C as

$$\phi(2) = (i \vee a) \wedge \neg s \wedge \neg o$$
$$\phi(t) = \top$$

$$\phi(4) = i$$
$$\phi(4') = \neg i$$
$$\phi(6) = a$$
$$\phi(6') = \neg a$$
$$\phi(9) = \top$$
$$\phi(9') = \bot$$

(a) making total   (b) making complete

Fig. 3. In (a), the procedure of making the automaton total is shown for state 2 of A (from Fig. 2). (b) illustrates the procedure of completion for the same state. The dashed parts are those that are inserted in the course of a transformation. "*" means that there is a transition for every element in the alphabet.

a constant. This is supported by the observation that the interface of a service tends to be very small in comparison to its number of states[1].

Figure 3 (a) illustrates the procedure for state 2 of the annotated automaton A in Fig. 2. The alphabet of A is $\{?a, ?i, !o, !r, !s\}$. State 2 has only three outgoing edges labeled with $?a$, $?i$ and $!r$. Therefore, it is not total and two new edges (with the missing labels $!s$ and $!o$) are added by the above algorithm. To avoid that $Match(A^{\phi(,\chi)})$ is changed by this modification, we set the Boolean formula of state 2 to $\phi(2) = (i \vee a) \wedge \neg s \wedge \neg o$.

The next transformation shows that every *extended* annotated automaton can be made complete.

To this end, transform $A^{\phi,\chi}$ as follows.

- If $(\bigvee_{q \in Q_0} \phi(q)) \neq \top$, insert an additional initial state q_0^* with successors in a trap state like in the previous construction, and replace χ with $\chi \wedge \neg q_0^*$. Let the annotation of q_0^* be $(\bigwedge_{q \in Q_0} \neg \phi(q))$.
- If, for some q and x, $\bigvee_{q':[q,x,q'] \in \delta} \phi(q') \neq \top$, insert an additional successor state q^* with successors in a trap state as in the previous construction, and replace χ with $\chi \wedge \neg q^*$. Let the annotation of q^* be $(\bigwedge_{q':[q,x,q'] \in \delta} \neg \phi(q'))$.

The procedure can at most insert $|Q| \cdot |C|$ states, so the costs in space and time are again linear. Observe that both transformations preserve determinism of the involved annotated automaton.

In Fig. 3 (b), we illustrated the transformation of state 2 of the annotated automaton A in Fig. 2. State 2 has only one $?a$-labeled edge leading to state 4. Since the annotation $\phi(4) \neq \top$ we add an edge labeled with $?a$ to a new state $4'$. State $4'$ has for every label a transition to the trap state t. With the remaining successor states of state 2 is proceeded in the same manner.

The purpose of the last transformation is to align the interfaces (the alphabet) of two (extended) annotated automata. This transformation consists just of the replacement of the alphabet C by some superset C'. In fact, $Match(A^{\phi,\chi})$ does not change if the alphabet C is replaced with any superset $C' \supseteq C$ as there is no edge in A with a label in $C' \setminus C$. The only issue here is that the modified interface

---

[1] In a project conducted under participation of an enterprise, we analyzed several real WS-BPEL specifications. In average, such a service had about a dozen message types whereas the service itself had some 100,000 states.

typically turns a total automaton into a partial one as no state has successors for the new messages in C'. However, the transformations presented above solve this problem. For this reason, we shall assume throughout the remainder of this paper that the involved annotated automata range on exactly the same alphabet.

5   Set Operations on Annotated Automata

In this section, we present approaches for the implementation of basic set operations on deterministic extended annotated automata. The result will again be a deterministic extended annotated automaton. Each subsection is devoted to one of the operations.

Complement

In this section, we aim at constructing an extended annotated automaton $A'^{\phi',\chi'}$ from a given extended annotated automaton $A^{\phi,\chi}$ such that, for every service automaton S operating on the alphabet C_A, $S \in Match(A^{\phi,\chi})$ if and only if $S \notin Match(A'^{\phi',\chi'})$.

Thanks to the concepts coined in the previous section, the operation can be implemented trivially. All we need to do is to negate the global formula χ.

Theorem 1 (Negation). *Let $A^{\phi,\chi}$ be a total and complete deterministic annotated automaton. Let S be a service automaton with $C_S = C_A$. Then $S \in Match(A^{\phi,\chi})$ if and only if $S \notin Match(A^{\phi,\neg\chi})$.*

Proof. As the matching relation is the same for $A^{\phi,\chi}$ and $A^{\phi,\neg\chi}$, the claim reduces to: An assignment γ satisfies the Boolean formula χ if and only if γ does not satisfy $\neg\chi$ which is obvious.   □

Given a total and complete extended annotated automaton, negation can be executed in constant time. If an arbitrary deterministic extended annotated automaton is given, we can still execute negation in linear time which is then consumed for the transformation into a total and complete automaton.

A nice feature of our construction is that a double application of negation yields the original automaton. While this is trivial for total and complete automata, it is actually true also for partial or incomplete automata as long as we allow ourselves to

– remove a state q if χ implies $\neg q$;
– remove an edge $q \xrightarrow{x}_\delta q'$ if $\phi(q)$ implies $\neg x$.

Both transformations are safe in the sense that they do not change $Match(A^{\phi,\chi})$. They can undo the effect of the transformations in the previous section.

Intersection

An intersection operation on annotated automata has already been proposed in previous work [24,21]. Although we basically apply the same ideas, we present the approach again as we use a more general setting: we use extended automata instead of plain annotated automata and we permit arbitrary Boolean annotations rather than negation-free formulas only.

The idea of implementing intersection is to construct the product automaton known from classical automata theory. A product automaton implements the idea that both constituents run in parallel, in every step executing transitions with the same label.

We annotate states in the product automaton with the conjunction of the annotations of the constituents. The global formulas are connected by $\wedge$. The only just technical difficulty is that the global formulas of the input automata A and B range over completely different alphabets (Q_A resp. Q_B) than the product automaton (having $Q_A \times Q_B$ as its set of states). As the meaning of a proposition q in a global formula is that q is "touched" by the matching relation, it is natural to replace a proposition q_A in χ_A with $\bigvee_{q_B \in Q_B}[q_A, q_B]$. Likewise, a proposition q_B in χ_B should be replaced with $\bigvee_{q_A \in Q_A}[q_A, q_B]$. Other than this, the construction is straightforward.

Definition 8 (product automaton). *Let A^{ϕ_A, χ_A} and B^{ϕ_B, χ_B} be extended annotated automata with $C_A = C_B$. Then the* production automaton $P = A \times B$ *is the extended annotated automaton defined as follows:*

- $Q_P = Q_A \times Q_B$;
- $C_P = C_A(= C_B)$;
- $Q_{0P} = Q_{0A} \times Q_{0B}$;
- $[q_A, q_B] \xrightarrow{x}_{\delta_P} [q'_A, q'_B]$ *if and only if* $q_A \xrightarrow{x}_{\delta_A} q'_A$ *and* $q_B \xrightarrow{x}_{\delta_B} q'_B$;
- *for all* $[q_A, q_B] \in Q_P$, $\phi([q_A, q_B]) = \phi(q_A) \wedge \phi(q_B)$;
- $\chi_P \equiv \chi_A^* \wedge \chi_B^*$.

In the last item, χ_A^* and χ_B^* are obtained from χ_A and χ_B as explained in the text above.

The construction of a product automaton preserves each of the properties deterministic, total, and complete. Figure 4 shows an example for the construction of a product automaton, applied to non-total and incomplete automata.

Theorem 2 (Intersection). *Let A^{ϕ_A, χ_A} and B^{ϕ_B, χ_B} be extended annotated automata with $C_A = C_B$. Then $Match(A \times B) = Match(A) \cap Match(B)$.*

The result actually holds also for non-total and incomplete extended automata. For better readability, however, we support our claim only for the total and complete case.

Proof. (Sketch.) Let A and B (and in consequence P, too) be total and complete extended annotated automata. By Cor. 1, the matching relations $\varrho_{S,A}$, $\varrho_{S,B}$, and $\varrho_{S,P}$ all exist and the local annotations are satisfied by every service. Let S be a service with $C_S = C_A$. There is a transition in P if and only if there are corresponding transitions in A and B. Using an inductive argument along the lines of Def. 4, we can show the following relations between the various matching relations which all exist by Cor. 1.

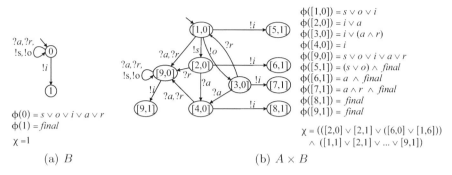

$\phi(0) = s \vee o \vee i \vee a \vee r$
$\phi(1) = final$

$\chi = 1$

(a) B

$\phi([1,0]) = s \vee o \vee i$
$\phi([2,0]) = i \vee a$
$\phi([3,0]) = i \vee (a \wedge r)$
$\phi([4,0]) = i$
$\phi([9,0]) = s \vee o \vee i \vee a \vee r$
$\phi([5,1]) = (s \vee o) \wedge final$
$\phi([6,1]) = a \wedge final$
$\phi([7,1]) = a \wedge r \wedge final$
$\phi([8,1]) = final$
$\phi([9,1]) = final$

$\chi = (([2,0] \vee [2,1] \vee ([6,0] \vee [1,6]))$
$\wedge ([1,1] \vee [2,1] \vee \dots \vee [9,1]))$

(b) $A \times B$

Fig. 4. In the left part, we show another extended annotated automaton. The right part depicts the product automaton $A \times B$ where A is the automaton in Fig. 2. The global constraint of A is $\chi = 2 \vee 6$ for A. For simplifying the resulting automaton, we could have set all propositions in local constraints to *false* where no corresponding edge is present. This modification is safe as no service automaton having such a transition can match. States $[5,1]$, $[6,1]$, and $[7,1]$ would be removed this way.

- If $[q_S, q_A] \in \varrho_{S,A}$ then there exists a q_B such that $[q_s, [q_A, q_B]] \in \varrho_{S,P}$;
- If $[q_S, [q_A, q_B]] \in \varrho_{S,P}$ then $[q_S, q_A] \in \varrho_{S,A}$.

From these facts, we may conclude that, for some q_B, truth of proposition $[q_A, q_B]$ in the assignment to χ_P implies truth of proposition q_A in the assignment to χ_A and truth of $\bigvee_{q'_A \in Q_A} [q'_A, q_B]$ in χ_A^*. In the other case, if $[q_A, q_B]$ is false for all q_B in the assignment to χ_P, so is the assignment to q_A in χ_A and $\bigvee_{q'_A \in Q_A} [q'_A, q_B]$ in χ_A^*. Consequently, χ_A^* and χ_A always get the same values. Arguing symmetrically, χ_B and χ_B^* evaluate to the same value, too. In consequence, S satisfies χ_P if and only if S satisfies both χ_A and χ_B. $\square$

The costs for executing intersection are in the magnitude of $\mathcal{O}(|Q_A| \cdot |Q_B|)$ both regarding space and time.

Union

Given the two operations of intersection and complement from the previous subsections, the implementation of union is trivial using De Morgan's rule: $M \cup N = \overline{\overline{M} \cap \overline{N}}$.

Given the efficiency of our approach to complement and intersection, and further remembering that the operations preserve the properties of being total and complete, there is most likely no significantly more efficient realization of union.

In consequence, we have an implementation for union that costs $\mathcal{O}(|Q_A| \cdot |Q_B|)$.

Membership

The membership problem is a decision problem, Given a service automaton S and an extended annotated automaton $A^{\phi,\chi}$, we want to know whether $S \in Match(A^{\phi,\chi})$.

Following the lines of Def. 4, we need to compute the relation $\varrho_{S,A}$. This can be done in time $\mathcal{O}(|Q_S| \cdot |Q_A|)$, using a coordinated depth-first search through S and A. The evaluation of the local annotations can be done during this search and does not require extra costs. After having computed $\varrho_{S,A}$, the value of the propositions of the global annotation χ is determined, and the formula can be evaluated in linear time.

Hence, the overall costs for membership amount to $\mathcal{O}(|Q_S| \cdot |Q_A|)$.

Emptiness

Emptiness is the problem of checking whether, for a given deterministic extended annotated automaton $A^{\phi,\chi}$, $Match(A^{\phi,\chi}) = \emptyset$.

Checking emptiness is significantly more difficult than emptiness checks in previous approaches that involved plain annotated automata [16,18]. These approaches did not use negated propositions. Negation is, however, essential for implementing complement.

As a first result, we show that emptiness is at least decidable. We prove this by showing the following result.

Theorem 3 (Emptiness). *Let $A^{\phi,\chi}$ be a total and complete deterministic extended annotated automaton. If there is a service $S \in Match(A^{\phi,\chi})$, there is also a service $S^* \in Match(A^{\phi,\chi})$ which is a subautomaton of some automaton M which can be computed from $A^{\phi,\chi}$.*

The idea for building M is to separate the states of A into several copies. Each copy $[q_A, \beta]$ of a state q_A corresponds to a particular assignment β that satisfies the local annotation $\phi(q_A)$. Successors of $[q_A, \beta]$ are chosen such that state $[q_A, \beta]$ produces exactly the assignment β in the matching procedure. In addition, we take care that $[q_A, \beta]$ matches exactly with state q_A in $A^{\phi,\chi}$. Then, a given service automaton S that matches with $A^{\phi,\chi}$, can be transformed into the subautomaton S^* of M by restricting M to exactly those states and assignments that are actually occurring in S.

Proof. (Sketch.) Let $A^{\phi,\chi}$ be given. Construct M as follows. $Q_M \subseteq Q_A \times 2^{C \cup \{final\}}$ where $[q_A, \beta] \in Q_M$ iff β satisfies $\phi(q_A)$. $Q_{0M} = Q_M \cap (Q_{0A} \times 2^{C \cup \{final\}})$. $[[q_A, \beta], x, [q'_A, \beta']] \in \delta_M$ iff $[q_A, x, q'_A] \in A$ and $x \in \beta(q)$. It can be shown by induction that $\varrho_{M,A} = \{[q_A, \beta], q_A] \mid [q_A, \beta] \in Q_M\}$.

Assume now that $Match(A^{\phi,\chi}) \neq \emptyset$ and let $S \in Match(A^{\phi,\chi})$. It is our task to exhibit a service $S^* \in Match(A^{\phi,\chi})$ that is a subautomaton of M. We set $Q_{S^*} = \{[q_A, \beta(q_S)] \mid [q_S, q_A] \in \varrho_{S,A}\}$. $\delta_{S^*} = \delta_M \cap (Q_{S^*} \times Q_{S^*})$. $q_{0S^*} = [q_A, \beta(q_{0S})]$, for the unique q_A with $[q_{0S}, q_A] \in \varrho_{S,A}$ and $q_A \in Q_{0A}$. $\Omega_{S^*} = \{[q_A, \beta] \mid final \in \beta\}$.

For showing that S^* is well defined it suffices to show that all states in Q_{S^*} are reachable from q_{0S^*}. This can be verified by induction along the lines of the definition of $\varrho_{S,A}$ which proceeds along immediate successor states. For the same reason, a state $[q, \beta]$ in S^* contains successors for exactly the elements in β. Next it is obvious that S^* is a subautomaton of M. From this fact, we conclude that $\varrho_{S^*,A} \subseteq \varrho_{M,A}$. Thus, all local annotations are satisfied since they are satisfied in

M as well. Since every state in M occurs in only one pair of ϱ_{MA}, and all states in Q_{S^*} are reachable, we obtain $\varrho_{S^* A} = \varrho_{MA} \cap Q_{S^*} \times Q_A$. Consequently, S^* touches the same states of A as S, so χ evaluates to the same value for both S and S^*. This means that $S^* \in Match(A^{\phi,\chi})$. □

Theorem 4 (Complexity of emptiness). *The non-emptiness problem for total and complete extended annotated automata is NP-complete.*

Proof. Theorem 3 shows that it is sufficient to "guess" a subautomaton of the automaton M considered there. The size of M is linear in the size of A, so guessing can be done in polynomial time. Checking whether the guess is correct amounts to the membership problem for which we provided a polynomial solution in the previous subsection. Hence, non-emptiness is in NP.

For showing NP-hardness, we reduce the satisfiability problem SAT for Boolean formulas to non-emptiness. To this end, let ψ be a Boolean formula with n propositions. Consider an extended annotated automaton with $n + 1$ states and a singleton alphabet where one state is the initial one and all others correspond to the propositions of ψ. Insert transitions from the initial state to all other states. Use ψ as the global constraint, let the local constraints all be equal to $\top$. This means that $Match(A^{\phi,\psi})$ is non-empty if and only if ψ is satisfiable. □

Some time ago, NP-completeness was identified with "intractability". Recent results in several domains show, however, that NP-complete problems can very well be solved for many problem instances of practically relevant size. For the SAT problem, for instance, there exist solvers which deal with formulas having more than 1,000,000 variables.

Whether or not emptiness can be decided quickly, can only be stated on the basis of case studies using an actual implementation. We have to leave this investigation to future work. We see, however, such strong links between our emptiness problem and the Boolean SAT problem, that we believe that many of the sophisticated techniques used there can be adapted to our setting.

6   Applications Revisited

We motivated the implementation of set operations by three applications sketched in Sect. 2. Now, having seen our actual approach to set operations, we may add a few remarks to each approach.

Substituting a service

We proposed to reduce substitutability (defined as partner preservation) be reduced to the check $Partners(S_1) \cap \overline{Partners(S_2)} = \emptyset$. Execution of our approach amounts to executing a complement operation, an intersection operation, and an emptiness check, that is two efficient operations and an NP-complete decision. A previous approach in [21] can be implemented more efficiently, but only using plain annotated automata and negation-free local annotations. In addition, the original approach did not naturally provide a counterexample for non-substitutability which is the case in our approach. In fact, the emptiness check returns a service S^* in the case of non-emptiness which is a partner of the first, but not of the second given service.

In [18], we proposed a substitutability check using extended automata, again restricted to negation-free annotations. The construction used there is similarly complex as the one used here, and again the old approach did not naturally provide a counterexample facility.

Querying a set of behaviors

We proposed to represent behavioral properties by a finite representation of the set of services that satisfy the property, and to realize Boolean connectives as corresponding set operations.

The formalism of extended annotated automata is at least as expressive as standard automata which have been used in [24] for expressing behavioral properties. Hence, our approach is capable of dealing with many interesting properties. Moreover, it turns out that all constructions that correspond to Boolean connective can actually be efficiently implemented. For the final check whether a service has a correctly interacting partner that meets the given properties, only a single NP-complete emptiness check needs to be performed.

Navigating in a behavioral registry

We proposed to iteratively cut the search space (a set of services) into halves, using unions of the sets of partners of the given services.

The sets of correctly interacting partners of services can be represented as annotated automata and thus fit into our framework. Moreover, union can be implemented efficiently. In the case that the product automata constructions involved in the union operation yield too big automata, it may well be possible to overapproximate the result. Overapproximation means to represent a superset of the originally intended set of services for the sake of obtaining a smaller representation as extended annotated automaton. This idea needs to be detailed out in future work but it demonstrates that there are options that can be used in case of exploding results.

Concerning the actual search, only the initial step involved an NP-complete emptiness check. In the subsequent steps of the procedure sketched in Sect. 2, we may replace the emptiness check by the much more efficient membership check since emptiness returns a suitable example service. This way, we do not need to solve NP-complete problems repeatedly.

7   Conclusion

We provided constructions for the basic set operations on sets of service automata, represented as extended annotated automata. Negation, intersection, union, and membership can be implemented efficiently. The complexity of these operations is actually comparable to implementations of other sophisticated and successfully used representations of sets of objects, including finite automata [26] for representing regular languages (i.e., sets of words) or binary decision diagrams [27] for representing Boolean functions (i.e., sets of Boolean vectors). In all mentioned cases, intersection and union require costs in the product of the input sizes while negation is constant to linear.

Membership is solved in linear time in all approaches. It is remarkable that we match the complexity of these formalisms although the elements of our sets (service automata) are much more complex than the elements in the other approaches (words or Boolean vectors). To a large degree, the simplicity of negation, intersection, and union is due to the carefully chosen formalism of deterministic extended annotated automata.

The weakest link in our approach is certainly the emptiness check, which is intrinsically non-polynomial, unless P = NP. However, present-day technology for other NP-complete problems teaches us that NP-completeness is no longer a reason for resignation. Moreover, all sketched application scenarios involve only a single application of an emptiness check while the efficient operations are applied frequently. Nevertheless, the performance of the emptiness problem must be carefully evaluated in a forthcoming case study on realistic examples.

In the motivation part, we outlined the usefulness of our approach. We actually open the way to selecting a service from a registry by purely behavioral specifications. Using our techniques, a query language used for this purpose may freely use Boolean combinations of languages primitives. We also showed that substitutability investigations can return counterexamples in case of non-substitutability. We actually believe that there are more useful applications of set algebra on sets of service automata. Consequently, we shall devote some our our future efforts for looking into other behavioral problems and their potential to be expressed in terms of set algebra on service automata.

References

1. Papazoglou, M.: Agent-oriented technology in support of e-business. Commun. ACM 44, 71–77 (2001)
2. Hull, R., Benedikt, M., Christophides, V., Su, J.: E-services: a look behind the curtain. In: Proc. PODS, pp. 1–14. ACM, New York (2003)
3. Alonso, G., Casati, F., Kuno, H., Machiraju, V.: Web Services: Concepts, Architectures and Applications. Springer, Heidelberg (2003)
4. Gottschalk, K.: Web Services Architecture Overview. IBM Whitepaper, IBM developerWorks (2000), http://ibm.com/developerWorks/web/library/w-ovr
5. Ferrara, A.: Web services: a process algebra approach. In: Proc. ICSOC, pp. 242–251 (2004)
6. Rao, J., Kungas, P., Matskin, M.: Logic-based web services composition: From service description to process model. In: Proc. ICWS, pp. 446–453 (2004)
7. Fisteus, J.A., Fernández, L.S., Kloos, C.D.: Formal Verification of BPEL4WS Business Collaborations. In: Bauknecht, K., Bichler, M., Pröll, B. (eds.) EC-Web 2004. LNCS, vol. 3182, pp. 76–85. Springer, Heidelberg (2004)

8. Fu, X., Bultan, T., Su, J.: Analysis of interacting BPEL web services. In: Proc. WWW, pp. 621–630 (2004)
9. Farahbod, R., Glässer, U., Vajihollahi, M.: Specification and Validation of the Business Process Execution Language for Web Services. In: Zimmermann, W., Thalheim, B. (eds.) ASM 2004. LNCS, vol. 3052, pp. 78–94. Springer, Heidelberg (2004)
10. Fahland, D., Reisig, W.: ASM-based semantics for BPEL: The negative Control Flow. In: Proc. ASM, pp. 131–151 (2005)
11. Lohmann, N., Verbeek, H., Ouyang, C., Stahl, C.: Comparing and evaluating Petri net semantics for BPEL. Int. J. Business Process Integration and Management (in press, 2009)
12. Lohmann, N., Verbeek, H., Dijkman, R.: Petri net transformations for business processes - a survey. In: ToPNoC II. LNCS, pp. 46–63 (2009)
13. Lohmann, N., Kleine, J.: Fully-automatic translation of open workflow net models into simple abstract BPEL processes. In: Proc. Modellierung. LNI, vol. P-127, pp. 57–72. GI (2008)
14. Alves, A., et al.: Web Services Business Process Execution Language Version 2.0. OASIS Standard, April 11, 2007, OASIS (2007)
15. OMG: Business Process Modeling Notation (BPMN). Version 1.2, OMG (2008)
16. Lohmann, N., Massuthe, P., Wolf, K.: Operating guidelines for finite-state services. In: Kleijn, J., Yakovlev, A. (eds.) ICATPN 2007. LNCS, vol. 4546, pp. 321–341. Springer, Heidelberg (2007)
17. Stahl, C., Wolf, K.: Covering places and transitions in open nets. In: Dumas, M., Reichert, M., Shan, M.-C. (eds.) BPM 2008. LNCS, vol. 5240, pp. 116–131. Springer, Heidelberg (2008)
18. Stahl, C., Wolf, K.: Deciding service composition and substitutability using extended operating guidelines. Data Knowl. Eng. (2009) (accepted for publication)
19. Wombacher, A., Fankhauser, P., Mahleko, B., Neuhold, E.: Matchmaking for business processes based on choreographies. Int. Journal of Web Services Research 1, 14–32 (2004)
20. Wolf, K.: Does my service have partners? In: ToPNoC 2009. LNCS, vol. 5460, pp. 152–171. Springer, Heidelberg (2009)
21. Stahl, C., Massuthe, P., Bretschneider, J.: Deciding Substitutability of Services with Operating Guidelines. In: ToPNoC II. LNCS, pp. 172–191. Springer, Heidelberg (2008)
22. Kaschner, K., Lohmann, N.: Automatic test case generation for interacting services. In: ICSOC 2008. LNCS, vol. 5472, pp. 66–78. Springer, Heidelberg (2008)
23. van de Aalst, W.M.P., Lohmann, N., Massuthe, P., Stahl, C., Wolf, K.: Multiparty contracts: Agreeing and implementing interorganizational processes. Comput. J. (in press, 2009)
24. Lohmann, N., Massuthe, P., Wolf, K.: Behavioral constraints for services. In: Alonso, G., Dadam, P., Rosemann, M. (eds.) BPM 2007. LNCS, vol. 4714, pp. 271–287. Springer, Heidelberg (2007)
25. Mietzner, R., Ma, Z., Leymann, F.: An algorithm for the validation of executable completions of an abstract bpel process. In: Multikonferenz Wirtschaftsinformatik (2008)
26. Hopcroft, J.E., Ullman, J.: Introduction to Automata Theory, Languages and Computation. Addison-Wesley, Reading (1979)
27. Bryant, R.: Graph-based algorithms for Boolean function manipulation. IEEE Trans. Computers C-35, 677–691 (1986)

A Restructuring Method for WS-BPEL Business Processes Based on Extended Workflow Graphs

Thomas S. Heinze[1], Wolfram Amme[1], and Simon Moser[2]

[1] Friedrich Schiller University of Jena,
Institute of Computer Science,
07743 Jena, Germany
{T.Heinze,Wolfram.Amme}@uni-jena.de
[2] IBM Software Laboratory Böblingen
Business Process Solutions
71032 Böblingen, Germany
smoser@de.ibm.com

Abstract. Much research effort has been spent on the provision of analysis methods for business processes specified by means of Web Services Business Process Execution Language (WS-BPEL). Nevertheless, most approaches neglect conditional control flow, though running the risk of erroneous analysis results. In this paper, we present a restructuring approach for WS-BPEL processes, which helps to partly remedy conditional control flow. We therefore use a combination of workflow graphs and Concurrent Static Single Assignment Form. Based on the hybrid format, we are able to identify loops with static quasi-constant loop condition and transform them in such a way, that conditional control flow is replaced by unconditional control flow. Augmenting an existing analysis with the proposed restructuring then enables more precise results, as is shown for a compatibility analysis of WS-BPEL business processes.

1 Introduction

Automating business processes using an IT infrastructure has become increasingly important throughout the last years. Business process management technology has been proven to be a suitable platform for consolidating distributed information resources and thus promoting interoperability across cross-platform systems. Moreover, the concept of loosely coupled (web) services, where a service describes an individual distributed piece, has further augmented this technology field. The *Web Services Business Process Execution Language (WS-BPEL)* [1] offers a standards-based approach to build service-based business processes. A WS-BPEL process therefore implements one service by orchestrating other services. This can become problematic when two interacting services are implemented in a stateful way, which would be the case with long-running business processes. In this case, *syntactical compatibility*, i.e. that the WSDL interfaces of both services match, is not sufficient since both services also have to comply with a stateful interaction protocol in order to prevent invalid behavior at runtime, e.g.

U. Dayal et al. (Eds.): BPM 2009, LNCS 5701, pp. 211–228, 2009.
© Springer-Verlag Berlin Heidelberg 2009

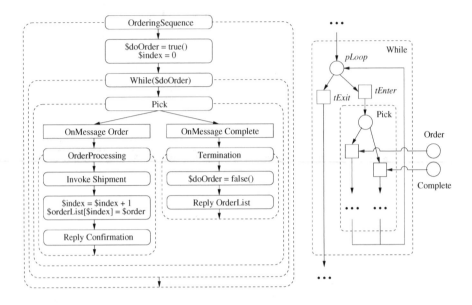

Fig. 1. OrderingSequence (left) and part of its Petri net model (right)

deadlocks. Such potentially dangerous situations should already be detected in a static fashion at design time. This so-called *control-flow verification* has been a major subject of research within the last years [2]. One subject of control-flow verification, the concept of *behavioral compatibility* of WS-BPEL processes, has been introduced in [3]. Two interacting business processes are behavioral compatible if and only if, for any case, both processes terminate properly, i.e., termination is guaranteed, there are no dangling messages, and deadlocks and livelocks are absent. Behavioral compatibility can theoretically be verified using various formalisms. In [3], Petri nets have been used, where *behavioral compatibility* becomes equivalent to the soundness property [4] of the composed Petri net representation of the two interacting services. However, using Petri nets as a formalism for process model analysis requires an abstraction which omits its data aspects, in order to allow for a feasible analysis (refer to the state space explosion problem). Thus, conditional control flow, i.e. conditions of branchings or loops, is often neglected in the Petri-net-based process representation. Instead, non-determinism is introduced, inducing a semantic gap between a process and its model and thereby provoking erroneous analysis results.[1]

A process fragment that can not be accurately analyzed, if conditional control flow is neglected, is shown on the left-hand side of Figure 1. The depicted activity is used as an example throughout this paper and may be part of a WS-BPEL process modeling an online shop. Therein, a customer can order multiple items by sending repeatedly Order, which triggers the shipment of the item and an acknowledgment message. Having transmitted all orders, the customer

---

[1] This approach seems to only weakly preserve properties like compatibility [5].

signals his orders to be finished by sending `Complete`, which is confirmed by message `OrderList`. In order to implement this business protocol, activity `OrderingSequence` embeds a loop whose execution is controlled by variable `doOrder`: If the value of `doOrder` is true, the loop is executed, and not otherwise. Initially, the value is set to true, and the loop will be executed. The `Pick` activity in the loop either activates `OrderProcessing`, in case `Order` is received, or sets the value of `doOrder` to false and replies `OrderList`, if `Complete` is received.

Consider a corresponding counterpart of `OrderingSequence` which successively sends and receives messages `Order`, `Confirmation`, `Complete`, `OrderList` in this order. Obviously, in reality, both process fragments are behaviorally compatible, i.e. the composed system of both processes would be deadlock-free. However, the use of Petri nets for the compatibility analysis and the associated abstraction from data aspects yields a different outcome: The loop in `OrderingSequence` is mapped to conflicting transitions $tEnter$ and $tExit$, modeling loop entry and exit non-deterministically, as shown in Figure 1. As Petri net semantics state to solve the conflict arbitrarily, the loop may be (re-)entered more than twice such that `OrderingSequence` still awaits messages after `Complete` has been received. Consequently, a deadlock may occur in this model and the analysis arrives at the result that both process fragments are not compatible.

Such erroneous situations can be avoided by defining a set of modeling rules for WS-BPEL processes. A scenario as described above is out of question, if the usage of interacting loops is restricted to such cases where the start of a loop iteration is immediately communicated to the partner processes, e.g. by a rule to define mandatory outgoing message activities at loop entry and exit, respectively. Similar rules have been defined to ensure the soundness property in [4]. Although they provide an easy to use criterion, erroneous situations cannot be ruled out in practice, since such rules are only best practices and it cannot be safely assumed that they are adhered to. Thus, we have chosen a different approach to reduce the number of failures in compatibility analysis. We propose to *restructure* WS-BPEL processes prior to their analysis. By the help of our technique, we are able to identify a special class of loops: loops with static quasi-constant loop condition, i.e. conditions where used variables are defined over constant values only. Loops belonging to this class can be transformed such that conditions are replaced with unconditional control flow. Thus, there is no need to use non-deterministic structures for the representation of these activities and this potential source of imprecision is avoided. The remainder of the paper is structured as follows: Section 2 introduces extended workflow graphs, which are used as our process representation format. The restructuring technique itself is presented in Section 3. This is followed by a discussion of related work in Section 4 and a conclusion in Section 5.

2   Extended Workflow Graphs

Precise analysis of business processes requires a representation format which is able to reflect the control flow as well as data aspects. A number of excellent representation formats have been established for this purpose, especially in the field of compiler construction. Often used formats include abstract syntax trees, control-flow graphs and program dependence graphs. The advantages of these formats, especially those of control-flow graphs, have already started to attract the area of business process analysis [6]. In our approach, we use extended workflow graphs as process representation format, driven by two major reasons: their similarity to control-flow graphs and the existence of a Petri net mapping [7].

Workflow graphs have been commonly used in the analysis of business processes [7,8], though also modeling control flow only. *Extended workflow graphs* can be seen as an enrichment of ordinary workflow graphs by a notation of process data. In principle, nodes in an extended workflow graph represent activities and edges connect nodes according to control flow and synchronization, i.e. links:

Definition 1 (Extended Workflow Graph). *An extended workflow graph is a directed graph $WFG = (N, E)$ with two special nodes $Start, End \in N$, such that the set of nodes $N = N_{Act} \cup N_{Split} \cup N_{Merge} \cup N_{Fork} \cup N_{Join}$ consists of*

- N_{Act}, *i.e. nodes to represent activities,*
- N_{Split}, *i.e. nodes to split the control flow in branchings or loops,*
- N_{Merge}, *i.e. nodes to merge the control flow in branchings or loops,*
- N_{Fork}, N_{Join}, *i.e. nodes to mark the begin and end of parallel sections.*

The set of edges E includes control flow and synchronization edges.

In Figure 2, the extended workflow graph for our sample process fragment is shown. Each basic activity is therein mapped to a single node (boxes), as done for activity `Reply Confirmation`. Sequences of basic activities are then represented by interconnecting these nodes one after another using control-flow edges. For the representation of the structured activities `While` and `Pick`, further nodes are added to indicate the divergence and convergence of control flow, i.e. *Split*, *Pick* (diamonds) and *Header*, *Merge* (trapezoids), respectively. Outgoing edges of the nodes that split the control flow are additionally labeled by corresponding events, in case of the event-driven branching, or by *True* and *False*, denoting the entry and exit of the loop. Finally, special nodes *Start* (circle) and *End* (filled circle) are used to designate begin and end of the process fragment.

Process data and its manipulation is encoded in extended workflow graphs by means of *Concurrent Static Single Assignment Form (CSSA-Form)*. CSSA-Form has been introduced in [9] for supporting the optimization of parallel programs. It has also been adapted for the analysis of WS-BPEL processes and is able to cover the more complex features of the language, e.g. links and dead path elimination (refer to [6] for details). Its key advantage is that each variable is defined exactly once, enforcing the direct representation of data dependences. In order to guarantee this property, variables are renamed in such a way, that each definition of a variable is assigned a unique name, typically denoted by subscripts: Variable v becomes values $v_1, \ldots, v_n$, one for each definition, and uses are adjusted accordingly. The single-assignment property is considered as static. Due to this, the definition of a variable inside a loop is regarded as single definition, although, the loop, and therefore the definition, may be executed more

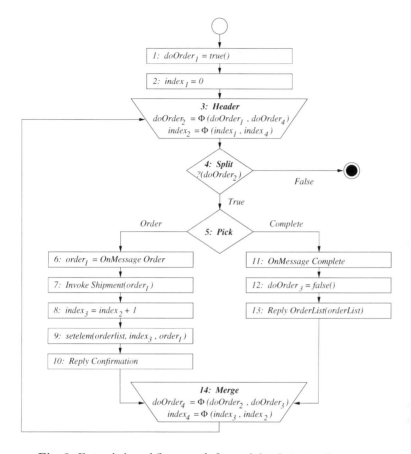

Fig. 2. Extended workflow graph for activity `OrderingSequence`

than once. Special handling is required if multiple definitions of the same variable reach a node via different branches or threads, i.e. at nodes of sets N_{Merge} and N_{Join}. In such cases, Φ-*functions* are inserted to merge the conflicting definitions.

Definition 2 (Φ-function). *A Φ-function for variable v is of the form $v = \Phi(v_1, \ldots, v_n)$, where n denotes the number of incoming control-flow edges of the node containing the function and thus the number of conflicting definitions of variable v at this node. The value of the function is one of its operands, depending on the actual control flow: v_i, if the i-th incoming edge denotes the branch taken at runtime or the thread whose associated operand is defined last.*

A further special function is required to support concurrent read and write access to variables. Often, such access to shared variables implies some kind of race condition, i.e. if a definition of a variable may reach uses in another thread depends on the interleaving of threads at runtime. In order to represent these race conditions, π-*functions* are introduced.

Definition 3 (π-function). *A π-function for variable v is of the form $v = \pi(v_1, \ldots, v_n)$, where n denotes the number of reaching definitions of variable v from the thread the function is located in and all other threads containing definitions of this variable. The value of the function is one of its operands v_i, selected non-deterministically.*

In the extended workflow graph of our sample in Figure 2, all used variables have been renamed in order to ascertain a (static) single definition of each variable, e.g. doOrder has become $doOrder_1, \ldots, doOrder_4$. Furthermore, several Φ-functions have been introduced in order to model the confluence of conflicting definitions at nodes $Header$ and $Merge$. As an example, the Φ-function $doOrder_2 = \Phi(doOrder_1, doOrder_4)$ in $Header$ models the confluence of the values the variable doOrder takes before loop execution starts ($doOrder_1$) and after a single iteration of the loop ($doOrder_4$). The latter value is thereby defined by means of another Φ-function in $Merge$, which merges the values of the variable on the two possible control-flow paths in the loop ($doOrder_2, doOrder_3$).

3   Restructuring Technique

Our restructuring technique can be applied to a loop only, if the conditional branch of the loop can be statically evaluated for each iteration. As a consequence, for each loop iteration, an execution or non-execution of the loop body can always be derived from the incoming values of the condition variables, i.e. variables used in the loop condition, directly. In our restructuring process we take advantage of this property, by replacing conditional branches with unconditional control flows to copies of the loop that are representing a certain state.

In principle, our restructuring technique can be seen as a simple unrolling process, that divides a given loop into multiple copies of its loop body, where each copy represents the execution of the loop for a certain assignment of values to variables, i.e. its state. In our terminology, such kind of copies are called *loop instances*. Conditional branches in these instances are redundant and are replaced by mapping the outgoing control flows of a loop iteration to instances that correspond to its outgoing assignment of variables. Figure 3 visualizes the main concept of our approach applied to the process fragment in Figure 1.

In our sample, the condition of the loop exactly matches with boolean variable $doOrder_2$, whose value is defined by a constant in each iteration of the loop. If the loop is entered for the first time, the value will be $true()$. This value will not change until the right branch, including the assignment $12 : doOrder_3 = false()$, is chosen for execution. As a consequence, we only need two copies of the loop body for restructuring. In one copy, which we call $Instance_{doOrder_2=true()}$, occurrences of the variable are replaced by the value $true()$, and in the other copy, which we call $Instance_{doOrder_2=false()}$, by $false()$.

In both copies, the conditional branch can be replaced by unconditional control-flow edges, such that in $Instance_{doOrder_2=true()}$ the loop is always executed and in $Instance_{doOrder_2=false()}$ the loop is exited immediately by jumping

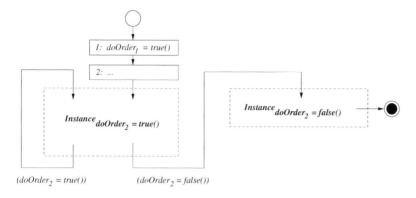

Fig. 3. Restructuring applied to the loop in `OrderingSequence`

to the end node. The control flow of the restructured process is completed by inserting edges to corresponding copies, depending on the assignment of $doOrder_2$. For example, since the value of $doOrder_2$ is not changed on the left branch in $Instance_{doOrder_2=true()}$, an edge connecting the branch to the header of this copy is introduced. On the other hand, the value is changed on the right branch to $false()$, and therefore this branch is connected to $Instance_{doOrder_2=false()}$.

3.1 Static Quasi-Constant Loop Condition

In a more formal sense, restructuring of loops using our technique can always be performed, if the conditional branch is defined by a static quasi-constant loop condition. *Static quasi-constant loop conditions* are characterized by the use of variables whose values are restricted to constants. As in other representation formats, each of the variables may be defined by multiple, differing constant assignments on varying control-flow paths, which are merged in CSSA-Form by the help of Φ-functions. In order to verify whether a given loop condition satisfies this property, definitions of the therein used variables have to be determined, as well as definitions of variables used in these definitions, and so on. If all definitions then represent assignments of constant values only, the condition under consideration obviously forms a static quasi-constant loop condition.

Loops in WS-BPEL processes are always defined by using block-oriented activities, i.e. `While`, `RepeatUntil`, and serial `ForEach`, because cyclic control flow defined by links is not allowed [1]. Furthermore, since any for or do-while loop can be transformed into an equivalent while loop [9], WS-BPEL activities `ForEach` (serial) and `RepeatUntil` can be transformed into a `While` activity as well. We are therefore able to restrict the presentation of our restructuring technique to loops specified by means of the block-oriented `While` activity without loss of generality. In particular, we define the class of loops with static quasi-constant loop condition to consist of those `While` activities whose conditional expression only relies on variables defined by constants, however, exclusive of interleaving definitions in different threads of a `Flow` activity. The latter is due to the fact that our technique is not applicable across the boundaries of threads, i.e. we are not "serializing" parallel threads which would be necessary in this case. Therefore, a condition which relies on variables defined by one or multiple constant assignments embedded in different threads, without any detectable synchronization by links, is not considered as static quasi-constant loop condition.

In principle, our restructuring technique can be applied to a single loop or to a complete WS-BPEL process. In both cases, we need to identify loops with static quasi-constant loop conditions, based on the variables used in the conditions. In extended workflow graphs, the retrieval of this kind of loop conditions is fairly simple, since relations of variable use and definition are directly reflected due to the encoding of WS-BPEL activities by means of CSSA-Form [9]. Identifying loops with static quasi-constant loop condition can therefore be done by traversing these relations. Starting with the variables used in the loop condition, definitions of variables are inspected to be either constant assignments, ordinary copy instructions, Φ-functions, or π-functions with one operand only, representing a parallel definition without interleaving definitions, i.e. race conditions. All other kinds of definitions are excluded for variables of static quasi-constant loop conditions. In case of a Φ-function, i.e. $v = \Phi(v_1, \ldots, v_n)$, the traversal continues with the definitions of the operands v_i. In the presence of an ordinary copy instruction, i.e. $v = v_1$, or a one-operand π-function, i.e. $v = \pi(v_1)$, the definition of the source or operand v_1 is further inspected, respectively.

As a second requirement of our technique, we furthermore have to guarantee during the verification process, that the inspected Φ-functions are only included in the header node of the loop under consideration, or in other merge nodes which are *postdominated* by this node.[2] This restriction does not tamper the applicability of our restructuring technique in most cases, since nodes contained in the body of a loop are always postdominated by the header node of the loop. If one of the Φ-functions is defined in the header node of another loop, this loop needs to be restructured first. We therefore propose a restructuring procedure, where loops are processed from the inside to the outside of a loop.

3.2 Loop Normal Form

An arbitrary loop can always be transformed into an unconditional loop, if it has been identified as a loop with static quasi-constant loop condition. In principle, the restructuring of a loop is performed in two steps. In a first step, the loop is converted into a normal form, and in a second step, the dividing of the normalized loop into copies of the loop is performed. In fact, the normalization of a loop is not absolutely necessary, but it eases requirements of the subsequent step.

One requirement for a valid loop transformation is, that a condition variable has exactly one definition on each incoming control-flow edge of the loop header node in an arbitrary iteration. However, this is not always the case for a loop

---

[2] A node n *dominates* a node m, if every path from *Start* to node m contains n. A node n *postdominates* a node m, if every path from node m to *End* contains n. A node n is said to *strictly* dominate or postdominate a node m, if $n \neq m$ and n dominates or postdominates m, respectively [9].

procedure $resolve(WFG = (N, E), merge, header)$ is
    $Subgraph = (N_{Subgraph}, E_{Subgraph}) \subseteq WFG$ such that $header \in N_{Subgraph}$
            $\wedge \forall n \in N_{Subgraph} \setminus \{header\} : n$ is strictly dominated by $merge$;
    foreach $e_i = (n_i, merge) \in E$ do
        $Subgraph_i$ = create copy of $Subgraph$;
        $substitutes_i = \emptyset$;
        foreach Φ-function $v = \Phi(v_1, \ldots, v_n)$ in $merge$ do
            let v_i be the operand associated to edge e_i;
            $substitutes_i[v] \rightarrow v_i$;
        end for;
        replace uses of variables in $Subgraph_i$ according to $substitutes_i$;
        let $successor_i$ be the copy of $successor(merge)$ in $Subgraph_i$;
        let $header_i$ be the copy of $header$ in $Subgraph_i$;
        $WFG = WFG \cup Subgraph_i$;
        $E = (E \setminus \{e_i\}) \cup \{(n_i, successor_i)\}$;
        update Φ-functions in $header_i$;
    end for;
    remove all unreachable nodes in WFG and adjacent edges;
    merge $header$ and copies of $header$ into single node;
    update CSSA-Form in WFG;
end.

Fig. 4. Algorithm for resolving a single merge node

with static quasi-constant loop condition. A condition variable may be assigned multiple values on different control-flow paths, which are merged by the help of Φ-functions. Determining and resolving these Φ-functions, except for those contained in the header node, is the purpose of the normalization process.

In extended workflow graphs, Φ-functions, that are relevant for normalization, can easily be identified by traversing the definitions of variables that are directly or indirectly used in the loop condition. If in this traversal, a Φ-function is observed, which is not contained in the loop header, this function is marked for resolving. It should be pointed out, that Φ-functions need to be processed in such a way, that no function which dominates another candidate function is resolved prior to the dominated function. This is mainly due to the prevention of unnecessary copying, but also required for the proper operation of the algorithm.

Actual resolving of Φ-functions is done in groups, broken down by the merge nodes that contain marked Φ-functions. In principle, a single merge node $merge$ is resolved by inserting multiple copies of the subgraph, which immediately starts after node $merge$ and ends at the loop header node, into the workflow graph. Thereby, one copy is created for each incoming control-flow edge of $merge$ and references to Φ-functions defined in node $merge$ are replaced by the operands associated to the corresponding edge. Eventually, operands of Φ-functions contained in the loop header node are updated accordingly. As a result, all control-flow paths which included confluent definitions of variables that had been merged in node $merge$ are now separated and explicitly modeled.

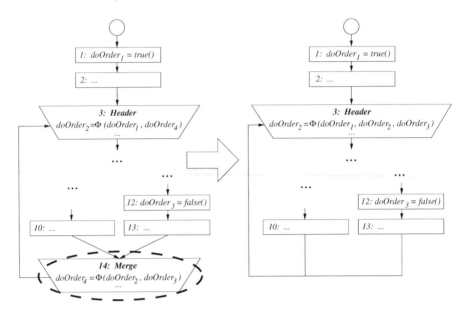

Fig. 5. Derivation of the normal form for the loop in OrderingSequence

In our normalization process, resolving a single merge node *merge* for a loop with header node *header* within an extended workflow graph *WFG* can be done by applying the algorithm denoted by procedure *resolve* shown in Figure 4. As a result, a copy of the subgraph *Subgraph* ⊆ *WFG*, which contains *header* and all nodes strictly dominated by *merge*, is created and inserted into the extended workflow graph for each incoming edge e_i of node *merge*. Uses of variables defined by a Φ-function are therein replaced by the operands associated to the respective edge e_i. Furthermore, Φ-functions in copies of header node *header* are appropriately updated. In a last step, all now unreachable nodes are removed from the extended workflow graph (in particular *merge*), copies of the header node are merged into a single header node, and CSSA-Form is re-established. Note that the algorithm will merely alter the target of edges e_i to *header* and update Φ-functions, if the successor of *merge* matches with *header*.

The normalization process for our example is shown in Figure 5. As can be seen, the performed transformation is quite simple, since we have only a single merge node (labeled 14), which needs to be resolved, and this node is immediately succeeded by the loop header (labeled 3). Therefore, in the normal form, the merge node is removed by connecting its predecessor nodes (labeled 10, 13) to the loop header node and updating Φ-functions' operands accordingly.

For establishing the correctness of the normalization process, first, we have shown in [10], that applying the resolving algorithm to a single merge node, which is postdominated by the header node, does not change the execution semantics of the process comprising the considered loop. An iteratively application of this proposition then shows the correctness of the overall transformation process.

3.3 Loop Instantiation

In the loop instantiation pass of our approach, the actual restructuring process, i.e. the replacement of conditional control flows by unconditional control flows, is performed.[3] In order to guarantee a proper transformation, two constraints must be hold. On the one hand, the restructuring must not change the execution semantics of the process containing the loop under consideration. Thus, the original and the modified process should be semantically equivalent. On the other hand, the technique needs to be effective, i.e. the condition of the loop should be resolved in the restructured process.

A simplified restructuring is possible, if the loop condition is directly related to the possible control-flow paths of the normalized loop. In principle, this can be achieved by replacing the split node containing the loop condition with multiple split nodes, which are positioned on each of the incoming control-flow paths of the loop header node. But, in CSSA-Form, such a transformation could change the execution semantics of the loop, since the copies of the split node are not dominated by the header node of the loop anymore. One possible solution addressing this problem would be to combine the repositioning of the loop condition with a simultaneously performed resolving of Φ-functions.

A Φ-function $v = \Phi(v_1, \ldots, v_n)$ in the loop header node is replaced by inserting assignments $v = v_i$ for each incoming edge of the loop header, assigning the operand associated to the edge to the variable defined by the Φ-function. In our approach, inserted assignments are grouped to so-called *set blocks* for each incoming edge. The loop condition by itself can then be safely repositioned by creating multiple copies of the split node containing the condition, each of them positioned directly after each set block. A nice side effect of this repositioning of the loop condition is, that we have derived a pattern which represents the execution of an arbitrary loop instance. This instance pattern is used afterwards as a kind of blueprint, when replacing the loop with its loop instances. Since the copied and inserted split nodes control whether an instance is entered or not during the instantiation process, we call them *instance guards*.

The result of this preprocessing step is shown for our sample in Figure 6. As can be seen, three set blocks (labeled 15, 16, 17) have been inserted into the extended workflow graph, one for the entry edge of the loop and one set block for each of the two possible control-flow paths in the loop. The inserted set blocks contain definitions of variables $doOrder_2$ and $index_2$, assigning the values associated to the respective edges. Furthermore, each set block is followed by an instance guard (labeled 18, 19, 20), i.e. a split node containing the loop condition. The instance pattern of our sample is depicted by a dotted frame.

As mentioned above, in the instantiation process, instance guards are permanently evaluated, and replaced by edges to corresponding loop instances, if evaluation results true, or by an edge to the exit node of the loop, in the other case. In particular, an instance in this process stands for an execution of the loop with respect to a certain assignment of condition variables. For such an

---

[3] It shall be noted that our technique is not restricted to loops, since it can be easily adopted to block-oriented conditional branchings as well [10].

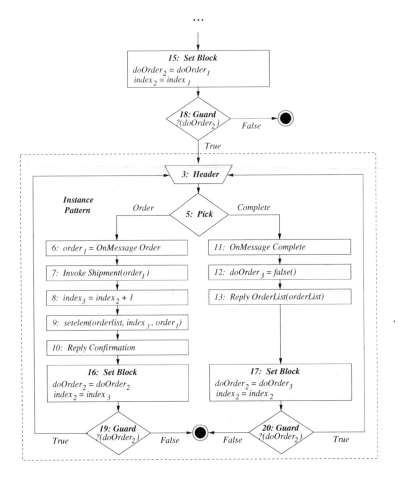

Fig. 6. Instance pattern of the loop in activity `OrderingSequence`

assignment of values $(v_1, \ldots, v_n)$ to condition variables $(var_1, \ldots, var_n)$, we can derive the corresponding loop instance by replacing all occurrences of condition variables $var_1, \ldots, var_n$ in the instance pattern according to the assignment.

A description of our instantiation algorithm for an extended workflow graph WFG and a normalized loop, given by its header node $header$ and exit node $exit$, is presented by procedure $instantiate$ in Figure 7. The algorithm initially derives the instance pattern of the loop by inserting instance guards and set blocks on each incoming edge of the loop header node $header$ (procedure $prepareInstancePattern$). Afterwards, instance guards are iteratively visited and processed until no further guard exist (main loop in procedure $instantiate$)[4]. Eventually, the algorithm terminates by removing all set blocks from the restructured extended workflow graph and re-establishing CSSA-Form.

---

[4] Iterative processing is accounted to the permanent generation of instances, which by themselves include new instance guards.

procedure $instantiate(WFG = (N, E), header, exit)$ is
    $prepareInstancePattern(WFG, header, exit)$;
    while ($\exists\ guard \in N$ such that $guard$ is instance guard) do
        $setblock = predecessor(guard)$ in WFG;
        let $[v_1, \ldots, v_n]$ be assignment of condition variables defined by $setblock$;
        if $(condition(guard) == true)$ then
            let $Instance_{[v_1, \ldots, v_n]}$ be the instance for assignment $[v_1, \ldots, v_n]$;
            let $instanceheader$ be the copy of $header$ in $Instance_{[v_1, \ldots, v_n]}$;
            if $(Instance_{[v_1, \ldots, v_n]} \subseteq WFG)$ then
                $WFG = WFG \cup Instance_{[v_1, \ldots, v_n]}$;
            end if;
            $E = E \cup \{(setblock, instanceheader)\}$;
        else $E = E \cup \{(setblock, exit)\}$;
        end if;
        remove $guard$ and adjacent edges from WFG;
    end while;
    remove set blocks from WFG and update CSSA-Form;
end.
procedure $prepareInstancePattern(WFG = (N, E), header, exit)$ is
    $split = successor(header)$ in WFG;
    foreach $e_i = (n_i, header) \in E$ do
        $guard_i =$ create copy of $split$ as instance guard;
        $setblock_i =$ create new node;
        foreach Φ-function $v = \Phi(v_1, \ldots, v_n)$ in $header$ do
            let v_i be the operand associated to edge e_i;
            add assignment $v = v_i$ to $setblock_i$;
        end for;
        $E = E \cup \{(n_i, setblock_i), (setblock_i, guard_i),$
                        $(guard_i, exit)$ with label $False$,
                        $(guard_i, header)$ with label $True\}$;
        $N = N \cup \{setblock_i, guard_i\}$;
    end for;
    remove $split$ and adjacent edges from WFG;
    remove all Φ-functions in $header$;
end.

Fig. 7. Algorithm for the instantiation of a loop in normal form

During the traversal of instance guards, a new loop instance is only created for an instance guard $guard$, if evaluation of its condition $condition(guard)$ yields true and an appropriate instance has not already been created. Such a loop instance is generated by creating a copy of the instance pattern and replacing therein all occurences of condition variables with their corresponding values.

Figure 8 contains the result of applying the instantiation algorithm to our sample, prior to restoring CSSA-Form. Instantiation has been performed in three iterations. In the first iteration, the instance guard from the entry of the loop was replaced by creating and inserting a new instance for the assignment $doOrder_2 = true$. This instance itself contained two further guards (refer to

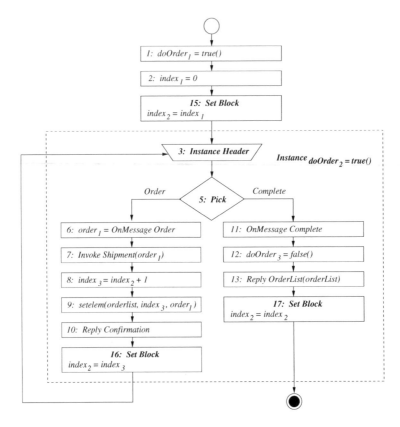

Fig. 8. Result of loop instantiation: Loop instance for assignment $doOrder_2 = true()$

the instance pattern in Figure 6), that have been processed in subsequent iterations. Thereby, since the evaluation of the left guard of this instance also resulted true, that guard has been replaced by an edge to the already created instance $Instance_{doOrder_2=true()}$. However, since the condition contained in the right instance guard yielded false, this guard has been replaced by an edge to the exit node of the loop. Thus, only a single loop instance has been created.

3.4   Restructured Loop

Figure 9 shows on the left-hand side the final restructured extended workflow graph, after set blocks have been removed and CSSA-Form is re-established, i.e. a Φ-function has been inserted to merge the confluent definitions of variable $index$. Since the compatibility analysis in [3] utilizes a Petri-net-based model, finally, the restructured extended workflow graph needs to be mapped to a Petri net. This can be done quite easily, because an already existing mapping of ordinary workflow graphs to Petri nets can be adapted for that purpose [7,10]. The resulting Petri-net-based model for our sample is shown on the right-hand

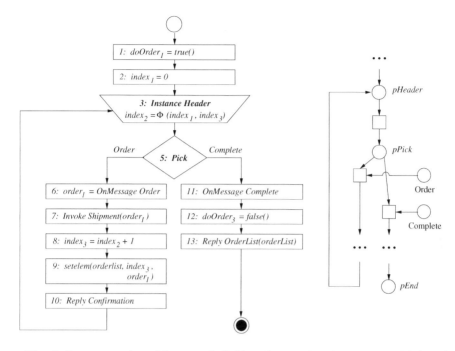

Fig. 9. Restructured workflow graph (left) and part of its Petri net model (right)

side of Figure 9. As expected, the non-deterministic conflict included in the original model could be avoided since the split node representing the conditional branch of the loop is now removed. Consequently, compatibility analysis of **OrderingSequence** and its counterpart, as described in Section 1, now yields the correct result: both activities are behavioral compatible in terms of [3].

As can be seen by this sample, we are able to restructure a loop with static quasi-constant loop condition such that the conditional branch is replaced by unconditional control flow. Therefore, modeling the control flow of the loop can now be done precisely without considering data aspects, i.e. data dependences of the loop condition. In general, any WS-BPEL process that only contains loops or conditional branchings with static quasi-constant condition can be restructured using our technique and the resulting extended workflow graph provides a precise model of the process with respect to control flow. For these reasons, we propose to use our restructuring technique prior to existing control-flow verification analyses. Any conditional branch defined by a static quasi-constant condition can thus be resolved beforehand and will not interfere with the analysis. However, any other conditional branch, not defined by a static quasi-constant condition, remains unchanged and could still provide a potential source of imprecision.

Effectiveness and correctness of the restructuring technique is shown in [10]. The proof of effectiveness is fairly simple, since it must be only shown that the algorithm terminates. However, this immediately follows from the fact that an upper bound for the number of instances is given by the number of combinations for possible values of condition variables, which is finite for a loop with static quasi-constant loop condition. For the proof of semantically correctness of our algorithm, we have shown that the set of possible executions of the original loop and the set of possible executions of the altered loop are identical. In fact, obviously, this claim holds via construction, but also can be proven formally using complete induction over loop instances.

4   Related Work

Most of current approaches for control-flow verification of WS-BPEL business processes do not consider process data, justified by the enabling of a feasible analysis [2]. In general, when modeling WS-BPEL with Petri nets, a common argument is to apply high-level Petri nets, if process data needs to be considered. In contrast to our technique, such an approach is restricted to processes of finite data domain, since an infinite data domain causes an infinite state space of the model. However, except for this restriction, process features using data of finite domain only, e.g. join conditions of links [1], can be mapped clearly and precisely using unfolding techniques for high-level Petri nets [5].

Advanced techniques for the analysis of behavioral compatibility have also been described in earlier works [6,11]. Both approaches emphasize the problems of ignoring data dependences when analyzing WS-BPEL processes and therefore propose the use of CSSA-Form as process representation format. In the first approach [6], a data-flow analysis allows to determine inbound messages which may reach a certain activity in a process. The thus discovered dependences can be used to improve the Petri net model of the process.

The second approach [11] provides an automated method to enhance the mapping of WS-BPEL processes to Petri nets. Therefore, data dependences of branching or loop conditions are analyzed such that conditions of boolean data domain are identified. Using data-sensitive Petri net patterns then allows to precisely map these conditions, conjoined with their data dependences, and thus improves the Petri net model. Other than the technique described in this paper, the method is also restricted to conditions of boolean data domain.

Another method for enhancing control-flow verification of WS-BPEL processes is proposed in [12]. The depicted approach addresses conditional control flow and its implications for a Petri-net-based analysis. Therefore the Petri net model is enhanced by means of effect places and predicate transitions. Process information that controls the translation procedure in this technique is derived by a simple simulation of the considered process. It is not clear to the authors, if the technique is effectively feasible in the presence of loops, since in this case all iterations of a loop have to be simulated. Furthermore, the addressed problem seems to be in fact a classical control-flow problem and therefore should be solvable without an extension of the underlying Petri net model, using a combination of control-flow expansion and dead code elimination.

In [13], a data-flow analysis is presented, which is able to determine reaching definitions of variables in WS-BPEL processes. The derived information is then used for process partitioning. If a process was split into fragments such that a data dependence exists between two fragments, a message exchange is required. In order to minimize the number of required message exchanges, the described analysis has to yield precise results. Therefore, the analysis respects dead path elimination [1] and provides a handling of complex data types [13]. In extended workflow graphs, reaching definitions are already reflected due to the use of CSSA-Form. However, since CSSA-Form provides an over-approximation of reaching definitions, it may be refined using the proposed analysis.

A method addressing classical control-flow graphs and data-flow analysis in general, is presented in [14]. The therein proposed approach identifies destructive merges, i.e. nodes of a control-flow graph where data flow information is lost due to the merging of differing incoming data flow facts. A subsequent restructuring of the graph, duplicating nodes in such a way that destructive merges can be eliminated, allows to increase the precision of the data-flow analysis. In a broader sense, our technique can be seen as an application of this method to business processes. However, the concept of loop instances, used in our technique, allows to restructure loops in a single pass, without the need for an additional constant propagation. Furthermore, the representation of process data in CSSA-Form simplifies parts of the restructuring enormously, e.g. the identification of static quasi-constant loop conditions, and also allows to handle parallelism.

5   Conclusion

In this paper, we motivated the use of a restructuring method for WS-BPEL business processes, prior to an existing compatibility analysis [3]. Due to the coarse representation of conditional loops or branchings by the use of non-determinism in the Petri-net-based model, the analysis is often prone to erroneous results. Therefore, resolving conditional control flow in WS-BPEL processes, whenever possible, allows for a more precise analysis. To this end, we introduced extended workflow graphs, as a combination of ordinary workflow graphs and CSSA-Form, in order to represent WS-BPEL processes without loss of precision.

Building on that, we were able to identify block-oriented loop activities with static quasi-constant loop condition, i.e. conditions where used variables are defined over constant values only. Furthermore, we provided a safe algorithm which enables the removal of such conditions by replacing conditional with unconditional control flow. Applying the Petri-net-based analysis of behavioral compatibility to such a restructured process results in more precise analysis results, since the use of non-deterministic structures within the model can be reduced.

Main subject of future work will be the evaluation of our restructuring technique with respect to a set of real-world WS-BPEL processes, in order to assess the practical relevance of our approach besides establishing its theoretical foundation as conducted within this paper. Unfortunately, we are currently not aware of such a "benchmark" for WS-BPEL and therefore require to firstly define a set of realistic business processes which can be used for that purpose.

Another direction for future work is the extension of our technique to all loops, whose conditions can be statically evaluated and removed by the help of control-flow restructuring, rather than processing static quasi-constant loop conditions only. In particular, we are interested in loops for which the values of some condition variables can only be restricted to certain ranges of values, or are even unknown at all. In principle, this can be done by introducing a more abstract notion of loop instances, i.e. instances where assignments of condition variables are described by means of intervals or predicates.

References

1. Alves, A., Arkin, A., Askary, S., Barreto, C., Bloch, B., Curbera, F., Ford, M., Goland, Y., Guízar, A., Kartha, N., Liu, C.K., Khalaf, R., König, D., Marin, M., Mehta, V., Thatte, S., van der Rijn, D., Yendluri, P., Yiu, A.: Web Services Business Process Execution Language Version 2.0. Standard, OASIS (April 2007)
2. van Breugel, F., Koshkina, M.: Models and Verification of BPEL (September 2006)
3. Martens, A., Moser, S., Gerhardt, A., Funk, K.: Analyzing Compatibility of BPEL Processes. In: Advanced Int. Conf. on Telecommunications and Int. Conf. on Internet and Web Applications and Services, p. 147. IEEE, Los Alamitos (2006)
4. van der Aalst, W.M.P.: Structural Characterizations of Sound Workflow Nets. Computing Science Report 96/23, Eindhoven University of Technology (1996)
5. Lohmann, N.: A Feature-Complete Petri Net Semantics for WS-BPEL 2.0 and its Compiler BPEL2oWFN. Techn. report 212, Humboldt University of Berlin (2007)
6. Moser, S., Martens, A., Görlach, K., Amme, W., Godlinski, A.: Advanced Verification of Distributed WS-BPEL Business Processes Incorporating CSSA-based Data Flow Analysis. In: 2007 IEEE Int. Conf. on Services Computing. IEEE, Los Alamitos (2007)
7. van der Aalst, W.M.P., Hirnschall, A., Verbeek, H.M.W(E.): An Alternative Way to Analyze Workflow Graphs. In: Pidduck, A.B., Mylopoulos, J., Woo, C.C., Ozsu, M.T. (eds.) CAiSE 2002. LNCS, vol. 2348, pp. 535–552. Springer, Heidelberg (2002)
8. Sadiq, W., Orlowska, M.E.: Analyzing Process Models Using Graph Reduction Techniques. Information Systems 25(2), 117–134 (2000)
9. Lee, J., Midkiff, S.P., Padua, D.A.: Concurrent Static Single Assignment Form and Constant Propagation for Explicitly Parallel Programs. In: Carter, L., Ferrante, J., Sehr, D., Chatterjee, S., Prins, J.F., Li, Z., Yew, P.-C. (eds.) LCPC 1998. LNCS, vol. 1656, pp. 114–130. Springer, Heidelberg (1999)
10. Heinze, T.S., Amme, W., Moser, S.: Resolving Conditional Branches in WS-BPEL Business Processes. Report, Friedrich Schiller University of Jena (to appear)
11. Heinze, T.S., Amme, W., Moser, S.: Generic CSSA-based Pattern over Boolean Data for an Improved WS-BPEL to Petri Net Mapping. In: Third Int. Conf. on Internet and Web Applications and Services, pp. 590–595. IEEE, Los Alamitos (2008)
12. Monakova, G., Kopp, O., Leymann, F.: Improving Control Flow Verification in a Business Process using an Extended Petri Net. In: 1st Central-European Workshop on Services and their Composition, vol. 438, 95–101. CEUR-WS.org (2009)
13. Kopp, O., Khalaf, R., Leymann, F.: Reaching Definition Analysis Respecting Dead Path Elimination Semantics in BPEL Processes. Report 2007/04, IAAS (2007)
14. Thakur, A., Govindarajan, R.: Comprehensive Path-sensitive Data-flow Analysis. In: 6th Int. Symp. on Code Generation and Optimization, pp. 55–63. ACM Press, New York (2008)

The Triconnected Abstraction of Process Models

Artem Polyvyanyy, Sergey Smirnov, and Mathias Weske

Business Process Technology Group
Hasso Plattner Institute at the University of Potsdam
Prof.-Dr.-Helmert-Str. 2-3, D-14482 Potsdam, Germany
{Artem.Polyvyanyy,Sergey.Smirnov,Mathias.Weske}@hpi.uni-potsdam.de

Abstract. Companies use business process models to represent their working procedures in order to deploy services to markets, to analyze them, and to improve upon them. Competitive markets necessitate complex procedures, which lead to large process specifications with sophisticated structures. Real world process models can often incorporate hundreds of modeling constructs. While a large degree of detail complicates the comprehension of the processes, it is essential to many analysis tasks. This paper presents a technique to abstract, i.e., to simplify process models. Given a detailed model, we introduce abstraction rules which generalize process fragments in order to bring the model to a higher abstraction level. The approach is suited for the abstraction of large process specifications in order to aid model comprehension as well as decomposing problems of process model analysis. The work is based on process structure trees that have recently been introduced to the field of business process management.

1 Introduction

Business process modeling is a well-established technique for designing and communicating how work activities are related to each other, and how these activities contribute to a business goal. To provide a common understanding of the language used, standard modeling notations are proposed, for instance, Business Process Modeling Notation (BPMN) [1], Event-driven Process Chains (EPC) [2], and Petri nets [3]. Business process models serve as a communication vehicle for different stakeholders, e.g., business analysts and software designers. Moreover, process models are used to analyze working procedures, to propose improvements, and even to provide a blueprint for a software realizing the process.

With the increasing complexity of services which companies provide to markets, business processes fulfilling these services are getting more and more complex, too. As a result, business process models often consist of dozens or even hundreds of nodes, making these models hard to understand. There is a dilemma: On the one hand, too much detail hampers the understanding of the overall process. On the other hand, this level of detail might be required for process analysis and for implementing the process in software.

There are two approaches to address the problem. Either different models serving different purposes are developed, or different models, catering to different process modeling needs, are generated from a detailed original model. If

U. Dayal et al. (Eds.): BPM 2009, LNCS 5701, pp. 229–244, 2009.

the former approach is followed, consistency of the models is a severe problem. Changes on one level need to be reflected on other levels as well, which is often done manually. Experience shows that due to model evolution on different levels of detail, the models become inconsistent quite soon. Therefore, we opt for the latter approach: We generate different process models from a given detailed model by introducing transformation rules. These rules abstract from details of a process model and provide abstracted models that non-technical stakeholders can understand. At the same time, any evolutionary changes will be taken into account, since effectively there is only one process model, and the others are generated from it on demand. Technically, the work is based on the program parsing technique, known from the compiler theory of sequential programs [4]. The method was introduced to the business process management community in the refined process structure tree (RPST) decomposition of workflow graphs [5].

While the results in this paper are of a conceptual and rather theoretical nature, they emerged from an industry project conducted with a large health insurance company, just like a previous study focusing on pattern-based process abstraction [6]. In this initial endeavor, we developed an automated abstraction control mechanism guided by the average execution time of tasks included in a model. The proposed technique attempts to first abstract from tasks which are rarely observed. Of course, in order to allow such an abstraction control, models must be additionally annotated with with the tasks' average execution times. The main limitation of the pattern-based approach is the problem of completeness, i.e., the necessity to have a full set of patterns which can support the abstraction of arbitrarily structured process models. The idea of abstraction control mechanisms is elaborated in [7], where an abstraction slider is presented.

The completeness of a set of reduction rules is a well-known problem in the analysis of Petri nets. Berthelot proposed a set of rules which can be repeatedly applied to reduce live and bounded marked graphs to a single transition [8,9]. Desel and Esparza, in [10], proposed a complete kit of reduction rules for free-choice Petri nets. In [11], Murata presents reduction rules which preserve the liveness, safeness, and boundedness properties. However, all the mentioned rules are incomplete when operating on models of an arbitrary structure. The limitations of the pattern-based abstraction and the impossibility of closing the gap by adapting the existing reduction rules have inspired this work. We define and utilize for abstraction purposes a notion of a process component which permits achieving completeness when handling a process model of an arbitrary structure.

The rest of the paper is organized as follows: The next section sketches the research field of business process model abstraction, its perspectives and its challenges. In section 3, we provide definitions and a basic corollary that form the basis for further discussion. The structural decomposition of process models into triconnected components is presented in section 4. In section 5, the components are used for process model abstraction, resulting in the triconnected abstraction technique. The paper closes with ideas on future steps and conclusions that summarize our findings.

2   Business Process Model Abstraction

This section discusses the research field of business process model abstraction (BPMA). The core aspects of BPMA are identified. Finally, we position the contribution area of BPMA to be addressed in the rest of the paper.

Business process analysts often attempt to capture every detail of handling a particular business case for inclusion in a process model, which leads to excessive numbers of modeling constructs and sophisticated model structures. In order to reduce the complexity and to allow for the faster investigation of process logic, we started to look for automated techniques to abstract, i.e., to simplify, process models. Abstraction is the result of the generalization or elimination of properties in an entity or a phenomenon in order to reduce it to a set of essential characteristics. Information loss is the fundamental property of abstraction and is its intended outcome. When modeling, business process analysts abstract from the complex reality by extracting important behavioral aspects of a process. In BPMA, we investigate problems specific to the abstraction of process model entities. The challenge lies in identifying what is a meaningful generalization of process logic aimed at removing certain characteristics while at the same time emphasizing others.

In BPMA, identified process fragments can be eliminated or replaced by concepts of a higher abstraction level which conceal, but also represent, the logic of the underlying fragments. In both cases, generalization as well as elimination, sophisticated handling mechanisms need to be proposed. We refer to such mechanisms as *abstraction steps*.

Control mechanisms combine atomic abstraction steps into *abstraction strategies*. One can envision manual strategies in which a user specifies tasks to be abstracted, semi-automated, or automated control mechanisms.

Any process abstraction methodology aims at ensuring certain properties of abstracted models. The properties should allow a semantic relation between the original and abstracted models. The key property we pursue in our approach of process abstraction is order preservation. An *order preserving abstraction* is an abstraction that ensures that neither new task execution order constraints can appear after abstraction, nor existing ones (except for generalized ones) go away. For instance, assume that task A should be abstracted. Let f_A be a process fragment affected in the abstraction step (f_A contains A). As a result of abstraction, fragment f_A gets replaced by task F. If task B belongs to f_A, information about execution order constraints between task A and task B is lost. However, an order preserving abstraction ensures that between any pair of tasks not in f_A, e.g., task C and task D, execution order constraints are preserved. Furthermore, an order preserving abstraction guarantees that execution order constraints between any task not in f_A, e.g., task E, and any task in f_A, task A or task B in our example, are the same as between task E and task F. In the end, an order preserving abstraction secures the overall process logic to be reflected in the abstracted model.

A business process model abstraction methodology is a compromised combination of requirements and techniques picked out from all of the discussed abstraction aspects. Usually, such a combination is guided by project specific use cases. This paper primarily contributes to the BPMA aspects of discovering fragments which are structurally suitable for abstraction and further performing abstractions. Effectively, we define a structural fragment type which is accepted as a unit of process logic abstraction, provide mechanisms for the discovery of a complete set of process fragments suitable for abstraction, and specify the algorithm which aims at abstracting from a given task in a process model by utilizing the discovered fragments.

3   Preliminaries

In this section, we introduce basic definitions. We start with a process model formalism adapted from [12] which is based on generic modeling concepts. A process model consists of a set of tasks and their structuring using directed control flow edges and gateway nodes that implement process routing decisions.

Definition 1. $P = (N, E, type)$ is a *process model* if $N = N_T \cup N_G$ is a set of nodes, where N_T is a nonempty set of tasks and N_G is a set of gateways; the sets are disjoint. $E \subseteq N \times N$ is a set of directed edges between nodes defining control flow. $type : N_G \rightarrow \{and, xor, or\}$ is a function that assigns a control flow construct to each gateway. (N, E) is a connected graph—a *process graph*. Each task $t \in N_T$ can have at most one incoming and at most one outgoing edge $(|\bullet t| \leq 1 \wedge |t\bullet| \leq 1)$, where $\bullet t$ stands for a set of immediate predecessor nodes $(\bullet t = \{n \in N | (n, t) \in E\})$ and $t\bullet$ stands for a set of immediate successor nodes $(t\bullet = \{n \in N | (t, n) \in E\})$ of task t. A task $t \in N_T$ is a *process entry* if $|\bullet t| = 0$. A task $t \in N_T$ is a *process exit* if $|t\bullet| = 0$. There is at least one process entry task and at least one process exit task. Each gateway is either a split or a join. A gateway $g \in N_G$ is a *split* if $(|\bullet g| = 1 \wedge |g\bullet| > 1)$. A gateway $g \in N_G$ is a *join* if $(|\bullet g| > 1 \wedge |g\bullet| = 1)$.

To be able to refer to parts of a process model, we define a process fragment. A process fragment is a connected part of a process model.

Definition 2. A *process fragment* $F = (N_F, E_F, type_F)$ of a process model $P = (N, E, type)$, where $N_G \subset N$ is a set of gateways of P, consists of a connected subgraph (N_F, E_F) of the process graph (N, E) of P and function $type_F$, which is a restriction of the function $type$ of P to a set $N_F \cap N_G$.

Within a process fragment, nodes can be classified in regard to their structural relation to the whole process model.

Definition 3. A node $n \in N_F$ is a *boundary* node of a process fragment $F = (N_F, E_F, type_F)$ in a process model $P = (N, E, type)$ if n is a process entry of P, a process exit of P, or there exist edges $e_i \in E_F$ and $e_j \in E \backslash E_F$ adjacent through n. A non-boundary node $n \in N_F$ of F is an *internal* node of F.

Boundary nodes of a process fragment can be distinguished as fragment entries and fragment exits based on the directions of incident control flow edges.

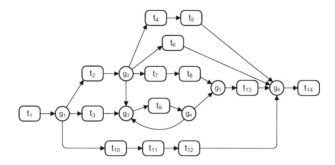

Fig. 1. A process model

Definition 4. Let $n \in N_F$ be a boundary node of a process fragment $F = (N_F, E_F, type_F)$ in a process model $P = (N, E, type)$, then:

- A node n is a *fragment entry* of F if all the incoming edges of n are outside of F ($\bullet n \subseteq N \setminus N_F$) or all the outgoing edges of n are inside of F ($n \bullet \subseteq N_F$).
- A node n is a *fragment exit* of F if all the outgoing edges of n are outside of F ($n \bullet \subseteq N \setminus N_F$) or all the incoming edges of n are inside of F ($\bullet n \subseteq N_F$).

Finally, we recognize a special class of process fragments—process components.

Definition 5. A *process component* $C = (N_C, E_C, type_C)$ is a process fragment with two boundary nodes: one fragment entry and one fragment exit.

This notion of a component was first introduced in [4] as a concept of a *proper subprogram*. A process component is a process fragment in which it is assured that if control flows through a fragment's edge, it has first entered the process fragment through the fragment entry and will subsequently leave the process fragment through the fragment exit.

Structurally, a process component is a self-contained block of process logic with strictly defined boundaries. Semantically, a process component can be addressed as a detailed specification of task execution scenarios. Hence, any process component can be formalized as a WF-net [13] of, potentially, an arbitrary structure. Therefore, in the triconnected abstraction approach, a process component is accepted as a unit of meaningful aggregation of process logic, i.e., detailed specifications get represented by a corresponding task concept. In the following sections, we discuss issues relevant to the identification and abstraction of process components in process models.

We require process models to be *structurally sound* [12], i.e., a process model should have exactly one process entry, exactly one process exit, and each process model node should be on a path from the process entry to the process exit. The prerequisite introduces a minimal correctness notion for process models—subjects for abstraction. Moreover, the stated structural requirement is crucial when it comes to the discovery of process components in process models. Figure 1 provides an example of a process model suitable for abstraction.

4   The Triconnected Decomposition

This section explains how to discover process fragments that relate to the notion of a process component as defined in section 3. First, we give the basic intuition inherent in the algorithm. Afterwards, we show the relation of the discovery process to the approach of SPQR-tree [14,15] decomposition. Finally, we discuss SPQR-tree fragments in the context of process models.

4.1   Basic Approach for Process Component Discovery

A search for a process component in a process model is guided by its definition (see Definition 5), which states that a process component is a process fragment with two boundary nodes. Boundary nodes are the nodes that connect the fragment to the model, i.e., if removed the fragment becomes disconnected from the model. Thus, in order to discover a process component, one must first look for a *separation pair*—a pair of nodes that disconnect a process fragment from the rest of the process model. For instance, gateways g_3 and g_4 disconnect task t_9 in the process model from Figure 1. Afterwards, the boundary nodes of the fragment need to be tested to give one fragment entry and one fragment exit.

A separation pair divides process model into two fragments. In order to find all fragments with two boundary nodes, the rationale of the described discovery step must be applied to each of the two fragments, resulting in a divide and conquer algorithm design. Each recursive thread terminates once the problem cannot be further subdivided, i.e., there is no separation pair in a process fragment.

The described algorithm is in fact the algorithm for the discovery of triconnected components in a graph. Connectivity is a property of a graph. It is known that a graph is k-connected if there exists no set of $k - 1$ elements, each a vertex or an edge, whose removal makes the graph disconnected (there is no path between some node pair in a graph). Such a set is called a separating $(k-1)$-set. Separating 1- and 2-sets of graph vertices are called cutvertices and separation pairs. 1-, 2-, and 3-connected graphs are referred to as connected, biconnected, and triconnected, respectively. Each recursive thread of the algorithm terminates once it encounters a triconnected component.

4.2   SPQR-Tree Decomposition

In order to discover process components, one can use SPQR-tree decomposition. SPQR-tree decomposition is a decomposition of an undirected biconnected multigraph induced by its split pairs aimed at identifying its triconnected components. A *split pair* is either a separation pair or a pair of adjacent nodes. Process models are connected, but not necessarily biconnected. For example, the process model from Figure 1 has cutvertex g_1. However, it is always possible to make a process model biconnected by adding a back edge connecting a process exit with a process entry. The requirement of structural soundness ensures that every process model has exactly one process entry and exactly one process exit.

The algorithm for the discovery of triconnected components of a graph was first proposed by Hopcroft and Tarjan in [16]. Later, Tarjan and Valdes in [4]

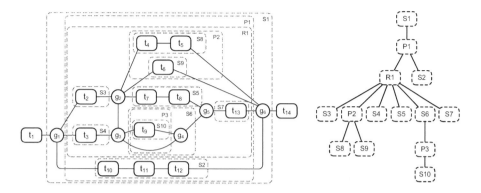

(a) Undirected process graph, triconnected components   (b) SPQR-tree

Fig. 2. SPQR-tree process model decomposition

applied the algorithm for sequential program parsing to obtain the *parse tree* (or *the tree of the triconnected components*). The tree was studied as SPQR-tree in [14,15]. [16,17,18] show the path towards a linear time complexity algorithm implementation of SPQR-tree decomposition. The decomposition results in triconnected components of four structural types, in the following using the SPQR-tree terminology, S, P, Q, and R types.

○ *Trivial case.* A split pair is a pair of adjacent graph vertices—a fragment consists of one edge—the Q-type fragment.
○ *Parallel case.* A split pair is a pair of adjacent graph vertices in k distinct edges ($k \geq 2$)—the P-type fragment.
○ *Series case.* A split pair is a pair of graph vertices giving a maximal sequence of vertices and consists of k nodes and k edges ($k \geq 3$)—the S-type fragment.
○ *Rigid case.* If none of the above cases applies, a fragment is a triconnected fragment—the R-type fragment.

SPQR-tree decomposition of the process model from Figure 1 is exemplified in Figure 2. Each process fragment corresponds to a triconnected component of the model and is defined by edges that are inside or intersect with a corresponding region visualized with a dashed line in Figure 2(a). Fragment names hint at structural fragment types, e.g., $P1$, $P2$, and $P3$ are all parallel case fragments. Boundary nodes of a fragment are the nodes incident with edges crossing the region borderline and are outside of the region.

Figure 2(b) shows an SPQR-tree that visualizes hierarchical fragment relations. Fragment $P1$ contains fragments $R1$ and $S2$ and is fully contained within fragment $S1$. Each SPQR-tree node represents a *fragment skeleton*, i.e., basic structure of a fragment and its relations with a parent and child fragments. Figure 3 shows fragment skeletons of SPQR-tree nodes from Figure 2(b). Boundary nodes are highlighted with a thick borderline, e.g., nodes g_1 and g_6 in fragment $R1$ (see Figure 3(b)). Each fragment skeleton can consist of edges of three types.

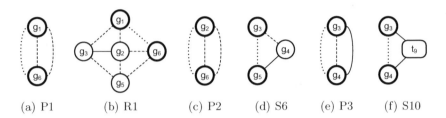

(a) P1   (b) R1   (c) P2   (d) S6   (e) P3   (f) S10

Fig. 3. SPQR-tree fragment skeletons

Original graph edges are drawn with solid lines, whereas dotted and dashed lines represent *virtual edges*. Each virtual edge is shared between two fragment skeletons and hints at a parent-child relation. An edge visualized by a dotted line shows a child relation of the fragment skeleton with another skeleton which contains the same virtual edge; a dashed line signals a parent relation. For instance, the fragment skeleton from Figure 3(f) contains one virtual edge (g_3, g_4), which hints at a child relation with another fragment skeleton that contains the same virtual edge—fragment skeleton $P3$ (see Figure 3(e)). In order to obtain the graph fragment given by fragment skeleton $P3$, one must "glue" it together with fragment skeleton $S10$ along virtual edge (g_3, g_4). Once the fragments are combined, the virtual edge is removed. In general, a graph fragment represented by an SPQR-tree node can be obtained by combining all its descendants.

SPQR-tree provides process model decomposition that ignores control flow edge directions. At this point, there has still been no distinction made between entry and exit boundary nodes; obtained fragments still cannot be classified as process components.

4.3   SPQR-Tree Fragments in the Context of Process Models

In this section, we examine fragments obtained after the SPQR-tree decomposition of a process model, i.e., edges of a process graph are directed and nodes distinct as tasks and gateways.

In general, an SPQR-tree can be rooted to any node. However, in the context of a process model it makes sense to root the tree to a node representing the fragment containing the deliberately introduced back edge (node $S1$ in Figure 2(b)). As a result, one obtains the structural hierarchical refinement of a process model.

Further observations are: Task nodes can only be present, but are not always necessarily present (see Figure 3(d)), inside of S-type fragments, while boundary fragment nodes are always gateways. The former property comes from the definition of the S-type fragment. Any sequence of nodes in a process graph can only be formed by task nodes embraced by gateways. Thus, any maximal sequence, also composed of one task (see Figure 3(f)), is recognized as the S-type fragment with two boundary gateways: one at sequence entry and another at sequence exit. This also means that other fragment skeletons are composed of gateways only, which testifies the latter property.

Until now, we have recognized sequences as S-type fragments. Q-type fragments stand for original process graph edges, e.g., the edge (g_4, g_5) of fragment skeleton from Figure 3(d). P-type fragments (see Figures 3(a), 3(c), and 3(e)) allow identification of block and loop structures within process models. The control flow of the process model from Figure 1 specifies fragments $P1$ and $P2$ as blocks and fragment $P3$ as a loop (there exists a back edge between boundary nodes g_3 and g_4). The fragment from Figure 3(b) is the triconnected fragment that explicitly defines what makes the process model graph-structured. There are no R-type fragments in a block-structured process model. A block-structured process model can be inductively composed based on sequence, block, and loop patterns (S-type and P-type fragments) [19].

Finally, we are ready to make the concluding proposition of section 4:

Theorem 1. *Any process fragment obtained after SPQR-tree decomposition of a structurally sound process model is a process component.*

Proof. Any process fragment obtained after SPQR-tree decomposition of a process model has two boundary nodes. A pair of boundary nodes of a process fragment is a split pair of the process model. Thus, it is necessary to show that one of the boundary nodes is a fragment entry and the other is a fragment exit.

First, we show that any boundary node of a process fragment induced by SPQR-tree decomposition is either a fragment entry or a fragment exit. All the edges incident with a boundary node are divided into two disjoint sets of those inside and those outside the fragment. Definition 4 states that a boundary node of a process fragment is a fragment entry or a fragment exit if either all the incoming or all the outgoing edges incident with the node are either the edges of the fragment or are outside the fragment. As explained above, any boundary node is a gateway. For any gateway, either a set of all incoming edges or a set of all outgoing edges consists of one element (see Definition 1). The relation of this one edge, either belonging to the process fragment or not, defines the relation of the whole set. Therefore, any boundary node can only expose the logic of a fragment entry or a fragment exit.

The rationale towards a formal proof of the "pure" logic of a boundary node of a process fragment can be approached as follows. Let $P = (N, E, type)$ be a process model, $F = (N_F, E_F, type_F)$ be a process fragment of P. Let us define auxiliary predicates:

○ $i : E \times N \rightarrow \{true, false\}$ is $true$ if $e \in E$ is the incoming edge of node $n \in N$, $false$ otherwise, and
○ $o : E \times N \rightarrow \{true, false\}$ is $true$ if $e \in E$ is the outgoing edge of node $n \in N$, $false$ otherwise.

One can now define predicates which check if a node $n \in N$ can be an entry of F—$canEnter$, or an exit of F—$canExit$:

○ $canEnter(n, F) = \exists e_1 \in E \backslash E_F \exists e_2 \in E_F : i(e_1, n) \wedge o(e_2, n)$,
○ $canExit(n, F) = \exists e_1 \in E_F \exists e_2 \in E \backslash E_F : i(e_1, n) \wedge o(e_2, n)$.

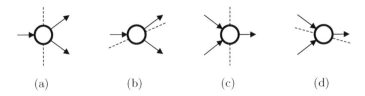

(a)   (b)   (c)   (d)

Fig. 4. All possible combinations for edge separation on internal and external fragment edges for a boundary gateway connecting three edges

In order to show that any boundary fragment node cannot at the same time expose entry and exit logic, one must show that the logical statements $canEnter(n, F) \models \neg canExit(n, F)$ and $canExit(n, F) \models \neg canEnter(n, F)$ hold. Hence, one must show that $canEnter(n, F) \land canExit(n, F)$ is a false statement on all interpretations which in a prenex normal form says:

$$\exists e_1 \in E \backslash E_F \exists e_2 \in E_F \exists e_3 \in E_F \exists e_4 \in E \backslash E_F : i(e_1, n) \land o(e_2, n) \land i(e_3, n) \land o(e_4, n)$$

If n is a split gateway, the statement might evaluate to *true* only if e_1 and e_3 are bound to the same edge. This, however, is impossible, as e_1 and e_3 belong to different sets which are disjoint: E and $E \backslash E_F$. The same rationale applies for a join gateway and edges e_2 and e_4. Therefore, a logical expression $canEnter(n, F) \land canExit(n, F)$ always evaluates to *false*, which proves the pure logic of any boundary node of F.

Figure 4 shows all possible combinations of internal and external fragment edges incident with a boundary gateway which connects three edges. The dashed line separates edges on fragment's internal and external edges. Regardless of a separation and a gateway type, control flow is only allowed to "penetrate" a fragment's boundary in one direction, either to enter or to leave a process fragment.

Finally, it is necessary to show that only one arrangement of boundary nodes is possible, i.e., one of the nodes is a fragment entry and the other is a fragment exit. We show this by contradiction; the settings of two fragment entries or two fragment exits are not possible under the correctness criteria imposed on a process model—a process model is structurally sound. Two cases can be reduced to one. For instance, in case of a fragment with two exits, one can discuss a two entry fragment formed by the edges outside the two exit fragment. A process fragment with two entries violates the requirement of a structurally sound process model which states that each node in a process model is on a path from a process entry to a process exit. Once we enter a two entry fragment, we never leave it. Any node of a two entry fragment cannot be on a path from the process entry to the process exit. Therefore, one of the boundary nodes must be a fragment exit. If the process entry and the process exit are the boundary nodes of a process fragment, the process entry is a fragment entry and the process exit is a fragment exit.   □

5 The Triconnected Abstraction

This section presents the triconnected abstraction. The approach is based on the decomposition technique described in section 4. First, we define abstraction rules. Afterwards, we combine the rules into the process model abstraction algorithm.

5.1 Abstraction Rules

The triconnected process model abstraction technique is founded on the idea of interchanging process fragments with process tasks of higher abstraction levels. In this section, we present abstraction rules that utilize process components obtained after SPQR-tree decomposition for this purpose. The approach assumes abstraction control mechanism that delivers collection of tasks to be abstracted in the process model.

Once a task to abstract is selected, it uniquely identifies the S-type fragment that contains the task and its structural relation within SPQR-tree. There can be seven types of SPQR-tree edges based on the types of adjacent nodes of S-, P-, and R-type; Q-type fragments are not considered. Edges of (S, S)-type and (P, P)-type are recognized as single fragments of S- or P-type, respectively. Edges are proposed as $(parent, child)$ pairs. Out of seven edge types, four connect S-type nodes: (S, P), (S, R), (P, S), and (R, S). The abstraction rules we propose operate within a single series case process fragment, or assume one of the four stated structural relations of an S-type process fragment.

Sequential (Q-Type) Abstraction. A task in a process model can be structured in a sequence with other tasks. We implement abstraction of this task by aggregation with one of its neighbors. Any maximal sequence of tasks is recognized within an S-type process component. Thus, the abstraction is performed locally, i.e., within one process component.

Figure 5 shows an example of a *sequential abstraction* performed inside of the S-type process component. The structure of the original process component is given on the left of the figure. The component is a maximal sequence of three tasks. The example ignores boundary gateway logic, which can be either split or join. In the case task A or C should be abstracted, selection of the neighbor task to aggregate with is obvious—it is task B. However, if task B trig-

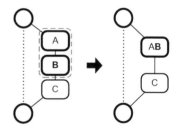

Fig. 5. Sequential abstraction

gers abstraction, the selection is delegated to the abstraction control mechanism. If structural process model generalization is of interest, abstraction control mechanism can allow nondeterministic task choice. In the example, task A is selected to be aggregated with task B, the corresponding process fragment is enclosed in the region with a dashed borderline and constitutes a single Q-type component.

The process component structure on the right of Figure 5 is the output of the sequential abstraction step. As a result, tasks A and B are aggregated into one task AB that semantically corresponds to the activity of first accomplishing task A and then task B. The process component keeps its structural type—the S-type. Sequential abstraction preserves SPQR-tree structure.

S-Type Abstraction. A maximal sequence of tasks in a process model can consist of one task. The situation might occur in the original model or be a result of the prior application of sequential abstractions. This task can be structured in a sequence with process components of P-type or R-type. Within SPQR-tree, such structural relations are captured by (S, P)- or (S, R)-type edges. If it is necessary to abstract the task, aggregation with a neighbor component is performed to result in S-type abstraction.

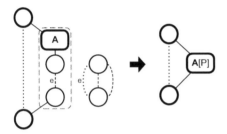

Figure 6 shows an example of S-type abstraction. Task A is designed for abstraction (highlighted with a thick borderline on the left in Figure 6). Task A has no neighbor task—sequential abstraction is not possible. However, the task is in a sequence with the P-type component to form the abstraction fragment in the region enclosed by the dashed borderline. The result of S-

Fig. 6. S-type abstraction

type abstraction is given on the right of the figure. Abstraction results in task $A[P]$, which semantically corresponds to the activity of first accomplishing task A and then performing a process fragment captured by the P-type component. S-type abstraction results in SPQR-tree transformation. The branch representing the abstracted component gets removed. Abstraction leads to a restructuring of the S-type component that contained the task which triggered abstraction. However, the component retains its type—the S-type.

S-type abstraction is presented by means of a structural relation of an (S, P)-type edge in SPQR-tree. The procedure for an (S, R)-type edge is analogous. In the example, the boundary gateways of the abstracted component are reduced. In general, if a boundary gateway of an abstracted component is shared with some other process component, it must be preserved in the abstracted model.

P-Type Abstraction. Sequential and S-type abstractions tend to generalize S-type components into simple components. A *simple component* is a S-type component composed of a single task (see Figure 3(f)). Simple components are structured by (P, S)- or (R, S)-type edges in SPQR-tree. If a task from a simple component is selected for abstraction and its parent component is a P-type component, *P-type abstraction* is performed. The task is aggregated with some other child component of the parent component. The selection of the child component to aggregate with is carried out by the abstraction control mechanism.

Figure 7 shows an example of P-type abstraction. Task A is selected for abstraction. The task is highlighted with a thick borderline and is the only task of the simple component (shown on the left of Figure 7). The simple component is the child component of

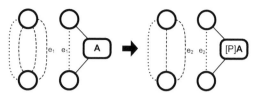

Fig. 7. P-type abstraction

the P-type component. It shares virtual edge e_1 with its parent. The result of the P-type abstraction step is given on the right of the figure. Two child components of the P-type component are aggregated into one simple component that contains task $[P]A$. This task semantically corresponds to the execution of two abstracted branches following the type of the boundary gateways. The obtained simple component shares virtual edge e_2 with the parent P-type component.

P-type abstraction results in SPQR-tree transformation. The branch that represents the abstracted component is completely removed. The number of child components of the parent parallel component is reduced by one. If the P-type component initially contains two branches, abstraction results in a single branch. Afterwards, the boundary gateways must be reduced if they do not specify any routing logic, i.e., have single incoming edge and single outgoing edge. In such a case, the P-type component node is further reduced in the SPQR-tree to represent a single task within the next level parent component.

R-Type Abstraction. A task intended for abstraction can be contained in a simple component within a process model that is a child of a R-type component. Such a structural relation is specified by a (R, S)-type edge within SPQR-tree. R-type abstraction is proposed to handle this situation. As a result of the R-type abstraction the task is aggregated with the whole parent component.

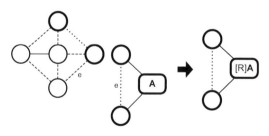

Fig. 8. R-type abstraction

Figure 8 shows an example of R-type abstraction. Task A is selected to be abstracted. The task is highlighted with a thick borderline and is the only task of the simple component on the left of the figure. The simple component is the child of the R-type component (the same component as in Figure 3(b)). The simple component shares the virtual edge e with its parent and corresponds to fragment $S7$ from Figure 2. The result of R-type abstraction step is given on the right of Figure 8. The abstraction results in the aggregation of the whole parent R-type component into a simple component that has task $[R]A$ and boundary gateways of the R-type component. The task semantically corresponds to the execution of the whole rigid component.

R-type abstraction results in SPQR-tree transformation. The abstracted R-type component gets replaced by a simple component. The branch of the R-type fragment is completely removed. Similar to *P*-type abstraction, the boundary gateways can be skipped to further reduce the resulting simple component.

5.2 Abstraction Algorithm

Section 5.1 presented four abstraction rules. The rules cover all possible structural relations of a task in a process model. In this section, we organize them into a procedure that handles a single abstraction step of a task. As input, the algorithm obtains a process model, its SPQR-tree decomposition, and a task to abstract. As output, the algorithm delivers a process model with the specified task abstracted. Algorithm 1 formalizes the procedure in pseudo code.

The algorithm orchestrates abstraction rules and attempts to aggregate a minimal number of tasks at each abstraction step; empirical insights for the proposed solution were obtained in [6]. In line 1, the component *c* which contains task *a* is identified—it is a *S*-type component. If *c* is not a simple component (line 2), then either it has a neighbor task (line 3) or a neighbor component (line 4) that can be aggregated with task *a*. Otherwise (line 5), abstraction of task *a* depends on the parent component of *c*. If *c* is the root component of SPQR-tree, then *p* consists of a single task *a* and there is nothing else to abstract (line 6). Otherwise, get the parent component of *c*—component *cp* (line 7). If *cp* is a *P*-type component (line 8) or a *R*-type component (line 9), then *P*-type abstraction or, respectively, *R*-type abstraction is performed.

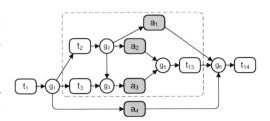

Fig. 9. An abstracted process model

Algorithm 1 provides a formal relation between an original model and an abstracted one. The triconnected abstraction is the order preserving abstraction. Figure 9 shows the abstraction example of the process model from Figure 1. In the example, a collection of tasks selected for abstraction caused process

---

Algorithm 1. The Triconnected Abstraction

TriAbstraction(ProcessModel *p*, SPQRtree *t*, Task *a*)

1. *c* := component of process model *p* from SPQR-tree *t* containing task *a*
2. **if** *c* is not a simple component **then**
3.   **if** *a* has neighbor task in *c* **then** perform *sequential abstraction* of *a*
4.   **else** perform *S-type abstraction* of *a*
5. **else** // *c* is a simple component
6.   **if** *c* is the root component in *t* **then** *p* is already abstracted to one task **return**
7.   *cp* := get a parent component of *c* in SPQR-tree *t*
8.   **if** *cp* is *P*-type component **then** perform *P-type abstraction* of *a*
9.   **if** *cp* is *R*-type component **then** perform *R-type abstraction* of *a*

components $S2$, $S5$, $S6$, $S8$, $S9$, $S10$, $P2$, and $P3$ to get abstracted. These tasks can be t_6, t_8, t_9, t_{10}, and t_{12} (see Figure 1). After abstraction, aggregating tasks a_1, a_2, a_3, and a_4, highlighted with grey background in the figure, conceal the process logic of abstracted components. For instance, task $a1$ is the abstraction of two branches: one composed of tasks t_4 and t_5, and the other of a single task t_6. The type of gateway g_2 specifies the behavioral relation of both branches inside the abstracted task. Task a_1 can be derived using a single P-type abstraction step triggered by task t_6 or by a series of sequential then P-type abstractions if first triggered by either t_4 or t_5. The only R-type component of the process model, shown in the region enclosed by the dashed borderline in Figure 9, is not abstracted. An algorithmic step aimed at abstracting any of the tasks contained within the region will cause the whole component to aggregate into one task.

6   Conclusions

In this paper, we investigated how the SPQR-tree decomposition of process models can help the task of process model abstraction, in particular the discovery of structurally meaningful process model fragments and their aggregation. We defined abstraction rules based on the notion of a process component and proposed their arrangement in the algorithm.

The triconnected abstraction technique defines structural model transformations and can be generalized to any process modeling notation which uses directed graphs as the underlying formalism. Limitations of the triconnected abstraction technique come from restrictions on process model structure. Process models must be free of self-loop structural patterns (should have no cutvertices), and must contain no "mixed" gateways with multiple incoming and multiple outgoing edges (should decompose onto process components). The limitations described above can be overcome by a preprocessing step which transforms mixed gateways into a sequence of first a join, then a split. Alternatively, one can generalize abstraction mechanisms to operate with the RPST decomposition [5].

While the results have proven very useful to our project partner, the abstraction mechanisms only take into account the structure of a business process. In particular, the user of the abstraction might decide that certain activities need to be present in several or even all abstractions. In this case, the application of the mechanisms introduced in this paper needs to be restricted. Therefore, studies regarding the methodology of abstractions need to complement the more technical studies reported in this paper. In future works, we also plan to investigate multiple entry multiple exit components; this should allow further decomposition of rigid case fragments in process models. Theorem 1 gives promising insights into the problem of RPST computation, which we plan to develop in the following work. A promising research direction is to look into how the triconnected abstraction technique can be employed for decomposing problems of process model verification—process model behavior analysis, and which model properties are preserved by the abstraction rules.

References

1. OMG: Business Process Modeling Notation, Version 1.2 (January 2009)
2. Keller, G., Nüttgens, M., Scheer, A.: Semantische Prozessmodellierung auf der Grundlage "Ereignisgesteuerter Prozessketten (EPK)". Technical Report 89, University of Saarland (1992)
3. Petri, C.: Kommunikation mit Automaten. PhD thesis, Institut für instrumentelle Mathematik, Bonn, Germany (1962)
4. Tarjan, R.E., Valdes, J.: Prime Subprogram Parsing of a Program. In: Proceedings of the 7th Symposium on Principles of Programming Languages (POPL), pp. 95–105. ACM, New York (1980)
5. Vanhatalo, J., Völzer, H., Koehler, J.: The Refined Process Structure Tree. In: Proceedings of the 6th International Conference on Business Process Management (BPM), Milan, Italy, September 2008, pp. 100–115 (2008)
6. Polyvyanyy, A., Smirnov, S., Weske, M.: Reducing Complexity of Large EPCs. In: Geschäftsprozessmanagement mit Ereignisgesteuerten Prozessketten (MobIS: EPK), Saarbruecken, Germany (November 2008)
7. Polyvyanyy, A., Smirnov, S., Weske, M.: Process Model Abstraction: A Slider Approach. In: Proceedings of the 12th IEEE International Enterprise Distributed Object Computing Conference (EDOC), Munich, Germany (September 2008)
8. Berthelot, G.: Checking Properties of Nets using Transformation. In: Advances in Petri Nets 1985, London, UK, pp. 19–40. Springer, Heidelberg (1986)
9. Berthelot, G.: Transformations and Decompositions of Nets. In: Advances in Petri nets 1986, London, UK, pp. 359–376. Springer, Heidelberg (1987)
10. Desel, J., Esparza, J.: Free Choice Petri Nets. Cambridge University Press, New York (1995)
11. Murata, T.: Petri Nets: Properties, Analysis and Applications. Proceedings of the IEEE 77(4), 541–580 (1989)
12. Weske, M.: Business Process Management: Concepts, Languages, Architectures. Springer, Heidelberg (2007)
13. Aalst, W.: Verification of Workflow Nets. In: Azéma, P., Balbo, G. (eds.) Application and Theory of Petri Nets, Berlin, Germany, pp. 407–426. Springer, Heidelberg (1997)
14. Battista, G.D., Tamassia, R.: Incremental Planarity Testing. In: Proceedings of the 30th Annual Symposium on Foundations of Computer Science, FOCS (1989)
15. Battista, G.D., Tamassia, R.: On-Line Maintenance of Triconnected Components with SPQR-Trees. Algorithmica 15(4), 302–318 (1996)
16. Hopcroft, J.E., Tarjan, R.E.: Dividing a Graph into Triconnected Components. SIAM Journal on Computing 2(3), 135–158 (1973)
17. Fussell, D., Ramachandran, V., Thurimella, R.: Finding Triconnected Components by Local Replacement. SIAM Journal on Computing 22(3), 587–616 (1993)
18. Gutwenger, C., Mutzel, P.: A Linear Time Implementation of SPQR-Trees. In: Proceedings of the 8th International Symposium on Graph Drawing (GD), London, UK, pp. 77–90. Springer, Heidelberg (2001)
19. Liu, R., Kumar, A.: An Analysis and Taxonomy of Unstructured Workflows. In: van der Aalst, W.M.P., Benatallah, B., Casati, F., Curbera, F. (eds.) BPM 2005. LNCS, vol. 3649, pp. 268–284. Springer, Heidelberg (2005)

Granularity as a Cognitive Factor in the Effectiveness of Business Process Model Reuse

Oliver Holschke, Jannis Rake, and Olga Levina

Technische Universität Berlin, Fachgebiet Systemanalyse und EDV
FR 6-7, Franklinstr. 28-29, 10587 Berlin, Germany
{oliver.holschke,jannis.rake,olga.levina}@sysedv.tu-berlin.de

Abstract. Reusing design models is an attractive approach in business process modeling as modeling efficiency and quality of design outcomes may be significantly improved. However, reusing conceptual models is not a cost-free effort, but has to be carefully designed. While factors such as psychological anchoring and task-adequacy in reuse-based modeling tasks have been investigated, information granularity as a cognitive concept has not been at the center of empirical research yet. We hypothesize that business process granularity as a factor in design tasks under reuse has a significant impact on the effectiveness of resulting business process models. We test our hypothesis in a comparative study employing high and low granularities. The reusable processes provided were taken from widely accessible reference models for the telecommunication industry (enhanced Telecom Operations Map). First experimental results show that Recall in tasks involving coarser granularity is lower than in cases of finer granularity. These findings suggest that decision makers in business process management should be considerate with regard to the implementation of reuse mechanisms of different granularities. We realize that due to our small sample size results are not statistically significant, but this preliminary run shows that it is ready for running on a larger scale.

Keywords: Business Process Model Reuse, Reuse Economics, Process Granularity, Design for Reuse, Reuse of Non-Code Artifacts, Experiment.

1 Introduction

Reuse of concepts is a basic human psychological mechanism. Humans draw on memory and knowledge represented in the brain, i.e. activation of neurons as a reaction to specific input patterns, but also represented in other media, e.g. scripts and other visual models. From an evolutionary perspective humans have been very successful using reuse mechanisms. Efficient reuse of patterns, e.g. "Escape when predator is spotted!" played a central role. It is a fundamental psychological activity. Therefore the range of application areas is very wide: learning, training, knowledge management, classification and others [1].

For specialized tasks such as designing and managing IS systems in enterprise contexts, reuse mechanisms as well are very attractive because of the potential

U. Dayal et al. (Eds.): BPM 2009, LNCS 5701, pp. 245–260, 2009.

economic benefits conveyed as time-savings, qualitative improvements and economies of scale. There are many examples in system design and development that focus on reuse drawing upon various artifacts settled in different enterprise contexts, e.g. code reuse [2-6], component reuse [7, 8], reuse of (industrial) reference models (e.g. enhanced Telecom Operations Map (eTOM) [9] and Supply Chain Operations Reference Model (SCOR) [10]), reuse of architectural decision topics based on a reference model [11], design patterns [6, 12-15], reuse of conceptual models [8], and reuse of IS as distributed functionality in the form of, e.g. multi-tenancy platforms and Software-as-a-Service (SaaS) [16].

Regarding the size and complexity of many enterprises, design artifacts on a conceptual level (as opposed to, e.g. code) gain importance as their higher abstraction, also referred to by the term *granularity*, allows decision makers to cope with new requirements and exert control on the enterprise (artifacts of higher abstraction/ granularity are useful due to limited cognitive capacities of decision makers and astronomic cost when controlling *all* parameters in an enterprise). The reuse of proven conceptual models, such as methodological patterns or business process models may provide an efficient means for successfully exerting that control.

However, successful reuse depends on various parameters, such as the variability of the problem domain, the availability and applicability of a reusable design artifact, and the human cognitive activities that are performed when identifying and, most importantly, *adapting* an artifact for a present managerial challenge. Particularly the latter is dependent on the size of informational building blocks – i.e. a specific granularity – represented in the reusable artifact, which must be sufficiently aligned with the designer's cognitive structures in order to facilitate an economic reuse process. Too fine granular representation of a reusable artifact may overwhelm the respective user exacting high adaptation costs. Too coarse granular representation may fail to exert necessary control of underlying relationships. While different granularity levels of models all have their target audiences who can efficiently apply them within their respective context, the question remains what the optimal granularity (band) for conceptual models may be, so that decision makers can successfully exploit the benefits of model reuse when managing the enterprise to flexibly react to changes.

We explore these relationships by conducting two studies that make use of conceptual models of different granularities. We believe that, given equal problem complexities, equal similarity distances between problem and reuse artifact, and equal reuse mechanisms (determined by the cognitive abilities of the designers), the model granularity affects the outcome of the design (under reuse) process depending on how well the designer's cognitive reuse capabilities can cope with the respective model granularity. Our findings are early indications of how the granularity of conceptual models should be designed so that the IS management of enterprises employing conceptual models as powerful tools can be improved.

Related works also concerned with behavioral aspects of business process management in general and comprehension and complexity in particular, exist to a limited degree. Impacts of modularity on business process understandability are evaluated and discussed in [17]. [18] present experimental findings on decomposition of UML models, considering the *Wand and Weber Good Decompositions Conditions* [19], showing that better decomposition increases *actual* understanding of the models.

However, although these works are concerned with different levels of granularity relating to human understandability, issues of reusing artifacts of different granularity have not been explored yet. The rest of this article is organized as follows. Section 2 introduces the concept of granularity and its quantitative measurement. In Section 3 the granularity concept is integrated into the design process under reuse and propositions regarding granularity as an important factor are derived. Section 4 describes the two studies that we have conducted, and discusses the results. Section 5 summarizes the results and concludes with the limitations of this study and a brief outlook on further research work.

2 Granularity: Concept and Quantification

2.1 Granularity as a Concept of Human Cognition

The term granularity has been discussed in various research areas such as Granular Computing, Cognitive Informatics, Pattern Classification, and Conceptual Modeling. Granularity is a fundamental concept in human cognition and deals with the construction, interpretation, and representation of *granules*. A granule is a clump of points (objects) drawn together by indistinguishability, similarity, proximity or functionality [20]. Granules are the result of a granulation process. It is a process that involves dividing some universe into subsets or the grouping of individual subjects into clusters. Granules can be viewed as subsets of the universe, which may be either fuzzy or crisp [1, 21]. Once granulation has been performed, it is necessary to label granules. This can be done by classification, i.e. assigning a name to a granule such that an element in the granule is an instance of the named category.

A *partition* of a universe U is a collection of nonempty and pair-wise disjoint subsets of U whose union is U. Each subset in a partition is also called a block. In the granulated view, *partitions*, being elements of the partition of U, are the basic building blocks and are called elementary granules. They are the smallest nonempty subsets that can be defined, observed or measured. From elementary granules, larger granules can be constructed by taking unions of elementary granules [1]. Since partitions are nonempty, they may have a cardinality bigger than 1. The parts or blocks of the partitions are countable, but not observable because they cannot be differentiated.

When applying the granularity concept to the context of enterprise systems and their management and design, the whole enterprise including the entirety of business processes, data structures, employees and other artifacts to conduct the business may be regarded as the universe U. This generally is the viewpoint of an enterprise designer/architect. Naturally, when incorporating relationships to other enterprises, those artifacts also become part of the designer's discourse and therefore enlarge the universe. To simplify matters we refrain from including B2B relations in the universe. In the enterprise context different partitions π_i can be related to the different design viewpoints that designers and developers in the organization may take. The structuring of the whole enterprise into coarse service domains for instance, may be seen as a coarse-granular partition π_1. In contrast, the design of a specific end-to-end business process with all the required data structures – this happening on a much finer level – may be regarded as a fine-granular partition π_2. A conceptual visualization of different granularities of partitions π_1 and π_2 is presented in Fig. 1. The granularity of different partitions is an important characteristic of design tasks that reside on a specific level of granularity as it may affect how designs are planned and developed, and how efficiently available design artifacts may be reused.

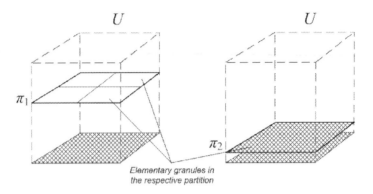

Fig. 1. Granularity of partition π_1 (left) and partition π_2 (right) of a universe U

2.2 Measuring Granularity

In order to apply the granularity concept as a factor in reuse-oriented design tasks, a quantitative measure has to be associated to granularity. For measuring the granularity of a partition the Shannon entropy measure can be used as a basis [1, 22]. With respect to a partition $\pi = \{A_1, A_2, ..., A_m\}$ the probability distribution is

$$P_\pi = \left(\frac{|A_1|}{|U|}, \frac{|A_2|}{|U|}, ..., \frac{|A_m|}{|U|} \right) \tag{1}$$

where $|\cdot|$ denotes the cardinality of a set [1]. The Shannon entropy function of the probability distribution is defined by

$$H(\pi) = H(P_\pi) = -\sum_{i=1}^{m} \frac{|A_i|}{|U|} \log \frac{|A_i|}{|U|} \tag{2}$$

According to [1] the following function G can be used as a measure of granularity for a partition π.

$$G(\pi) = \sum_{i=1}^{m} \frac{|A_i|}{|U|} \log |A_i| \tag{3}$$

In contrast to the entropy function, for two partitions π_1 and π_2 with $\pi_1 \succ \pi_2$ (i.e. π_1 is coarser than π_2, and/or, π_2 is a refinement of π_1, compare Fig. 1) we have now $G(\pi_1) \geq G(\pi_2)$. The coarsest partition $\{U\}$ has the maximum granularity value log $|U|$, and the finest partition $\{\{x\}\,|\,x \in U\}$ has the minimum granularity value 0. We can use the measure G as defined here to distinguish different granularities of different reuse-based process design tasks.

3   Granularity as a Factor in Effective Process Model Reuse

In order to analyze the effect of granularity on business process model reuse effectiveness we can develop the following propositions: Given potentially reusable business process artifacts of different granularities for system development tasks and given a possible quantification measure for the granularity of the various available information pieces in system development situations,

- If a reuse approach is chosen, then – independent of the granularity of the reuse artifact and the design task – the effectiveness of the modeling approach will be higher compared with the non-reuse approach.
- If two reuse approaches, that differ in granularity of design task and reuse artifact, are compared, and let all other experimental parameters be controlled, then there will be a difference between reuse effectiveness.

The literature review pointed to experience level of the designer, the phase in the reuse process, the nature of the reuse artifact, and the context under which the reuse is conducted [23] as factors relevant to understanding the impact of granularity in the context of designing enterprise systems/architectures. Our comparative study varies

- The granularity of the design tasks and reusable business process artifacts:
  - High granularity business process model in the form of process domains (Study 1) vs.
  - Low granularity business process model in the form of a business process modeled in BPMN (Business Process Modeling Notation [24]) (Study 2)

4   Study 1: Process Domain Reuse

4.1   Task, Reuse Artifact and Participants

In the first study the granularity in the design task was set to a relatively coarse level. The participating teams were asked to design process domains for the whole enterprise rather than a particular fine granular workflow (see Study 2). Process domains are clusters of business processes, functionality and/or data processing capabilities that share a certain similarity (other clustering criteria are possible). Structures of this type and granularity have been defined e.g., as industrial reference models [25]. Here, one group of participants was provided with the eTOM model (Levels 0 and 1, see Fig. 2) [9] as reusable artifact (Treatment group). The other group of participants did not receive this artifact (Control Group). The problem description for both groups consisted of data about the application landscape and the individual applications' functional and data processing characteristics. Based on this information participants had to design a process domain model.

The participants in the experiment were graduate or master students majoring in computer science, computer engineering, and business mathematics respectively. All students had completed a semester-long Systems Analysis lecture incl. tutorials in which several modeling methods, such as business process and class modeling for enterprise design, were presented and practiced in various business contexts. Twenty-three people participated in the study. Treatment groups (6) and Control groups (5) were randomly selected. Students were used in this study for several reasons. The concept of cognitive granularity relates to basic human psychological activities. Further, the student subjects in this study are likely to be future enterprise or software architects, at least be involved in process design activities of IT-supported businesses. Findings from this sample would at least have applicability in design situations involving inexperienced enterprise/software architects or consultants. Participants had 90 minutes to perform the task and deliver the designed process domain model.

4.2   Control of Experimental Setup

We controlled the experimental parameters a) problem complexity, b) the similarity distance between the reuse artifact and the presented problem, and c) the adaptation techniques, in order to isolate as far as possible granularity as a factor in business process reuse effectiveness.

To quantify the problem complexities we used the McCabe complexity defined as $M = e - n + 2p$ (M is the McCabe measure, e is the number of edges in a graph, n is the number of all nodes, and p is the number of individual control graphs in the overall graph structure) [26]. We considered both design tasks (Study 1: high vs. Study 2: low granularity) as essentially going through the cycle of steps 1. Identify a design element, 2. Classify this element by certain attributes, 3. Match it to an element in the reuse artifact, 4. Make decision whether to keep or omit it and 5. Enforce that decision. The problem description consists in both tasks of 32 essential informational elements: In study 1 it consists of the number of applications in the whole enterprise and their functional attributes (to be arranged in process domains); in study 2 it consists of the actual steps, decision logic, and data objects in the business process (to be designed). McCabe complexity in both tasks is $M = 4*32-5*32+2*32 = 96$.

Also in both studies the similarity distances between the task and the reusable artifact had to be controlled and kept equal. Otherwise the treatment group in the study that is provided a reuse artifact closer to the solution of the task would have an advantage. We used the syntactical similarity as the measure to determine the "distance" between task and reuse artifact as defined in [27]. Structural similarity and semantic similarity are also important dimensions to define overall similarity of business process models, but for our purpose syntactical similarity sufficed. For the study conducted with high granularity artifacts, structural considerations are not the center of attention, as process domains in the form of procedural clusters have to be identified. The study conducted with low granularity does consist of a reusable business process with a specific sequentiality, but the sequence does not essentially deviate from the structure demanded by the task. Potential semantic dissimilarities between task and reuse artifact can also be overlooked as the language in both artifacts was set understandable and allocable to each other for bilingual participants (German/English). The most outstanding differences we chose between tasks and reuse artifacts, therefore, reside on the syntactical level. Out of possible syntactical differences we decided the reuse artifacts foremost to have extraneous model elements, i.e. model elements in the reuse artifact that exceed the required functionality in the given task. As many repositories of reusable artifacts, such as logical information models, e.g. UBL [28], OAGIS etc. intend to cover as many problem cases as possible that could occur, the assumption that reuse artifacts will bear ample extraneous functionality in many specific modeling tasks, seems plausible. In both studies the reuse artifacts exceeded the design task by 77% - 81% of extraneous information.

By far the most difficult parameter to control is the applied adaptation technique when adapting the reuse artifact to fit the task. We tried to control these as best as possible by choosing 23 students of similar academic background. It can be assumed that there are certain similarities between styles of thinking regarding business process modeling as they have studied this topic in one of our courses. Nevertheless problem solving strategies may deviate between individuals. Moreover, all participants are human, i.e. the brain structures available for problem solving, at least on a most basic level, are the same. The actual usage of available conceptual models, i.e. application of reuse patterns from a cognitive perspective, still remains outside of our scientific realm, but will eventually have to be included. Here, we assume all participants to possess very similar adaptation techniques to modify reusable artifacts. We did not vary the factor of user experience here. At most the students' performance may be compared to novice modelers in enterprises. Experienced modelers are expected to behave differently.

4.3 Independent Variable

The independent variable for both of our studies was the presence or absence of a reuse artifact. One group of participants (Treatment group, 6 teams with two members each, one had three) was given a common standardized process domain model for the telecommunication industry. It was taken from the enhanced Telecom Operations Map (eTOM) [9] and shows the first two levels of the process hierarchy. The second group (Control group, 5 teams with two members each) did not receive this model. The partitioning π_1 and its corresponding granularity of the problem space, and the reusable process domain model respectively, are presented in formula (4) and Fig. 2.

The granularity here is 1,982 with 23 granules ($m = 23$) as the process domains and an average cardinality of one granule, $|A_i|$, of 96 (i.e. assuming an average of three processes of cardinality 32 per domain). We consider $G(\pi_1)$ in study 1 as high granularity. Both reuse artifacts include extraneous elements of the same proportion (see subsection above).

$$G(\pi_1) = \sum_{i=1}^{m} \frac{|A_i|}{|U|} \log |A_i| = 1{,}982 \, , \tag{4}$$

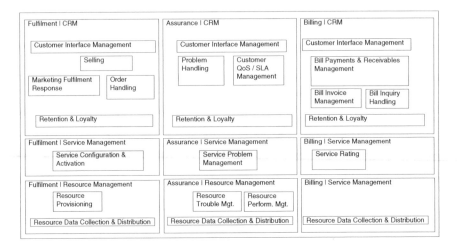

Fig. 2. Reusable eTOM process domain model provided to the treatment group in study 1

4.4 Dependent Variables

Overall, the dependent variable consisted of the solution to the problem and how effectively this was done. We are interested in the main characteristic of how effective the reuse of the design artifact is, given a design task and granularity. Various features to characterize modeling effectiveness have been proposed. In [5] some relevant variables have been defined, e.g. Reusing Unanticipated Components, Reducing Locating Time, Snowball Effects of Deliveries, and Knowledge Augmentation. References [29, 30] classify semantic, syntactic and pragmatic qualities of designed conceptual models. We define effectiveness by a group of variables according to [6] who regard the reuse task essentially as an information retrieval task, which therefore can be measured by recall and precision, and according to [23], who identify extraneous information in reuse artifacts that can impair the resulting design models.

We therefore define reuse effectiveness collectively by three variables:

- Recall: the relevant elements in a resulting list (here, the final designed business process model) in relation to all potentially relevant elements (all those available in the reuse model, as our reuse artifact fully covers the design task).
- Precision: the relevant elements in a resulting list (here, the final designed business process model) in relation to the size of the resulting list (here, the final designed business process model). Precision is a measure for how "clean" the final model is.
- Extraneous: those elements of the reuse artifact that were carried over into the final solution, but which were not required by the original design task. Too many carried over extraneous elements are an indicator of distorted design models, therefore models of inferior quality.

Fig. 3. Exemplary result of a design team that was provided with the reuse artifact

4.5 Operational Hypothesis Study 1

Our operational hypothesis for study 1 is as follows. Given the described reuse artifact and the design task of granularity $G(\pi_1)$, i.e. *high* granularity, and controlling the experimental parameters, we expect the treatment group to significantly perform better in terms of reuse effectiveness than the control group. Our null hypothesis therefore is that the treatment group is expected to perform worse compared to the control group. We will see if this null hypothesis can be falsified. In the following we can analyze our results and relate them to our hypothesis.

4.6 Results and Discussion

In Fig. 3 an exemplary result of the process domain design is depicted. As can be seen in Table 1 and 2 we can compare the control and treatment group in three effectiveness variables and analyze whether the treatment effect was significant or not. We conducted an independent samples t-test to analyze the equality of means. For Recall the Levene test showed that equal variances could not be assumed, therefore no student t-test could be performed. The t-test for unequal variances showed a significance of .019. This value is below .05, therefore we can assume that the average improvement of recall in the treatment group cannot be attributed to random alone. Similar for precision, the t-test for unequal variances showed that the precision improvement is unlikely to have occurred due to chance alone (significance is .039 < .05). In the variable extraneous no significant effect in the treatment group could be observed.

A general weakness in the control groups could be observed. The control groups were not able to / or not willing to advance to a more fine granular level of modeling. The process domains that were identified remained on a very coarse level, even though plenty of information was provided on the application level. Control groups also showed a tendency to cluster integration and mediation systems. This was perceived as a self-contained functionality and seen as important with regard to the business process. The actual functions and sub-processes that were required and to be clustered in domains were mostly neglected.

The overall result of study 1 is that in two out of three effectiveness variables a significant improvement could be observed. The null hypothesis can therefore be partly falsified.

Table 1. Group statistics of Study 1 (high granularity)

	Treatment[a]	N	Mean	Std. Deviation	Std. Error Mean
Recall	0	5	0,215	0,034	0,015
	1	6	0,577	0,261	0,106
Precision	0	5	0,393	0,040	0,018
	1	6	0,664	0,239	0,098
Extraneous	0	5	0,40	0,548	0,245
	1	6	2,50	2,588	1,057

[a] 0: Control 1: Treatment

Table 2. Independent samples test of Study 1 (high granularity)

	Levene's Test		t-Test for Equality of Means							
								95% Confidence Interval of the Difference		
	F	Sig.	t	df	Sig. (2-tailed)	Mean Difference	Std. Error Difference	Lower	Upper
Recall	-	-	-3,4	5,21	0,019	-0,362	0,108	-0,635	-0,088
Precision	-	-	-2,7	5,33	0,039	-0,271	0,099	-0,522	0,021
Extraneous	3,360	0,100	-1,8	9	0,111	-2,100	1,189	-4,790	0,590

5 Study 2: Business Process Model Reuse

5.1 Experimental Setup and Relation to Study 2

The 23 participants of study 1 were randomly reassigned to new treatment (5 teams) and control groups (6 teams) for study 2. The task of study 2 was to design a business process model of low granularity under reuse of an available artifact (Treatment group) and without reuse (Control group). The provided reuse artifact is depicted in Fig. 4 and Fig. 5. The partitioning π_2 and its corresponding granularity of the problem space, and the reusable business process model respectively, are presented as follows:

$$G(\pi_2) = \sum_{i=1}^{m} \frac{|A_i|}{|U|} \log |A_i| = 0, \tag{5}$$

with all granules as the business process elements and an average cardinality of one granule of 1 (i.e. assuming that the granularity of this partitioning of the problem is as low as zero). We consider $G(\pi_2)$ in study 2 as low granularity. The reuse artifact

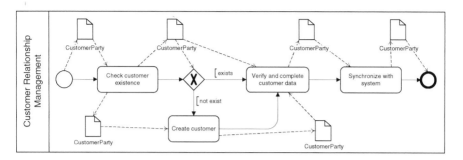

Fig. 4. Provided reusable process model of low granularity in study 2

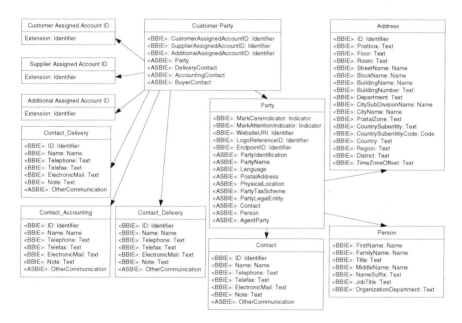

Fig. 5. Reusable Data model provided to participants in study 2

includes extraneous elements of the same proportion as in study 2 (see subsection above). The granularity difference between study 1 and study 2 can clearly be seen:

$$1,982 = G(\pi_1) > G(\pi_2) = 0, \tag{6}$$

All remaining experimental parameters were controlled as in study 1.

5.2   Operational Hypotheses

Our operational hypotheses for study 2 are as follows.

- Given the described reuse artifact and the design task of granularity $G(\pi_2)$, i.e. *low* granularity, and controlling the experimental parameters, we expect the treatment group to significantly perform better in terms of reuse effectiveness than the control group. Our null hypothesis therefore is that the treatment group is expected to perform worse compared to the control group. We will see if this null hypothesis can be falsified.

- Considering the study results of the treatment groups, i.e. the reuse groups, of both studies, we expect a significant difference between the reuse effectiveness which can be attributed to the varied artifact granularity. Our null hypothesis therefore is: Comparing the final solutions of both treatment groups does not show any significant differences in reuse effectiveness. We will see if our hypothesis can be falsified.

5.3 Results and Comparative Analysis of both Study Results

The group statistics of the results of study 2 and independent samples t-test are presented in Table 3 and Table 4 respectively.

We compared control group and treatment group in three variables. For the variables recall and precision the Levene test for equality of variances showed that equal variances could be assumed (significance of .118 > .100 and .200 > .100 respectively). Therefore the student t-test could be performed. Student t-test showed for recall a significance of .718. This value is > .05, therefore we cannot rule out that the recall differences in the treatment group have been due to chance alone. For precision however, student t-test showed a significance of .030 which is below .05. The precision drop in the treatment group can therefore be attributed to the effect of reusing an available process model which did not perfectly match the task. The excess information provided within the reuse artifact, i.e. the extraneous information, very likely led to the distortion of the resulting model. This can also be seen in the obvious difference between means of the variable Extraneous (0.00 vs. 8.40). Even though the t-test for unequal variances did not show sufficient significance, the amount of extraneous information that the treatment group was provided is evident. We can therefore not falsify our first null hypothesis: reuse performed worse precision-wise.

We also compare the treatment groups of studies 1 and 2 to test our second null hypothesis, as presented in Table 5 and Table 6. Only for Recall a significant impact due to granularity could be observed. It seems that in our experiment Recall in the high granularity case is significantly lower than in the low granularity case leaving all

Table 3. Group statistics of Study 2 (low granularity)

	Treentment[a]	N	Mean	Std. Deviation	Std. Error Mean
Recall	0	6	0,938	0,038	0,016
	1	5	0,919	0,124	0,055
Precision	0	6	0,972	0,068	0,028
	1	5	0,779	0,170	0,076
Extraneous	0	6	0,00	0,000	0,000
	1	5	8,40	9,072	4,057

[a] 0: Control 1: Treatment

Table 4. Independent samples test of Study 2 (low granularity)

	Levene's Test		t-Test for Equality of Means						
								95% Confidence Interval of the Difference	
	F	Sig.	t	df	Sig. (2-tailed)	Mean Difference	Std. Error Difference	Lower	Upper
Recall	2,98	0,118	0,373	9	0,718	0,020	0,053	-0,1	0,139
Precision	1,91	0,200	2,564	9	0,030	0,193	0,075	0,023	0,363
Extraneous	-	-	-2,07	9	0,107	-8,400	4,057	-19,66	2,864

Table 5. Group statistics of comparative study (high vs. low granularity)

	Treatment[a]	N	Mean	Std. Deviation	Std. Error Mean
Recall	H	6	0,577	0,261	0,106
	L	5	0,918	0,124	0,055
Precision	H	6	0,665	0,240	0,098
	L	5	0,779	0,170	0,076
Extraneous	H	6	2,50	2,588	1,075
	L	5	8,40	9,072	4,057

[a] H: Reuse at high granularity L: Reuse at low granularity

Table 6. Independent samples test of comparative study (high vs. low granularity)

	Levene's Test		t-Test for Equality of Means						
								95% Confidence Interval of the Difference	
	F	Sig.	t	df	Sig. (2-tailed)	Mean Difference	Std. Error Difference	Lower	Upper
Recall	1,944	0,179	-2,67	9	0,026	-0,342	0,128	-0,631	-0,522
Precision	0,946	0,356	-0,89	9	0,393	-0,115	0,128	-0,405	0,175
Extraneous	-	-	-1,41	4,54	0,224	-5,900	4,192	-17,01	5,210

other parameters controlled. The mapping of high granularity concepts under reuse of an available process domain model seemed to be a harder task than in the case when low granularity concepts are involved, even though the problem complexity in both studies was controlled. Our null hypothesis, that there would be no significant impact on modeling effectiveness at different granularities, was therefore partly falsified.

6 Conclusion

We have conducted a comparative study in order to test our hypothesis on the impacts of information granularity on the effectiveness of business process model reuse. We designed an experiment in which experimental parameters such as problem complexity, problem-reuse artifact similarity, and adaptation techniques were controlled as best as possible to focus on information granularity as the only factor affecting different modeling outcomes under reuse behavior. Our experimental results show that there is a difference in one out of three defined modeling effectiveness variables, i.e. Recall, partly falsifying our null hypothesis (no significant impacts induced by different granularities).

However, limitations to our study have to be recognized. The number of perceived concepts in both studies was probably not the same. Despite our effort to compensate the higher number of concepts (all the fine granular data fields) of study 2, by providing the same high number of applications in study 1, leading to the same McCabe complexity, this can be criticized. In a following experiment the number of granules in all treatment groups, independent of low or high granularity, should be kept exactly the same to fully isolate cognitive granularity as the only affecting factor. In our approach this aspect has not been fully considered yet. The assumption of equal adaptation techniques of participants can also be criticized. This assumption may be too simplifying. In our experiment we assumed a matching activity and a keeping and/or omitting activity (of modeling elements) only. The complexity of the human brain activity is surely much higher as we implied here, and also the decision mechanisms and computation costs between individuals may vary immensely. These topics are still under research in the areas of cognitive modeling, neural networks and others. Lastly we realize that due to our small sample size results are not statistically significant, but this preliminary run shows that the experimental set-up is ready for running on a larger scale.

In future research work the aspects identified above should be considered. Moreover, recall and precision measures could be extended to the dimension of granularity, i.e. which concepts of what granularity can be retrieved from reuse artifact with what success, to eventually gain more fine-grained knowledge of cognitive reuse processes in business design work. Also, granularity concepts should be investigated considering the different types of assistant artifacts, e.g. compare reference models with patterns and other knowledge transferring approaches.

References

1. Yao, Y.: Probabilistic approaches to rough sets. Expert Systems 20, 287–297 (2003)
2. Frakes, W., Kang, K.: Software reuse research: status and future. IEEE Transactions on Software Engineering 31, 529–536 (2005)
3. Frakes, W.B., Terry, C.: Software Reuse: Metrics and Models. ACM Comput. Surv. 28, 415–435 (1996)
4. Fischer, G.: Cognitive View of Reuse and Redesign. IEEE Software 4, 60–72 (1987)
5. Ye, Y., Fischer, G.: Supporting Reuse by Delivering Task-Relevant and Personalized Information. In: International Conference on Software Engnineering (ICSE 2002). ACM, Orlando (2002)

6. Purao, S., Storey, V.C., Han, T.: Improving Analysis Pattern Reuse in Conceptual Design: Augmenting Automated Processes with Supervised Learning. Information Systems Research 14, 269–290 (2003)
7. Szyperski, C.: Component Software: Beyond Object-Oriented Programming. Addison-Wesley, New York (1998)
8. vom Brocke, J., Buddendick, C.: Reusable Conceptual Models – Requirements Based on the Design Science Research Paradigm. In: Chen, H., Olfman, L., Hevner, A., Chatterjee, S. (eds.) Design Science Research in Information Systems and Technology (DESRIST 2006), Claremont, CA (2006)
9. Kelly, M.: Enhanced Telecom Operations Map (eTOM) - The Business Process Framework. TeleManagement Forum (2007)
10. Supply-Chain Council: Supply Chain Operations Reference-model Version 8.0. Supply-Chain Council, Inc. (2006)
11. Zimmermann, O., Gschwind, T., Küster, J.M., Leymann, F., Schuster, N.: Reusable Architectural Decision Models for Enterprise Application Development. In: Overhage, S., Szyperski, C.A., Reussner, R., Stafford, J.A. (eds.) Third International Conference on Quality of Software Architectures, Software Architectures, Components, and Applications (QoSA 2007), pp. 15–32. Springer, Medford (2007)
12. Buckl, S., Ernst, A.M., Lankes, J., Schneider, K., Schweda, C.M.: A Pattern based Approach for constructing Enterprise Architecture Management Information Models. Internationale Tagung Wirtschaftsinformatik. Universitätsverlag Karlsruhe, Karlsruhe (2007)
13. Coad, P., North, D., Mayfield, M.: Object Models: Strategies, Patterns, Applications. Prentice-Hall, Englewood Cliffs (1996)
14. Gamma, E., Helm, R., Johnson, R., Vlissides, J.: Design Patterns: Elements of Reusable Object-Oriented Design. Addison-Wesley, Reading (1995)
15. Zdun, U., Hentrich, C., Dustdar, S.: Modeling process-driven and service-oriented architectures using patterns and pattern primitives. ACM Transactions on the Web 1(3) (2007)
16. Sun, W., Zhang, X., Guo, C.J., Sun, P., Su, H.: Software as a Service: Configuration and Customization Perspectives. In: IEEE Congress on Services Part II. IEEE, Los Alamitos (2008)
17. Reijers, H.A., Mendling, J.: Modularity in process models: Review and effects. In: Dumas, M., Reichert, M., Shan, M.-C. (eds.) BPM 2008. LNCS, vol. 5240, pp. 20–35. Springer, Heidelberg (2008)
18. Burton-Jones, A., Meso, P.: How good are these UML diagrams? An empirical test of the Wand and Weber good decomposition model. In: 23rd International Conference on Information Systems, Barcelona, pp. 15–18 (2002)
19. Wand, Y., Weber, R.: A model of systems decomposition. In: Tenth International Conference on Information Systems, Boston, MA, pp. 41–51 (1989)
20. Zadeh, L.A.: Towards a theory of fuzzy information granulation and its centrality in human reasoning and fuzzy logic. Fuzzy Sets and Systems 19, 111–127 (1997)
21. Yao, Y.: A partition model of granular computing. In: Peters, J.F., Skowron, A., Grzymała-Busse, J.W., Kostek, B.z., Świniarski, R.W., Szczuka, M.S. (eds.) Transactions on Rough Sets I. LNCS, vol. 3100, pp. 232–253. Springer, Heidelberg (2004)
22. Shannon, C.E., Weaver, W.: The Mathematical Theory of Communication. University of Illinois Press (1963)
23. Parsons, J., Saunders, C.: Cognitive Heuristics in Software Engineering: Applying and Extending Anchoring and Adjustment to Artifact Reuse. IEEE Trans. Software Eng. 30, 873–888 (2004)

24. Object Management Group: Business Process Modeling Notation Specification, Version 1.0 (2006)
25. Fettke, P., Loos, P.: Classification of reference models: a methodology and its application Information Systems and E-Business Management 1, 35–53 (2003)
26. Latva-Koivisto, A.M.: Finding a complexity measure for business process models. Helsinki University of Technology, Helsinki (2001)
27. Ehrig, M., Koschmider, A., Oberweis, A.: Measuring Similarity between Semantic Business Process Models. In: Fourth Asia-Pacific Conference on Conceptual Modelling (APCCM 2007). Australian Computer Society, Inc., Ballarat (2007)
28. Bosak, J., McGrath, T., Holman, G.K.: Universal Business Language v2.0. OASIS (2006)
29. Lindland, I., Sindre, G., Sølvberg, A.: Understanding quality in conceptual modeling. IEEE Software 11, 42–49 (1994)
30. Moody, D.L., Sindre, G., Brasethvik, T., Sølvberg, A.: Evaluating the quality of process models: Empirical testing of a quality framework. In: Spaccapietra, S., March, S.T., Kambayashi, Y. (eds.) ER 2002. LNCS, vol. 2503, p. 380. Springer, Heidelberg (2002)

Artifact-Based Transformation of IBM Global Financing

Tian Chao[1], David Cohn[1], Adrian Flatgard[1], Sandy Hahn[2], Mark Linehan[1],
Prabir Nandi[1], Anil Nigam[1], Florian Pinel[1], John Vergo[1], and Frederick y Wu[1]

[1] IBM Research, 19 Skyline Drive, Hawthorne, New York 10598
[2] IBM Global Financing, 1 North Castle Dr., Armonk New York 10504-1785
`{tian,dcohn,flatgard,hahn,mlinehan,prabir,anigam,pinel,`
`jvergo,fywu}@us.ibm.com`

Abstract. IBM Global Financing (IGF) is transforming its business using the Business Artifact Method[1], an innovative business process modeling technique that identifies key business artifacts and traces their life cycles as they are processed by the business. IGF is a complex, global business operation with many business design challenges. The Business Artifact Method is a fundamental shift in how to conceptualize, design and implement business operations. The Business Artifact Method was extended to solve the problem of designing a global standard for a complex, end-to-end process while supporting local geographic variations. Prior to employing the Business Artifact method, process decomposition, Lean and Six Sigma methods were each employed on different parts of the financing operation. Although they provided critical input to the final operational model, they proved insufficient for designing a complete, integrated, standard operation. The artifact method resulted in a business operations model that was at the right level of granularity for the problem at hand. A fully functional rapid prototype was created early in the engagement, which facilitated an improved understanding of the redesigned operations model. The resulting business operations model is being used as the basis for all aspects of business transformation in IBM Global Financing.

Keywords: Business Process Management, Business Artifacts, Business Entities, Business Design, Business Architecture, Service-Oriented Architecture.

1 IBM Global Financing

IBM Global Financing [6] (IGF) is the largest IT financier in the world. This division of IBM has operated for more than 25 years and boasts an asset base of nearly $38 billion, with 125,000 customers in more than 50 countries. For the last three years, it has financed over $40 billion in IT assets per year. IGF is responsible for 9% of the total profit of the IBM Corporation.

IGF business operations span the full, end-to-end process of financing hardware, software and services in the IT industry. Once a financing opportunity is identified by

---

[1] The "Business Artifact Method" is offered commercially by IBM as the Business Entity Life Cycle Analysis capability pattern in the SOMA method [1].

U. Dayal et al. (Eds.): BPM 2009, LNCS 5701, pp. 261–277, 2009.

a sales team, IGF handles all aspects of the financing deal including validation, contracts, negotiations, pricing, coordination with 3rd party suppliers, offer letters, and credit verification. IGF is also responsible for lease management, end-of-lease processing and finally, asset recovery, resale and disposal. The resale of refurbished assets accounts for a large portion of IGF revenue.

2 Business Context and Challenges

By mid-2008, IGF worldwide business operations reflected 25 years of organic growth, with many country organizations evolving and adapting to local business and legislative forces. This effectively established each country as a "silo" with their own unique processes that acted as an inhibitor to global integration and became a major annoyance for IGF's global clients who complained that, "It's like we're doing business with different IGF companies around the world. Working with country unique financing processes is an unnecessary cost to us."

Multiple variations of processes meant that the same business function, (e.g. pricing, supplier management, asset recovery) was performed differently by the various geographies ("Geos"). A prime example of geographic variation in the IGF process was how financing deals were priced. Over time, individual Geos developed deep local expertise and frequently captured the expertise in tools and utilities which were local in their scope and use. Opportunities to reuse business functions, and hence achieve economies of scale, were missed. Some countries did not utilize any technology to support pricing. Over a period of many years, the same local teams created ad hoc utilities to price deals using general purpose software such as spreadsheets. The result was that people in local geographies were highly invested in and reliant on their local processes and tools, exacerbating the organizational barriers to change.

In 2006, IBM initiated an internal transformation effort to become the world's premier Globally Integrated Enterprise (GIE) [9]. Not surprisingly, IGF was challenged with globally integrating their operations in support of IBM's GIE strategy. In a GIE, business processes are integrated across the global operations of the enterprise and specific functions are flexibly sourced, often at 'centers of excellence', based on a variety of factors such as skills, availability, location, and cost.

Simultaneous with IBM's transformation to a GIE, the IT industry underwent a significant and continuing shift over the last few decades, characterized by fewer large financing deals and a concomitant, steady decline in the average size of leasing contracts. IGF naturally saw analogous changes to the size of their financing arrangements as well. As the number of large financing deals decreased and the volume of leasing requests increased, the need for efficiency and cost control grew in importance. The emergence of increasing numbers of smaller, nimbler competitors placed additional pressure on IGF to streamline and integrate their operations.

The variations in IGF's business operations from one country to the next made it very difficult to measure the performance of the business in a consistent way. This led to two undesired outcomes. First, it was hard to compare the operational efficiency of one geo to another. In addition, it was difficult to roll up metrics from across the entire business (i.e. all countries) and produce a coherent and consistent view of how the whole IGF business was performing over time.

Further complicating the challenge were typical organizational change factors. The individual geos had operated autonomously for many years. They were reluctant to change their operations to achieve global integration. Giving up control of their processes and depending on other geos to provide critical functions carries risk and requires trust that the end-to-end, globally integrated operation design is achievable.

It is important to note that IGF had significant process modeling efforts under way at the time they started considering the artifact-centric approach. They were employing Lean and Six Sigma ("Lean Sigma") methods [10] in areas outside the scope of standardization, as well as traditional process decomposition approaches. The main challenges they were experiencing were that the modeling efforts were too detailed, complex, and time-consuming, which prevented the key business stakeholders from effectively designing a standard, end-to-end operation. It also hampered IGF's ability to identify the key business operations and challenges and, most important, inputs and outputs across the end to end process. In the language of one of their stakeholders, the artifact method "Gave us a tangible model (in terms of touch and feel) to work with". It was at the right level of granularity to support effective collaboration among the stakeholders to re-design the business operation. The process decomposition methods produced numerous, detailed functional process fragments, but failed to support the creation of a single, business level, end-to-end process model.

IGF executives also stated that the business was moving fast and they could not understand why it took so long to build and deploy applications (i.e. there was a long path from requirements to deployment). In the words of an IGF transformation executive, the artifact method enabled them to "get an execution mentality" that was not present with "process mapping".

3   IGF's Business Strategy

IGF senior management developed a simple strategy to address the challenges and opportunities described above. Additional resources and increased funding were unavailable and as result, a change to the business model was imperative. They decided to standardize the end-to-end process for their client financing operations, but to do so in a way that allowed for variations as demanded by legislation and other business considerations through business rules (discussed below). Such a global standard process needed to have a number of important characteristics in order to succeed:

1. Fluent in describing financing operations at a high level of abstraction yet capture the essential elements of the business operations across 50+ countries.

2. A way of defining a standard operation, with variations specified within the scope of a task.

3. Support for moving easily between the global standard and local instances of the business operations
4. Expressed in a way that the business executives could easily work with as they designed the new business operation. Has to be the "right" level of granularity
5. Serve as an intuitive structure for identifying, organizing and measuring Key Performance Indicators (KPIs)

Historically, the team utilized methods, such as hierarchical process decomposition and Lean Sigma to document processes, identify opportunities for improvements and to try and deliver an integrated and aligned end-to-end process definition. These techniques were focused on sequencing activities, but country variations permeated the education, training and execution of independent processes. The General Manager of IGF Client Financing created a Single Operating Model team, a small matrix group of subject matter experts focused on global process standardization and simplification that was realized through the Business Artifact Method.

4   The Business Artifact Method for Business Process Design

We employed the artifact method, a business transformation method developed at IBM Research [6] to address IGF's business process management challenges. Given a specific purpose, the artifact method provides a modeling approach that identifies functional chunks (tasks) that are consistent with the purpose. This right-sizing of tasks is accomplished by focusing on affecting a meaningful change to the artifact. Activity-centered approaches tend to produce models that capture the details of how an activity is performed. By contrast, we focus on results that are produced and recorded on the artifact.

The goal of the project was to create a global standard business operation model (BOM) with appropriate local variation to accommodate the legislative and business requirements of individual geographies. There were two primary drivers of the transformation effort: 1) the need to lower costs through standard operations end to end across the globe and 2) IGF's multi-national customers wanted a common customer experience with the same "look and feel" in all countries. The redesigned operations had to efficiently support a high volume of smaller deals and improve customer satisfaction.

The Research team met with the IGF transformation team twice prior to launching the project. The initial meetings, which lasted a total of 3 hours, were exchanges of information and an initial assessment that the artifact method indeed had the potential to address the challenges faced by the IGF business transformation team. The Research team described the basic artifact method and discussed past applications of the artifact method. Three of these case studies are thoroughly documented here [3].

We exited the initial meetings with a firm commitment from the IGF executive leadership team to pursue an artifact-based solution, although some of the key IGF executives were quite skeptical about the approach. What followed next was an in-depth, 3 day workshop with 3 objectives:

1) Ensure the artifact-centric approach would meet the client's needs for a professional business modeling method for standardization, including segmentation approaches for client/partner types, channel types and deal types.
2) Explore how the artifact approach would lead to faster results in system configuration and deployment in multiple countries.
3) Validate that the artifact approach enabled the rapid creation of a workflow prototype.

The workshop was attended by 5 members of the Research team and by ~15 executives and senior business leaders from IGF who represented most aspects of their global, end-to-end business process.

The workshop was a breakthrough in gaining IGF executive leadership to understand and buy into the artifact-centric approach. Three factors were critical to achieving buy-in. First, their business transformation executives were strong, vocal champions of the artifact method. They were responsible for inviting and ensuring the attendance of a large team of leaders from IGF who provided critical, deep subject matter expertise during the workshop. More subtly, the general impression (as evidenced by many comments during informal discussions) by Global Process Owners (GPOs), project team members and stakeholders was that the method brought clarity, simplicity and order to their complex business operations. Interestingly, there were numerous occasions where people from different parts of the IGF business had moments of insight into each other's part of the business. The method gave the IGF stakeholders a way to think about their end-to-end business that allowed them to communicate much more effectively.

A second critical success factor was that the Research team was well staffed and had the right set of skills for the task at hand. The key roles included "the artifact lead", who was the expert on the artifact method and led all the discussions in the workshop. "The modeler" was dedicated to running the software tool that captured the results of the workshop discussions. "The rules expert" captured the business rules and mapped them to the artifact-based process model. "The metrics expert" focused on identifying Key Performance Indicators (KPIs) and associated metrics. "The organizer" kept the workshop focused on core activities, managed all logistics and kept the workshop and the teams on track. Finally, there was "the prototyper", who was responsible for the creation of a "fast-path prototype", which leads to the third critical success factor.

Experience has shown that modeling is not always a successful technique for wrestling with business problems. Often, the models themselves are difficult to understand and manipulate by business people. The artifact method is supported by tools that mitigate this problem through the creation of a "fast-path" rapid prototype. In simple terms, once a BOM is captured in our modeling tool, we can create a prototype application, literally at the push of a button. While the prototype is not aesthetically sophisticated, it does reflect the functionality of the business operation with a high degree of integrity. Instead of working with a graphical model, we were able (after one day) to show a functioning prototype. As has been our experience with most clients, the prototype immediately helps ground the conversations, enhances communication among workshop attendees and gives the client a strong and clear sense how the redesigned "to-be" business operation will behave. One comment from a workshop participant is that it was "a tangible model that we could touch and feel". The prototype enabled the business people to validate the operation model in a way that simply reviewing a diagram could not. Certain parts of the automatically generated prototype, such as the user interface, are not expected to be used beyond this model validation phase and therefore were not extensively tested. However, the core of the application generated from the BOM can serve as the harness that drives or observes the operations, as described in the section on the future of the IGF transformation.

The workshop included a focused effort to define what was meant by "global standard". The "unit of standardization" was defined to be the task in the BOM. As mentioned earlier, a task affected business-sensible change to an artifact. The tasks also provided the structure for specifying business rules. The business rules, in turn, served as the mechanism to capture "local variations" in the business process which were approved by the tax and legal compliance organizations. The variations from country to country are quite significant, e.g. some countries were required to operate as a bank, with all the associated myriad of regulatory requirements. Other dimensions of variability were legal and tax requirements. Coming into the engagement, IGF knew that they could not create a "one size fits all" model of their business operation, but they did not have a fundamental approach that allowed for capturing, representing and governing allowable variations. An artifact-based model augmented with business rules provided a manageable solution.

The two month engagement started out with the results from the workshop viz. one business artifact, some key states in its lifecycle, and a few business rules. The lifecycle of this business artifact, the Deal, was fleshed out in detail and along the way two other business artifacts were identified. In constructing the lifecycles of all the three business artifacts, a great deal of deliberation went into consolidating the fine-grain process steps into tasks that affected meaningful changes to the business artifact they worked on. These tasks could be thought of as "process fragments" required to transition the business artifact from one state to the next. Once we had a satisfactory operation designed, the subject matter experts provided the business rules about the information each task needed to start working and about constraints on the lifecycle processing steps. Through these work sessions, the detailed information content, lifecycle processing, and business rules of the three business artifacts were assembled.

During the course of the design work outlined above, numerous communication and collaboration meetings were held with the 15 key stakeholders representing the business, the geographies and the functions. These were meetings the business execs found useful because the artifact centric approach ensured they were conducted in business language and at a level of granularity such that the stakeholders could understand and contribute to the design. As one business exec said after the first such meeting, "I dreaded coming today because I thought I'd be subjected to conference room walls filled with flow charts that I'd be expected to understand and then engage in meaningful design discussions. The approach you've shown us was infinitely more productive."

In addition to the stakeholder meetings, the core team met 25 times in total, usually two days per week over the course of the 2 month engagement. These meetings were focused on executing the artifact based methodology to create the business operational model (BOM) and process fragments that included:

- o Identification of roles for the various tasks.
- o Key Performance Indicators and metrics. Each metric was associated with relevant parts of the BOM that will be instrumented to emit transaction events. The output was an Excel spreadsheet used to create the observation model.
- o Business rules modeled with the SBVR syntax associating constraints to different processing paths in the BOM. The output was a SBVR (Semantics of Business Vocabulary and Business Rules) model.
- o CRUDE matrix (Create, Read, Update, Delete, Execute), defining information access, organized by each role player in each task in the BOM, captured in an Excel spreadsheet.

When the design was near completion another prototype was run, reviewed by the team and additional elements were identified and resolved. This iterative method of "design – rapid prototype – design" gives confidence to the business design team that they have addressed the business problem and to the IT team who no longer must interpret the design.

5   The IGF Business Operation Model

The key to the artifact method is to identify the Business Artifacts whose lifecycles represent the essential elements of the business process. When done properly, these lifecycles present a clear picture of the business and help develop insight on operating and improving the process. By the end of the first day of the initial workshop, the IGF team agreed that the primary goal of their business was to create and complete business deals with clients. They understood that although there was no specific document which fully defined each deal, it was a well-defined notion, or *abstraction*, and we defined the **Deal** artifact (Figure 1) as the overall arrangement with a client.

During the second phase of the artifact method, the team structured the lifecycle of the Deal artifact. They agreed that a Deal was *Created* when a sales person identified a client need that could lead to a financing opportunity. The deal remained in a *Draft* state while the client's credit was checked; while pricing was determined; while terms and conditions were specified, and while other administrative tasks were performed. It was possible for the Deal to become *Failed* if, for example, the client's credit was bad, but most Deals were eventually *Offered* to the client. These might become *Expired* if the client did not act, or *Lost* if the client rejected the deal. However, if the client *Signed* the deal, it became *Active*. For *Active* deals, IGF worked on Ordering, Accounting, Billing, etc. Clients could ask that an *Active* deal come to an *Early End* or return to *Draft* state, but most often, they were *Completed*. Thus, the diagram below summarized the lifecycle of the Deal artifact.

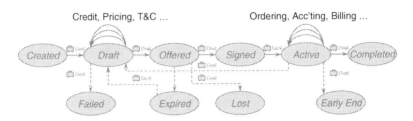

Fig. 1. The Deal artifact

The IGF team noted that although the lifecycle of the Deal artifact represented much of what IGF did, it omitted certain important aspects. For example, when Ordering was performed on an *Active* Deal, a separate part of IGF placed that order, received the items, made certain that the vendor was paid and handled the accounting. This led to the identification and definition of the **Supplier Invoice** artifact (Figure 2), so named because the document containing much of this artifact's information was the supplier invoice. The figure below shows that the Ordering action on a Deal creates the Supplier Invoice artifact which then goes through its own three-state lifecycle.

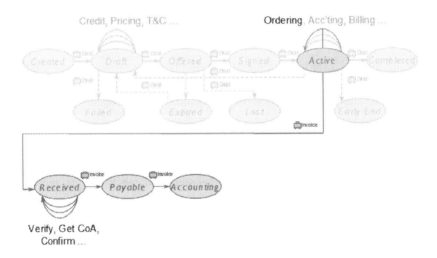

Fig. 2. The Supplier Invoice artifact

One final aspect of IGF that even the foregoing does not cover is the eventual disposition of hardware assets which clients no longer need. This led to the definition of the **Asset** artifact (Figure 3) which comes into existence when hardware has been delivered by the supplier. This is shown below as occurring when the Supplier Invoice lifecycle becomes *Payable* since at this point IGF takes title to the asset. Assets remain *Active* until they are *Concluded*, *Returned* or *Sold*. After this, the physical asset is no longer held by IGF and the Asset artifact becomes *Inactive*. Thus, IGF's complete Business Operations Model is as shown below:

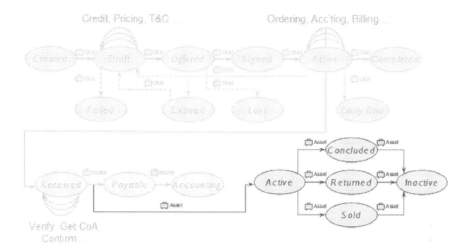

Fig. 3. The Asset artifact

This diagram is a significant simplification of the business operations model (BOM) produced during the two-month design effort. That model depicts ~65 individual tasks that define the work of the organization. However, the 18 artifact states shown here have remained in the detailed BOM which IGF continues to use to manage their operations.

A key aspect of the IGF global model development process was the identification and analysis of business rules that govern aspects of the BOM. An example business rule defines the information that must be available before the price quotation step:

"At minimum, a deal must have agreement type, credit reference number or dummy reference number for the GRMG rating, billing type, Machine-Type-Model (MTM), list and net prices, financial types, equipment source, term, payment frequency, payment timing, payment method, planned install date."

The rules were captured in English in a spreadsheet, organized by processing step (task) in the BOM. Rules constrain BOM processing steps. The rule given above says that price quotation cannot be done if the required information is not available. Documenting the rules in detail had several benefits: (a) it exposed many requirements on the BOM; (b) it helped stakeholders from the previously "siloed" IGF groups understand the business drivers that impact their activities; and (c) it created credibility that the BOM model met the requirements of those groups.

Each rule is associated with a single business artifact type, and executes against individual instances of that artifact at run time. The rules may reference any attributes both of the instance and of other artifacts that are associated with the artifact type in the information model.

Most rules constrain individual processing steps in the artifact lifecycle, but there is a mechanism for some rules to constrain multiple processing steps that have common semantics by labeling those steps with a common name. For example, if the lifecycle has several ways to gain approval of a Deal, and some rules constrain all those ways, the rules can be associated with all the ways via common use of the verb "approve".

Formally capturing the rules exposed systematic variations on two dimensions. Geography formed one dimension. For example, one country might require specific additional information as input to price quotation. Another dimension is the kind of business deal, characterized mostly by deal value and complexity. Many business rules that make sense for large, complex deals are overkill for small deals limited to predefined choices. Structuring the rules by processing step, then by deal kind, and ultimately by geography is a technique that both addresses the systematic variation and helps manage a large number of rules in a natural and comprehensible way.

Capturing the rules in a spreadsheet was easy and natural for the IGF team. Ideally, a further step would have converted the unstructured English statements to a more formal format such as the Structured English used by the OMG's *Semantics of Business Vocabulary and Business Rules* (SBVR) [8] specification. Benefits of formalizing the rules could include automated consistency checking among the rules, and direct execution of the rules in the generated application. Some promising work has been done in this area [5], but the existing prototype tool needs extension in order to support all the IGF needs.

One of the essential requirements for IGF's business standardization and transformation effort is to provide the end-to-end visibility to the business process performance [4]. Thus, it is critical to identify the right KPIs that can provide the measurements in an end-to-end view. Moreover, the KPI design approach described hereafter ensures that KPIs are an integral part of the artifact-based business design from the very beginning and not an add-on or afterthought. Throughout the transformation effort, KPI design and business operation design teams worked in tandem.

The first step in KPI design was to analyze the vision and strategy of IGF and identify an initial set of KPIs. What is measured is what gets done. Therefore, it is extremely important to identify the KPIs that are aligned with the organizational vision or strategy in the design phase. First, the objective statements of the IGF vision and top level business goals were analyzed and categorized in terms of the four Balanced Scorecard (BSC) perspectives: Financial, Customer, Process, and Learning & Growth to identify an initial set of KPIs from the IGF vision. Next, the KPIs were described using a KPI matrix template, categorized by each of the BSC perspectives and organized in two ways: 1) by user roles or persons/stakeholders who use or need the KPIs to design a role-based dashboard view (Figure 4), and 2) by dimensions (Figure 5) – a set of qualifying or grouping criteria to view KPIs on business

Deal by User (Deal level)	Business Support Op Exec	Pricing Exec	Credit Exec	Sales Exec	Accounting Exec
Process Perspective					
Total number of Deal (by an attribute)	X	X	X	X	X
Turnaround Time (TAT) of a Deal (E2E)	X			X	X
Average Turnaround Time (ATAT) of a Deal	X			X	X
Financial Perspective					
Win rate	X	X		X	X
Customer Perspective					
TotalNewClient	X	X	X	X	X
Learning & Growth Perspective					
Productivity_rate	X				

Fig. 4. KPIs Matrix Organized by User Roles/Stakeholders. This matrix is illustrative, and presents a small subset of all KPIs and GPOs.

Deal by Dimension (Deal level)	Deal size etc.)	Deal class (ratecard, non ratecard, rollout, etc.)	Channel (BP/ non BP)	Customer Type (Sector/ GB)
Process Perspective				
Total number of Deal (by an attribute)	X	X	X	X
Turnaround Time (TAT) of a Deal (E2E)	X	X		
Average Turnaround Time (ATAT) of a Deal	X	X		
Financial Perspective				
Win rate	X	X	X	X
Customer Perspective				
TotalNewClient	X	X	X	X

Fig. 5. KPIs Matrix Organized by Grouping Dimensions. This matrix is illustrative, and presents a small subset of all KPIs and grouping dimensions.

dashboard and/or On-Line Analytical Processing (OLAP) reporting to drill down on lower-level data.

The draft list of KPIs was then reviewed with IGF business process owners to prioritize into a set of agreed-upon high priority KPIs. The focus here was to ensure these high priority KPIs are accurate, clear, well-defined, and simple to understand. The goal was to identify a small number of "good" KPIs, not numerous, difficult to measure, metrics. To that end, each KPI is described in precise definition using a KPI Template, comprising several attributes: name, description, scope or business artifact the KPI is created for (e.g. Deal, Supplier Invoice & Asset), definition, source or referencing data in the business artifact and the exact computation expressions to calculate the KPI, grouping dimensions, users or stakeholders, and usage example. KPIs are defined with an end-to-end view in mind and not biased or sub-optimized for a specific functional area such as pricing, credit, or proposal. Please see an example of a KPI described in the KPI Template below.

In addition to displaying business performance on a dashboard, KPIs can also be used for evaluating and triggering real-time alerts that warrant immediate attention and corrective actions, thus providing feedback into the business process for continuous improvement. For example, a credit overdue condition can be defined based on "turn around time for Credit" (TAT4Credit) KPI (Figure 6), and an alert will be triggered if the KPI value exceeds a predefined target duration.

KPI Name	TAT4Credit
Description	Credit Request Turnaround Time (TAT)
Scope	Deal
Definition	Elapsed time between Credit Request was started and completed
Source / Calculation Expression	*Data source:* Deal Id, Credit Id, Credit_Request_Start_Timestamp, Credit_Request_Complete_Timestamp *Expression:* Credit_Request_Complete_Timestamp - Credit_Request_Start_Timestamp
Objective/Benchmark	TBD <Value objective specified in SLA or known data value as a baseline>
Dimensions	See KPI Matrix
Users	Credit Exec, Sales Exec
Usage Example	Find out longest, shortest Credit Request Turnaround Time, and calculate the average Credit requests TAT in a geo, for a deal class, a channel, and a customer type.

Fig. 6. Detailed example of a single KPI, Credit turnaround time (TAT)

The final step in the KPI design was to link the KPIs to the business operation model. Once identified, the high priority KPIs needed to be reviewed early on with the process team to ensure all source data needed to calculate KPIs and create the grouping dimensions are properly identified and defined in the business operation model. More important, the review uncovers any data that have not yet been accounted for in the business operation model. This integrated view of both the business processes and KPIs solidifies the overall design and avoids any surprises and potential rework when measurement systems are put in place.

6 Impact on the Business

The BOM continues to be used as a living document that resulted in significant changes to the way IGF operates and governs their business. Since the BOM captured a radically simplified yet right-sized model of the IGF operation, it was easy to adapt it to changing business conditions. Subsequent to the artifact centric design, dozens of work sessions with global teams were held to explain the global standard model and to gain awareness of potential country variations. Activities are currently underway to review and obtain consensus on the implementation of country level variations to the standard process. The artifact centric model, developed during the engagement, was subjected to many implementation planning challenges, organizational challenges and technical challenges from various departments of the IGF unit over the course of the ensuing 4 months. The model was subjected to careful scrutiny and analysis from a wide range of representative stakeholders in the IGF organization. It essentially stood intact and with minor changes and enhancements from the initial design that emerged from the engagement. The artifact centric model has become recognized as the fundamental, standard business operations model for the IGF Client Financing business.

One significant result of the BOM has been its use in identifying tasks that are global in nature and the assignment of GPOs to these tasks, resulting in increased accountability in the fundamental transformation of the IGF organization. As the name implies, GPOs "own" tasks of a global nature and are responsible for defining, implementing and governing their tasks. There are approximately 65 tasks with associated GPOs. Prior to our project and the creation of the BOM, IGF had identified GPOs. However, the BOM model greatly clarified the role and scope of responsibility of each individual GPO and GPOs were given much greater authority. Governance of the end-to-end IGF process is accomplished through the GPOs, who have authority and responsibility for standardizing their sub-processes and for approving deviations to the established standards. One way the GPOs control their sub-process is by issuing policies. Examples include the recording of Machine/Type/Models and serial numbers, revenue recognition upon shipment, accruals, price quotes, and clip levels for deal size. A member of the IBM CIO's Enterprise Process Framework team that is responsible for assessing process maturity, determined that Client Financing was well ahead of its peers in terms of process definition and Global Process Owner engagement.

The BOM was used to identify new GPO roles, most notably an end-to-end GPO who is responsible for ensuring the standardization and integrity of the full operation and has authority to add or redesign the tasks in the BOM. The end-to-end GPO is also accountable for the interlock of inputs and outputs across the process, from opportunity management through to end of lease returns. The BOM was color-coded to help visualize and communicate GPO ownership and scope of responsibility and authority.

The BOM is increasingly recognized as the lynchpin of IGF Client Financing process standardization. The BOM links to job roles, business rules, descriptions of the business operations, KPIs, Sarbanes-Oxley control points and training materials (Figure 7).

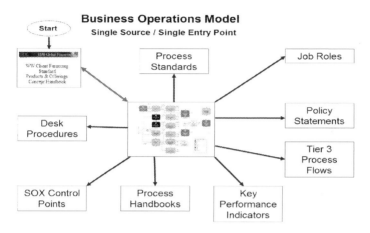

Fig. 7. Notional diagram of the Business Operations model and associated elements of the IGF operational specification

The General Manager of Client Financing recognized the results by saying,

> "Team, *congratulations* on achieving the first milestone in establishing global process standards. Global Process Owner sign-off on the initial set of standards establishes the baseline from which to build and refine global consistency in how we run the Client Financing business. In just two months as a Model Driven Business Transformation team, you set an aggressive target, established a date, scoped the deliverables... and delivered!

> The Business Operations Model and 60 global process standards defined below establishes the baseline, enabling us to expand globally, while allowing for additional deal types with approved country, legal, tax and regulatory requirements. A job well done!"

The VP of IGF Client Financing stated

> "The MDBT[2] initiative has brought structure and clarity to our transformation work and will also bring the work flows to life for our teams in the field and the international Centers of Excellence."

---

[2] MDBT is an acronym for Model-Driven Business Transformation, the IBM Research internal project name for the Business Artifact approach.

7 The Future of the IGF Transformation

The IGF BOM is a blue print of their Client Financing business operation. As discussed earlier, artifact tooling supports the creation of an executable prototype application with a mouse click.   The generated application enables role based manipulation of the 3 key business artifacts (Deal, Supplier Invoice & Asset) through their complete lifecycle.   Although shallow-by-design, the application in essence covers the end to end IGF operation, and more important, it is at the level desired by the business to track the performance of their key objectives. However, the reality of the IGF IT landscape is that it is comprised of both strategic and legacy applications developed by functional areas to serve very specific functional needs. These applications have grown over the years with varied technology bases. Some are customized off the shelf products, while others are home-grown.   Each serves a definite functional need and is indispensable to the business.

This is a very common situation in most large enterprises where the end-to-end processes are served by fragmented IT applications. End-to-end visibility to the operations is practically impossible to achieve. It is likely to be highly time consuming, bordering on impossible, to piece together the holistic metric information from each of the applications and as such unlikely to keep up with the business decision making cycles needed in today's highly dynamic business world.

Interestingly, the "application" generated from the BOM does provide the end-to-end perspective so eagerly desired by the business. But the reality is that we also have a tremendous and continuing investment in existing IT systems. To bridge the two, the concept of the business operation *harness* was born. The following figure explains the concept in the context of IGF. Needless to say, the same concept can easily be applied to other end-to-end business processes.

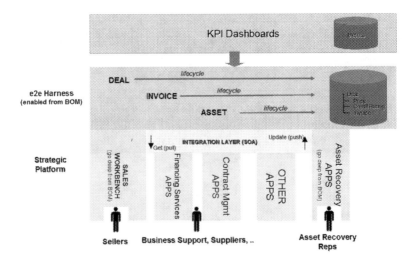

Fig. 8. End-to-end Harness solution architecture

The *e2e Harness* in the middle Figure 8 is the umbrella application enabled (generated) from the BOM. It is responsible for managing the lifecycle and data content of the 3 key business entities spanning the entire IGF business operations. The *harness* application will operate in two modes. The *observation* mode delegates processing to the legacy applications and periodically receives updates on the artifact by either proactively polling the legacy application (pull) or exposing a service for the legacy application to call (push). In the *application* mode, the *harness* is transformed into a full fledged application, managing the *artifacts* and the workflows around it.

The *Strategic Platform* (Figure 8) is the set of strategic functional applications that fulfill logical portions of the *harness*. In the *observation* mode, the *harness* delegates processing to these applications. For IGF, International Customer Financing System, Global Contract Management System and Global Asset Recovery Services are examples of existing applications that are triggered from the *harness* at the appropriate time with the relevant business artifact information. While the application is running or in some cases after it has completed processing, updates to the business artifacts are relayed back to the *harness* to update its own repository. For example, when the *Deal* is ready to be priced in the *harness* application, the Financial Services application is "kicked off". The application runs its course; a price is calculated and reported back to the *harness*. The *harness* updates the *Deal* with the pricing information, marking the milestone as completed and triggers the next set of processing activities.

In other cases, where the functionality cannot be mapped to an IT system, the *harness* goes "deep". This means that those portions of the *harness* will be developed into a full-fledged application implementing the process workflow. An example from IGF is the concept of a Sales's Workbench as shown in the bottom left of the figure. The *Deal* lifecycle starts with the sales cycle which hitherto has been carried out manually by the business. The *harness* can be expanded to provide the full workflow management for the sales team. As soon as the sales cycle is complete, the *harness* reverts back to the *observation* mode, delegating control to an existing application.

The top layer shows the KPI dashboards and metrics running off *harness* data. The KPIs are defined as a function of the BOM. The *harness* is the implementation of the BOM and the *harness* data reflects the lifecycle status and content of the 3 key business entities. Thus it is possible to get the complete end-to-end view of the entire IGF business through the *harness,* although most of the actual work is happening via the legacy applications.

In summary, a *harness*-like shallow application provides tremendous advantage to drive the end-to-end visibility the business really cares about. *The harness* also provides a single point of entry for all lines of business people so operations can be standardized. It also provides a blueprint to map current IT infrastructure supporting the business and further, a framework to evaluate future IT spending.

8   Conclusions, Discussion and Lessons Learned

IGF is a complex, global business operation. Re-designing the business model is a daunting task because of the inherent complexity and the sheer global scale of the business which spans over 50 countries. The IGF executive leadership team developed a strategy to standardize and integrate their operations globally.

Traditional approaches such as process decomposition were tried, but were impeded by the rich variations that are inherent in IGF's business operations. Cultural, organizational change and governance issues further slowed the progress of the teams. To address these issues, IGF executive leadership team utilized the Business Artifact Method to redesign their business.

The Business Artifact Method starts with identifying the key artifacts of the business, the "entities" that the business processes. For IGF, the key artifacts are the "deal", "supplier invoice" and "assets". The design process is centered on the artifacts and tracing the lifecycles of the artifacts as they are processed by the business. By contrast, traditional business process design methods focus on activities, with of the guidance provided by a focus on artifacts.

The business artifact provides the structure for capturing additional design elements of the operation including business rules, key performance indicators, and vocabulary. We were able to reduce the time and effort associated with modeling by addressing all design elements during BOM development. The IGF team also commented on the value of having an integrated model and how well the method facilitated communication among IGF stakeholders.

The Business Artifact method was credited with a number of intangible benefits by the IGF team. Of paramount importance is that the method yields an intuitive perspective on complex business operations that is focused on the primary purposes of the business. The method keeps the redesign team focused on the "right" level of granularity of business operations, while establishing a vocabulary and method which enhances stakeholder communication and consensus building. The BOM captures the business level requirements of the redesigned operation.

One of the more tangible results is the ability to generate working code directly from the BOM. This was important for two reasons. First, the "fast-path" rapid prototype allowed business process owners to easily comprehend how a BOM will behave when implemented. The ability to touch and operate the prototype grounded the semantics of the BOM, facilitating understanding and communication. Second, the generated code can serve as the skeletal code for a final application that realizes the BOM, greatly accelerating the development process. Because the BOM is used in generating a software solution, there is very low impedance between the intent of the business and the actual implementation.

The BOM defined 220 business rules, allowing us to manage country variation due to legal and regulatory requirements. This definition enabled the creation of 60 global process standards, being deployed across all geographies within IBM Global Financing. The BOM enables IGF to deploy the standards to all countries, resulting in significant cost savings. Other approaches had no vehicle to capture approved variations in local and regulatory country deviations to the standards.

We identified numerous enhancements to the tools which we used to capture the BOM. As detailed in section 5, the fundamental concept behind business artifacts are the information and life cycle of the artifact, and how the two relate to each other. We plan to investigate new tool features which improve a practitioner's understanding how each task creates, accesses, modifies and deletes information in the business artifact. Additionally, we see a need to help practitioners easily capture, visualize and comprehend the full life cycle of a single business artifact, including business rules, and KPIs. The tool needs to support large teams that work independently on parts of the BOM and then integrate their results.

The business artifact method was found to be an innovative technique for redesigning complex business operations. The redesigned BOM is being used as the structural backbone and single source, pivotal reference point of the IGF Client Financing business operation.

Acknowledgement

The authors gratefully acknowledge the thoughtful feedback provided by the Rick Hull and the anonymous reviews of our paper.

References

1. Arsanjani, A., Ghosh, S., Allam, A., Abdollah, T., Ganapathy, S., Holley, K.: SOMA: A method for developing service-oriented solutions. IBM Systems Journal 47(3), 377–396 (2008),
   `http://researchweb.watson.ibm.com/journal/sj/473/arsanjani.html`
2. Shapiro, B.P., Rangan, V.K., Sviokla, J.V.: Staple Yourself to an Order. Harvard Business Review (July-August 2004) (Originally published in 1992)
3. Bhattacharya, K., Caswell, N.S., Kumaran, S., Nigam, A., Wu, F.Y.: Artifact-centered operational modeling: lessons from customer engagements. IBM Systems Journal 46(4), 703–721 (2007)
4. Hammer, M.: The 7 Deadly Sins of Performance Measurement [and How to Avoid Them]. Sloan Review, 26–27 (April 2007)
5. Linehan, M.H.: SBVR use cases. In: Bassiliades, N., Governatori, G., Paschke, A. (eds.) RuleML 2008. LNCS, vol. 5321, pp. 182–196. Springer, Heidelberg (2008),
   `http://www.cs.manchester.ac.uk/ruleML/presentations/session5paper1.ppt`
6. IBM Global Financing,
   `http://www.ibm.com/services/us/financing/index.html`
7. Nigam, A., Caswell, N.S.: Business artifacts: An approach to operational specification. IBM Systems Journal 42(3), 428–445 (2003)
8. Object Modeling Group (OMG). Semantics of Business Vocabulary and Business Rules (SBVR), `http://www.omg.org/spec/SBVR/1.0/`
9. Palmisano, S.: The Globally Integrated Enterprise, Foreign Affairs (May/June 2006)
10. Wedgewood, I.: Lean Sigma: A Practitioner's Guide. Prentice Hall, Englewood Cliffs (2006)

Instantaneous Soundness Checking of Industrial Business Process Models

Dirk Fahland[1], Cédric Favre[2], Barbara Jobstmann[4], Jana Koehler[2], Niels Lohmann[3], Hagen Völzer[2], and Karsten Wolf[3]

[1] Humboldt-Universität zu Berlin, Institut für Informatik, Unter den Linden 6, 10099 Berlin, Germany
fahland@informatik.hu-berlin.de
[2] IBM Zurich Research Laboratory, Säumerstrasse 4, 8803 Rüschlikon, Switzerland
{ced,koe,hvo}@zurich.ibm.com
[3] Universität Rostock, Institut für Informatik, 18051 Rostock, Germany
{niels.lohmann,karsten.wolf}@uni-rostock.de
[4] EPF Lausanne, 1015 Lausanne, Switzerland
barbara.jobstmann@epfl.ch

Abstract. We report on a case study on control-flow analysis of business process models. We checked 735 industrial business process models from financial services, telecommunications and other domains. We investigated these models for soundness (absence of deadlock and lack of synchronization) using three different approaches: the business process verification tool Woflan, the Petri net model checker LoLA, and a recently developed technique based on SESE decomposition. We evaluate the various techniques used by these approaches in terms of their ability of accelerating the check. Our results show that industrial business process models can be checked in a few milliseconds, which enables tight integration of modeling with control-flow analysis. We also briefly compare the diagnostic information delivered by the different approaches.

1   Introduction

Various studies [1] show that many business process models contain control-flow errors such as deadlocks. Such errors obstruct the correct simulation, code generation and execution of these models. Therefore, detecting and removing control-flow errors becomes crucial in view of the increasing popularity of these use cases. Preventing errors by using a restricted, for example a purely block-oriented modeling language is rarely an option because a model typically needs to reflect the real causal process structures present in an enterprise.

In this paper, we are interested in checking business process models for the classical notion of soundness [2,3], which entails the absence of deadlocks and lack of synchronization, which are explained in more detail below. Our interest in soundness is motivated by an increased need in creating business process models not only for documentation purposes, but for an input into a translation and code generation process where, e.g., WS-BPEL code is generated. Soundness is necessary to translate a process modeled in a graph-based language, such as UML Activity Diagrams or BPMN, to WS-BPEL in a way that preserves the execution semantics *and* the structure of the process.

U. Dayal et al. (Eds.): BPM 2009, LNCS 5701, pp. 278–293, 2009.

This use case requires a process to be checked in a relatively short amount of time, say 500 ms or less, because checks are to be performed on each major modification, that is, at least on each save operation on the process model. Moreover, entire libraries of up to several hundred processes have to be checked when models are exchanged between modeling tools. Short response times make it possible to integrate control-flow analysis tightly with modeling such that errors are found at the earliest possible time, which would allow the user to relate an error to the latest change in the model. Furthermore, use cases such as code generation from models also require that an analysis produces sufficient diagnostic information to allow the user to locate and repair the detected errors.

A variety of techniques for checking soundness exists in the literature. They differ in their completeness, worst-case complexity, and quality of diagnostic information returned. Most techniques can be easily combined to optimize performance. The most flexible technique is state space exploration. It is most likely applicable to other similar use cases, such as checking a relaxed notion of soundness or checking more expressive languages supporting OR-joins and other advanced synchronization constructs. But state space exploration suffers from the state space explosion problem, i.e., the fact that the number of reachable states can be exponential in the size of the process model. On the other hand, many business process models have a simple structure, for instance, they are sequential to a large extent, hence they do not necessarily have a large state space.

At the onset of our project, it was not clear from the literature how large the state spaces of control-flow models of realistic business processes are and hence which additional techniques are needed to check their soundness as fast as required by our use case. It was completely open whether such a check can be performed in the required time and in such a way that sufficient diagnostic information is obtained. In addition, given the variety of available approaches, it was unclear which would be the most suitable techniques.

In this case study, we investigated three approaches implemented in three different tools as outlined in Fig. 1:

1. The Petri net model checker LoLA [4], from which we used CTL model checking with partial order reduction.
2. The business process verification tool Woflan [3], which uses a mixture of Petri net analysis techniques, most notably structural Petri net reduction and S-coverability analysis, as well as a form of state space exploration based on coverability trees.
3. The process validation technique used in the IBM WebSphere Business Modeler, which combines SESE decomposition [5] with heuristics and state space exploration.

The data set for our case study was a large collection of process libraries available in the IBM WebSphere Business Modeler tool. The first two approaches required a translation of these models into Petri nets, whereas for the third approach, the models were translated into workflow graphs.

We obtained the following results: Based on the 735 process models that we analyzed, soundness of industrial business process models can be decided in a few milliseconds per process. Although many processes are simple enough that state space exploration alone would be sufficient to decide soundness, this method is not sufficient

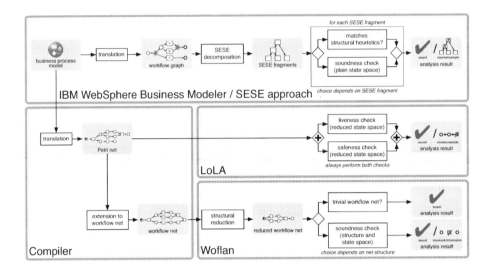

Fig. 1. Three different approaches and tools to check soundness

in general. However, all three approaches perform similarly fast, meeting the above-mentioned performance requirements. This implies that one can focus on different requirements such as the quality of the returned diagnostic information when deciding for a soundness-checking technique. Our study also shows that there is a high percentage of unsound models, confirming the need for better tool support for execution-aware modeling.

Previous studies [6,7,3] on checking soundness or the similar notion of *EPC soundness* of realistic business process models concentrate on error findings and error prediction. These studies do not report runtimes for the analysis. Mendling [8] reports an average analysis time of 1.8 secs and maximal time of 142 secs for checking the EPC soundness of 604 processes. His technique of using structural reduction rules that operate directly on the process model does not find all violations of soundness. A post-processing with state space exploration is not included in these runtimes. The same set of processes was also checked for relaxed soundness [9] with a reported runtime of 46 secs per process on average [8,1]; however, no maximal times are reported. Recent work [10] extends control-flow analysis to more advanced synchronization constructs such as OR-joins and cancelation regions, but so far no empirical results have been reported. A preliminary and incomplete version of the SESE decomposition technique that used heuristics only, but did not include state space exploration, was partially evaluated on a different set of data [5].

The remainder of this paper is organized as follows: In Sect. 2, we discuss the data used in this study, their translation to workflow graphs and Petri nets, and the notion of soundness. Sections 3, 4, and 5 present the three approaches together with the results they achieved on the data. In Sect. 6, we review the results in a comparison of the three approaches and draw conclusions.

2 Selecting the Empirical Data and Preparing the Case Study

2.1 Sampling the Process Data

We scanned a large set of real-world data available to the IBM team for our practical validation of the soundness-checking approaches and tools. These data mostly resulted from modeling activities in customer projects within a SOA context, i.e., processes were captured with the final goal of implementing them in a Service-Oriented Architecture. The models covered various industry domains such as financial services, automotive, telecommunications, construction, supply chain, health care, and customer relationship management. We also looked at large collections of reference processes that were created for the insurance and banking domain by users who explored different modeling styles, i.e., different ways of capturing data and control-flow at varying level of granularity. All models were available in the IBM WebSphere Business Modeler tool represented in a language that currently combines elements from UML Activity Diagrams and the Business Process Modeling Notation (BPMN), but some of them had originally been created in other tools first and then imported into the IBM product.

It turned out that only some of the model collections considered are useful for our purposes. Many process models are in fact quite small, as good modeling practice suggests an appropriate structuring of processes into subprocesses, and are therefore not a challenge for our soundness-checking approaches. Others, in particular those created in other tools, might not have been created with the appropriate notion of soundness or might have been created by non-experts and consequently turned out to be syntactically incomplete and therefore flawed in such a way that it made no sense to consider them further. In the course of our experimental studies, we therefore reduced our initial test set of approx. 3000 models to 5 libraries of 735 different models in total from the insurance, banking, customer relationship, as well as construction and automotive supply chain domains. We completely anonymized the data in these models, e.g., task names would be replaced by enumerations $t1, t2, \ldots$, and named these libraries A, B1, B2, B3, and C. These anonymized libraries, which have been stripped off all semantics and represent only purely structural information, were the input for the tools LoLA, Woflan, and the SESE approach. Libraries B1, B2, and B3 partially overlap as they represent a series of models from the same domain created over a period of two years, in which a library changed to the next by adding more process models and refining all models with further detail. The number of 735 different processes therefore counts only the latest library in this series, which is B3 with 421 processes, together with the 282 processes from library A and 32 processes from library C.

Table 1. Static data

	A	B1	B2	B3	C
Avg. / max. number of nodes	14 / 46	17 / 69	16 / 67	18 / 83	27 / 118
Avg. / max. number of edges	33 / 127	29 / 147	31 / 202	33 / 195	33 / 145
Avg. / max. node inflow	2.52 / 13	1.76 / 15	1.90 / 69	1.86 / 27	1.84 / 4
Avg. / max. node outflow	1.03 / 8	0.94 / 13	0.99 / 15	1.05 / 30	1.83 / 4

Table 1 characterizes the data from our process libraries by measuring the number of nodes that represent tasks, subprocesses, gateways, start and end events, and the number of edges that represent control- and data-flow connections between nodes. The inflow and outflow numbers capture the branching degree that occurs in the models. Note that for libraries B1 and B2 the average outflow is smaller than 1, because many end events occurring in these models have outflow 0.

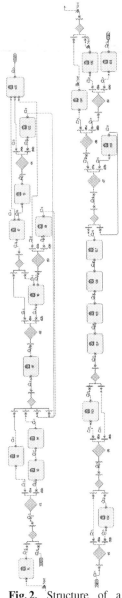

To illustrate such a process model, we show a typical average-sized example from library C in Fig. 2. We split the flow into two parts: the end of the left flow continues at the beginning of the right flow. This process model contains 21 tasks representing elementary, not further distinguished process steps, 16 gateways to encode XOR-splits and -merges, and 51 edges representing data- and control-flow connections. A task can have multiple incoming and outgoing edges that encode implicit AND-splits and -joins of the control and data flows. The example model also contains several cycles: There is a large cycle that spans almost the entire process and there are three smaller cycles within this large cycle – two of them are nested within each other, whereas the third occurs at the end of the process.

2.2   Translation into Workflow Graphs and Petri Nets

Data-flow constructs in the language of the current version of the IBM WebSphere Business Modeler are similar to UML activity diagrams. Here, we only consider explicit data-flow connections and no repositories, because each such connection implies a control-flow connection. Control-flow constructs are visualized in BPMN.

The translation of the process models into the format required by the soundness checkers focuses on the following modeling elements: start and end events, tasks, subprocesses, control flow, input and output sets, and gateways. Data flow is ignored during the translation, i.e., each explicit data-flow connection is replaced by a control-flow connection. Data flow connections from and to repositories were not considered at all. The current language supported by IBM WebSphere Business Modeler contains XOR- and AND-gateways as well as an OR-split, but no OR-join. The translation is well-known and therefore not repeated here; details are provided elsewhere [11].

A task can have multiple incoming and outgoing edges (inputs and outputs) that can be grouped into sets. Input and output sets of tasks are translated into gateway logic as illustrated

Fig. 2. Structure of a typical, average-sized process model

in Fig. 3. In Fig. 3, task A has inputs a, b grouped into one set and inputs c, d, e grouped into another set with the meaning that A can execute if it either receives a and b as input or $c, d,$

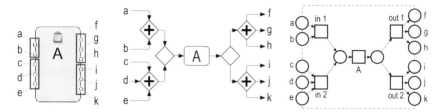

Fig. 3. Translation of a task with disjoint input and output sets (left) into the corresponding work-flow graph (center) and Petri net patterns (right)

and e. The output (sets) of task A are f, g, h and i, j, k. The presence of an input or output is expressed by placing a *token* on an edge between two nodes. Tokens move through the process as a task or gateway executes, taking the process from one state to another state in the usual way.

In the center of Fig. 3, we see the translation into a *workflow graph* [2,5], which is a control-flow graph containing only gateways and tasks. To the right, we see the resulting Petri net. In general, input and output sets can overlap, which would lead to *non-free-choice* Petri nets as a result of the translation [12]. However, none of the syntactically valid process models contained in our test set used overlapping inputs or output sets, i.e., the translation will only return free-choice nets in our case study. This makes it possible to benefit from fast analysis techniques for free-choice Petri nets, see for example Sect. 4. Furthermore, users of the tool can specify which input set activates which output set, but this information was not provided in any of the models. For the translation, we therefore assumed that each input set can potentially activate each output set. Two different translations into workflow graphs and Petri nets were implemented, although the Petri nets could also be directly obtained from the workflow graphs by a well-known construction [2]. The Petri net models are available at http://www.service-technology.org/soundness in PNML format.

2.3 Soundness

Figure 4 shows a workflow graph without any tasks as it occurs in the middle part of the process in Fig. 2 and to which we added a start and an end event. This process model contains a *lack of synchronization* error as well as a *local deadlock*, which are not so easy to spot in the first place.

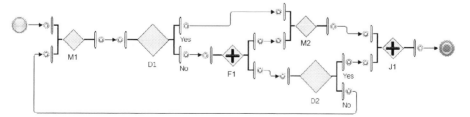

Fig. 4. Workflow graph with deadlock and lack of synchronization errors

A *local deadlock* is a reachable state *s* of the process that has a token on an incoming edge *e* of an AND-join such that each state that is in turn reachable from *s* also contains a token on *e*, i.e., the token is 'stuck' on *e*. A deadlock arises for example, if two alternative paths are merged by an AND-join or if an AND-join occurs as an entry to a cycle. In the example in Fig. 4, a deadlock occurs when a token travels the Yes edge leaving the XOR-split *D*1. Eventually, this token will reach the AND-join *J*1 via the upper incoming branch. However, no other token will ever arrive at the lower incoming branch of *J*1.

A reachable state *s* contains a *lack of synchronization* if there is an edge that has more than one token in *s*. If such an edge contained a task, it would be executed twice. A lack of synchronization arises for example, if two parallel paths are merged by an XOR-merge or if the exit of a cycle is an AND-split. In the example in Fig. 4, a lack of synchronization occurs when a token travels the No edge leaving the XOR-split *D*1. This token will activate the AND-split *F*1, which leads to a token reaching the XOR-merge *M*2 and another token traveling the cycle *D*2, *M*1, *D*1, *F*1. This can result in multiple tokens on the edge from *F*1 to *M*2.

A process model that has neither a lack of synchronization nor a local deadlock is said to be *sound*. This definition of soundness is equivalent to the classical definition of soundness in free-choice Petri nets [3]. There are other equivalent characterizations that are exploited by some of the tools used in our case study, see for example Sect. 5. Their formal treatment can be found elsewhere [2,3,13,5].

Table 2. Dynamic data

	A	B1	B2	B3	C
Processes in library	282	288	363	421	32
sound	152	107	161	207	15
unsound	130	181	202	214	17
Avg. / max. concurrency	2 / 13	8 / 14	16 / 66	14 / 33	2 / 4
Processes with >1000000 states	26	19	29	38	7
Processes with >1000000 states (only sound)	0	1	4	4	0
Avg. number of states (only sound, <1000000 states)	26	71	322	4911	680
Max. number of states (only sound, <1000000 states)	213	2363	28641	588507	8370

Table 2 summarizes the results of our analysis for the libraries. On average, only 46% of all process models are sound ranging from 37% for library B1 to 53% for library A. The table also shows the degree of concurrency that can be found in a process model, i.e., the maximum number of tokens that occur in a single reachable non-error state of the process. Row 5 shows the number of processes with more than one million reachable states, which include error states, and processes that have infinitely many reachable states such as the process shown in Fig. 4. To exclude those, we measured the size of the state space of each *sound* process, which is always finite, which still returned a few processes with more than one million states. The average values, however, suggest that such processes are rare.

3 State Space Verification with LoLA

LoLA [4] is a tool that decides numerous properties by an inspection of the state space of a given Petri net. For making state-space inspection feasible, it offers several state-space reduction techniques. The experiments were carried out with the current version of LoLA 1.11 [14].

Soundness as a model-checking problem. The process models have to be translated into Petri nets prior to the verification as sketched in Sect. 2.2. To verify soundness, LoLA works in two runs on the resulting Petri nets. In the first run, it checks for local deadlocks and in the second run for lack of synchronization.

A process has no deadlock iff a *final* state can be reached from every reachable state; a state is final iff each token has reached an end node. The latter can easily be expressed as a state predicate in LoLA. The former can be expressed as a CTL formula over this predicate and checked by LoLA directly. LoLA checks the property *on-the-fly*, i.e., while the state space is being generated. As soon as LoLA detects a violation, it stops and returns the violating state. Once an error state has been found, a reachability check is used to produce a trace to the error state.

LoLA has a switch that causes state-space generation to be stopped if an *unsafe* state is generated. A state is *unsafe* if a single place contains more than one token, which indicates a lack of synchronization in the original process model. This simultaneous check for lack of synchronization in the first run prevents that LoLA tries to generate an infinite state space and also optimizes performance for finite state spaces. If an unsafe state is found, a trace leading to it is returned immediately. However, the test for unsafe states cannot detect all lack of synchronization errors. Therefore, if no error has been detected during the first run, LoLA is invoked a second time on each net, this time explicitly checking for lack of synchronization.

Lack of synchronization, i.e., unsafeness of states, can be expressed in LoLA as the state predicate $\bigvee_{p \in P} m(p) > 1$, where P is the set of places of the Petri net. As this set can become very large, e.g., on our test data, a maximum of 275 places occurred, we simplified this predicate to optimize performance. We can assert by construction for several places in the Petri net that they cannot obtain more than one token unless a preceding place is also able to do so. In essence, only places that represent an XOR-merge or an exit of a cycle need to be considered. The resulting state predicate is checked for reachability by LoLA. If the predicate is satisfied, a lack of synchronization is identified and LoLA produces a trace to the error state.

We used *partial order reduction* [15] for the results of this paper. This technique suppresses insignificant orderings of concurrently enabled events. LoLA ensures that the property to be checked is preserved by the reduction.

For the example depicted in Fig. 2, LoLA detects a lack of synchronization in the first run, concludes that the net is unsound, and returns an error trace consisting of 36 states.

Experimental setup. After translating the process models into Petri nets with our compiler [11,16], we performed the two checks explained above. We ran the experiments on a notebook with a 2.16 GHz processor and 2 GB RAM. We set a bound of one million states for each net and classified a net as *intractable* if this bound was reached.

Table 3. Analysis statistics for LoLA

	A	B1	B2	B3	C
Intractable processes (no partial order reduction)	0	2	5	4	0
Avg. number of explored states (partial order reduction)	50.42	40.60	37.52	60.76	127.28
Max. number of explored states (partial order reduction)	187	1591	1591	6467	1469
Avg. length of error trace (partial order reduction)	30.24	10.81	12.12	11.21	53.17
Max. length of error trace (partial order reduction)	67	110	75	103	120
Analysis time for library (partial order reduction) [ms]	2680	2356	3184	3878	305
Analysis time for library (struct. reduced, partial order reduction) [ms]	2523	2192	3025	3575	275

Experimental results. Compared with the original models, the Petri nets that we obtained have about 5.5 times as many nodes and edges, see Table 1, which is due to the more fine-grained representation of the process logic in Petri nets as illustrated by Fig. 3. The largest net results from a process model in library C and has 558 nodes and 607 edges.

Without partial order reduction, not all nets could be analyzed, see row 1 of Table 3. When partial order reduction is used, there is no intractable process. In fact, the largest state space explored consists of only 6467 states. Only around 100 states need to be explored on average. During the experiments, LoLA never consumed more than 2 MB of memory, which allows for an unobtrusive verification process, which was not clear in advance. Table 3 summarizes the results.

In a variant of the experiment, we also applied structural Petri net reduction rules [17] to each Petri net before checking it with LoLA. These rules reduce the size of the net, while preserving soundness. The last row of Table 3 shows that structural net reduction hardly has any effect on the runtime. Note that these runtimes do not contain the time needed for structural reduction.

The longest error trace contains 120 Petri net states. When mapped to the original process model, this trace corresponds to a sequence of 40 tasks.

4   Soundness Verification with Woflan

Woflan [3] is a tool for verifying the soundness of business processes modeled as Petri nets. It poses syntactic restrictions on the Petri nets it can analyze, most notably, that each net must have a unique terminal place. Such a net is called a *workflow net*.

Preparing the input for Woflan. Only a few process models from our libraries have a unique terminal node, hence only a few of the resulting Petri nets would have a single terminal place and thus be workflow nets. However, a multi-terminal net N can be *extended* to a workflow net N' using the algorithm of Kiepuszewski et al. [13, Proof of Theorem 5.1]. This algorithm adds new edges to N that cause every terminal place of N to be marked in every run. It then synchronizes all terminal places of N by a final transition, which produces a token on a new unique terminal place. Kiepuszewski et al. [13] show that soundness is preserved by the extension assuming that the original net N is a free-choice Petri net. As we discussed in Section 2, our data set meets this assumption. It is also easy to see that the extension preserves unsoundness. Extending N only requires a depth-first search in N for each of its terminal places.

The tool Woflan. Woflan implements a complex algorithm [3] to check soundness. It uses various techniques from Petri net structure theory as well as state space exploration. If the workflow net is a free-choice net, which is the case in our experiments, Woflan's algorithm reduces to the following procedure (recall also Fig. 1):
(1) First, soundness-preserving structural reduction rules from Petri net theory [17] reduce the size of the input. If the resulting net is *trivial*, i.e., it has only one transition, Woflan immediately concludes that it is sound. (2) Otherwise, Woflan checks the *S-coverability* of the net [3] to exploit the following properties: (2a) A free-choice Petri net that is not S-coverable is unsound, and Woflan quits; the unsoundness can be caused by a deadlock or a lack of synchronization. (2b) A Petri net that is S-coverable has no lack of synchronization, but may contain a local deadlock [3]. (3) If step (2b) applies, Woflan searches for local deadlocks–in Petri net terms a *dead* or a *non-live* transition–by state space exploration, i.e., by constructing the net's coverability graph. The techniques underlying steps (2) and (3) have exponential worst-case complexity in the size of the net.
Woflan provides two kinds of diagnostic information in this setting: If step (2a) applies, it returns a list of places that are not S-coverable, i.e., that contribute to a deadlock or a lack of synchronization. If Woflan detects a deadlock in step (3), it returns a list of dead and non-live transitions that create this deadlock.

Experimental setup. We verified the workflow nets resulting from the translation with a command-line version of Woflan in a batch on a notebook with a 1.66 GHz processor and 2 GB RAM. We ran the experiments twice, the first time without applying structural reduction, the second time with. Aiming at *instantaneous* verification, we interrupted Woflan if the verification time exceeded 5000 ms. In these cases, we classified the process as *intractable* for the analysis.

Experimental results. Table 4 summarizes the results of our Woflan experiments. Our first analysis on the unreduced workflow nets was intractable for 46% of library A and for 19%-28% of libraries B1 to B3. The size of these nets corresponds to the numbers presented for LoLA in Sect. 3. Surprisingly, the analysis became intractable mostly when Woflan checked S-coverability–the technique's exponential worst-case complexity explains this observation. If S-coverability completed successfully, proving absence of deadlocks by state space exploration was tractable in all but 11 cases. Library C was analyzed completely and fairly quickly, see Table 4, row 4. The structure of its models seems to be more suitable for Woflan. We observed that without capping analysis after 5000 ms, Woflan's analysis frequently required between 15 min to more than 1 h per process.
In the second experiment, we let Woflan apply structural Petri net reduction rules prior to analysis, which on average reduced nets in size by a factor 5. The largest net, which resulted from a process in library B3, has 74 nodes and 232 edges. About a third of all models were reduced to the trivial workflow net, see Table 4, row 5. Thus, structural reduction alone identified 53% (libraries A and C) to 80% (libraries B) of all sound

Table 4. Analysis statistics for Woflan

	A	B1	B2	B3	C
1) Without structural reduction					
Intractable processes	129	54	77	119	0
due to S-coverability	129	53	74	112	0
due to state space exploration	0	1	3	7	0
Analysis time [ms]	860812	288218	429343	755875	2375
2) With structural reduction - no intractable processes					
Sound by structural reduction	81	79	134	162	8
Unsound by S-coverability	130	176	197	210	11
Processes that required state space exploration	71	32	32	49	8
Max. number of explored states	8	7	8	8	12
Analysis time per library [ms]	1120	1305	1795	2315	165
per process [ms], avg. / max.	3.97 / 20	4.55 / 40	4.94 / 91	5.50 / 1142	6.11 / 90

processes. Woflan classified about two thirds of the remaining nets as unsound by proving that a net is not S-coverable and free-choice. These nets constitute almost 100% of all *unsound* models. For example as Table 2 shows, library B3 has 213 unsound processes, out of which 210 are not S-coverable. Only for the remaining nets–between 9% (library B2) and 25% (libraries A and C) of the processes–was a state space of at most 12 states explored to complete the analysis. Woflan checks soundness of a process in about 4 to 6 ms on average, with a maximum runtime of less than 90 ms. The one exception in library B3 ran into the exponential worst-case complexity of the S-coverability check, see Table 4, row 10.

Interpreting Woflan's diagnostic information on the original process model is not trivial. For instance, in the workflow net that corresponds to the model of Fig. 4, Woflan reports *all* places to be not S-coverable, hiding the concrete source or location of the error.

We conclude that S-coverability checking alone does not sufficiently speed up the analysis for instantaneous verification of free-choice Petri nets. However, this technique becomes very powerful in combination with Petri net reduction rules. For up to 91% of our examples, soundness or unsoundness was proven alone by these two techniques. Only in the remaining cases, was a fairly simple state space exploration required.

5   The SESE Decomposition Approach

The SESE approach structurally decomposes a business process model into smaller fragments, for which soundness is analyzed by heuristics and state space exploration. If each fragment is sound, then the entire process is sound. The analysis is done on a workflow graph, which is obtained from the original process model as sketched in Sect. 2.2. The SESE approach combines the following three techniques.

State space exploration with SESE. The base technique for the SESE approach is state space exploration. Soundness of a workflow graph can be decided by checking that no explored state has more than one token on a single edge (lack of synchronization) and that each non-terminal state has a successor state (*global deadlock*). If a workflow graph has no lack of synchronization, then every local deadlock manifests itself eventually in a global deadlock in each execution. The workflow graph's state space is explored by depth-first search. The analysis terminates upon the first state that violates one of these two properties and returns a trace leading to this state. If there is no error, the entire state space must be explored.

SESE decomposition. To mitigate the state space explosion problem, we use a parsing technique called the *Refined Process Structure Tree (RPST)* [18]. The RPST decomposes a workflow graph into a hierarchy of *fragments* with a single entry and single exit (SESE) of control. A SESE fragment of a workflow graph is a subgraph that has a single entry node and a single exit node. Fig. 5 shows an example of a workflow graph that is decomposed into such fragments. Multiple end nodes can be handled by adding a unique dummy end node as shown in Fig. 5. Soundness is compositional with respect to SESE fragments, i.e., each fragment can be checked in isolation [5]. To verify the soundness of a fragment, each child fragment can be treated as a task (node) of the workflow graph.

The soundness of a SESE fragment can be checked using plain state space exploration. Because fragments are usually considerably smaller than the entire workflow graph, the input to the state space exploration is smaller, in turn resulting in smaller state spaces to be explored. The decomposition is done in linear time and the number of fragments is at most linear in the size of the workflow graph. The time to analyze an entire workflow graph is then dominated by the size of its largest fragment.

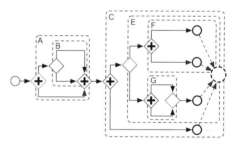

Fig. 5. Decomposition of a workflow graph using the Refined Process Structure Tree

The diagnostic information returned is a fragment showing the error as a trace relative to the fragment. This shows an error inside a smaller scope and shortens the error trace. Moreover, the checker can detect multiple errors at once, up to one per fragment. This includes 'unreachable' errors, such as a lack of synchronization in a fragment, e.g., in fragment *G* in Fig. 5, that cannot be reached by plain state space exploration because this fragment is obstructed by another deadlock earlier in the process, e.g., fragment *A* in Fig. 5.

Heuristics. In practice, many fragments have a simple structure that can be recognized as sound or unsound in linear time using structural heuristics [5]. For example, if a fragment contains only XOR-gateways, it is purely sequential and therefore sound. If a fragment contains at least one XOR-split, but no XOR-join it must be unsound. In this case, the XOR-split can be highlighted inside the highlighted fragment as diagnostic information. We implemented 14 heuristics, all of which can be evaluated based on a single count of the gateway types within a fragment. Only a fragment that does not match any of the heuristics becomes the subject of state space exploration; such a fragment is said to be *complex*. Therefore, heuristics are expected to speed up the analysis by bypassing the state space exploration.

Experimental setup. The SESE approach is implemented as part of the IBM Web-Sphere Business Modeler, in which we also conducted the experiments collecting results from the debugging console. The analysis time reported also includes the production of the regular error report in the tool.

We conducted three experiments to measure the impact of the SESE decomposition and the heuristics: First, we used plain state space exploration only. Second, we decomposed each process into its SESE fragments, and *all* fragments were then analyzed by state space exploration. In the third experiment, we used decomposition in combination with heuristics and state space exploration, i.e., state space exploration was applied only to complex fragments.

The analysis time is computed as an average over five runs. The overhead for loading the process models from the hard drive into memory was measured separately and factored out from the analysis time. The SESE experiment was conducted on a notebook with a 2 GHz processor and 3 GB RAM.

A process is *intractable* if more than 100000 states have to be explored. This threshold value is based on the experience that the time needed would otherwise exceed a value that is acceptable in the use case of instantaneous verification as described in Sect. 1.

Experimental results. Table 5 shows the results for the three experiments described above. For plain state space exploration, we observe that at most 6 out of 363 processes (all contained in library B2) are intractable, i.e., less than 2 percent. Analyzing library A, which contains no intractable process, only requires 490 ms.

When using the decomposition into fragments, we observe that there no longer is an intractable process. However, the analysis time of library B2 is dominated by one particular process which took 25 sec to be analyzed. All other processes took less than 1 sec each. SESE decomposition reduces the size of the input to state space exploration by an average factor between 1.5 and 4. The number of states that are explored for a particular process is the sum of the number of states explored for each fragment of the process. Table 5 shows that the number of states that have to be explored for a process on average reduces by up to a factor of 13.8 with respect to experiment 1. After decomposition, there is still a fragment that has 16403 states.

Library A shows that computing the decomposition does not always pay off: This library is analyzed faster without decomposition. The analyses of the other libraries, however, clearly benefit from the decomposition: Decomposition reduces the analysis time by a factor between 5 and 67 with respect to plain state space exploration.

In addition, we recorded the length of the error trace in both experiments. Error traces are notably smaller when they relate only to a fragment, rather than to the entire workflow graph. The error trace lengths were reduced by a factor of 4.7 on average. Note that the error trace using the decomposition into fragments starts at the start node of the fragment and not at the start node of the workflow graph. The decomposition allows us to detect multiple errors per process, at most one per fragment. For library B2, we measured an average of 1.55 and a maximum of 7 unsound fragments per unsound process.

Table 5. Experimental results for the SESE decomposition approach

		A	B1	B2	B3	C
1) State space exploration - reference						
Explored states per process (avg.)		42.8	826.4	1879.3	1508.1	149.7
Explored states per process (max.)		241	17176	28684	28688	2517
Intractable processes		0	2	6	5	0
Analysis time [ms]	library	490	30019	197670	135178	30019
	process (max.)	16	13186	76700	24624	62
2) Using the decomposition - no intractable processes						
Size reduction (workflow graph / largest fragment) (avg.)		4.1	3.8	3.9	4.7	2.7
Explored states	process (avg.)	52.4	31.6	86.7	38.4	61.7
	reduction w.r.t. exp. 1 per process (avg.)	1.0	13.8	13.4	10.2	1.5
	process (max.)	201	268	16534	311	356
	fragment (max.)	53	117	16403	68	120
Analysis time [ms]	library	1587	1359	35495	2446	447
	process (max.)	16	16	25286	32	32
3) Using the heuristics - no intractable processes						
Portion of fragments analyzed by heuristics		97%	97%	98%	98%	99%
Explored states	process (avg.)	6.0	2.3	3.2	2.5	10.1
	reduction w.r.t. exp. 2 per process (avg.)	28.3	22.9	78.4	29.6	34.1
	process (max.)	53	36	165	24	120
	fragment (max.)	53	36	165	20	120
Analysis time [ms]	library	1247	1390	1681	2303	318
	process (max.)	16	31	16	31	62

The third experiment shows that the heuristics speed up the analysis further. For all libraries, more than 97 percent of the fragments match some heuristic and only the remaining ones have to go into state space exploration. We noted that a process usually contains not more than one complex fragment, out of an average of 16 fragments per process. Only the largest process, which has 122 fragments, contains two complex fragments; no process contained more. The small number of complex fragments results in a reduction factor of up to 78.4 for the average number of states that were explored to analyze a process. The use of the heuristics reduces the analysis time of library B2 by a factor 21 with respect to experiment 2. For the other libraries, the differences in the analysis times is not significant. The maximum analysis times per process range from 10 to 62 ms.

6   Conclusion

We showed that different techniques can be used to check the soundness of industrial business process models reliably in fractions of a second.

For the *state space approach* using LoLA, we found that partial order reduction and on-the-fly verification are the essential factors for success. Although many processes could have been verified on a brute force state space, some state spaces exploded without the use of partial order reduction. While it was difficult to handle full state spaces, the exploration of erroneous state spaces up to the first error was efficient. Surprisingly, the prior application of structural Petri net reduction has only a minor impact on performance. This may be because many of the existing reduction rules address situations that also partial order reduction on the state space is dealing with.

In the *structural approach* using Woflan, we saw that the original models can easily be translated into the more restrictive notion of workflow nets with just one terminal node. Another observation was that the performance of Woflan can mainly be attributed to the structural Petri net techniques. In the few cases where Woflan had to explore a state space, this state space was rather small because of prior application of structural reduction. Here, structural reduction turned out to be beneficial as Woflan does not provide partial order reduction.

In the *decomposition approach* using SESE fragments, we learned that the approach did not suffer from severe state space explosion as the state space is only computed locally for a typically small fragment of the process model. Moreover, structural heuristics are sufficient to handle most of the fragments, which allows one to bypass state space exploration altogether.

While being similar in their performance, the three approaches chosen vary with respect to the diagnostic information they provide. The state space approach used by LoLA is able to return an error trace of manageable size that can be simulated or animated. The SESE approach can detect multiple errors in one analysis run and localizes each error in a particular, typically small, fragment of the original model. This also reduces the length of the error trace by a factor of 4.7 on average. Moreover, the approach can provide additional information depending on the heuristics applied. Woflan returns some Petri-net specific information that needs to be interpreted carefully before it can be shown to a business user.

Another notable difference between the three approaches is that Woflan is specifically built for checking soundness and the SESE approach is specifically designed to check soundness instantaneously, whereas LoLA is a generic model checker for Petri nets that could more easily be adapted to check other temporal properties of business processes.

We would like to point out that there are other promising algorithms to check soundness, especially polynomial-time algorithms exploiting the free-choice property [12]. We could not include those in our case study because we are not aware of available implementations.

Finally, note also that the various techniques could easily be combined in different ways. For example, one could apply SESE decomposition to break the model into smaller fragments, then use heuristics and structural Petri net reduction to quickly sort out sound fragments that have a simple structure, and then finally check the remaining fragments with state space exploration based on partial order reduction to obtain detailed localized error information.

Acknowledgements. We would like to thank Eric Verbeek for his substantial support in providing a Woflan version for our experiments. Dirk Fahland is funded by the DFG-Graduiertenkolleg "METRIK" (1324). Niels Lohmann and Karsten Wolf are supported by the DFG project "Operating Guidelines for Services" (WO 1466/8-1). Jana Koehler and Hagen Völzer were partially supported by the SUPER project (http://www.ip-super.org) under the EU 6th Framework Programme Information Society Technologies Objective (contract no. FP6-026850).

References

1. Mendling, J.: Empirical Studies in Process Model Verification. Trans. Petri Nets and Other Models of Concurrency (ToPNoC) 2, 208–224 (2009)
2. van der Aalst, W.M.P., Hirnschall, A., Verbeek, H.M.W(E.): An alternative way to analyze workflow graphs. In: Pidduck, A.B., Mylopoulos, J., Woo, C.C., Ozsu, M.T. (eds.) CAiSE 2002. LNCS, vol. 2348, pp. 535–552. Springer, Heidelberg (2002)
3. Verbeek, H.M.W.E., Basten, T., van der Aalst, W.M.P.: Diagnosing Workflow Processes using Woflan. Comput. J. 44(4), 246–279 (2001)
4. Wolf, K.: Generating petri net state spaces. In: Kleijn, J., Yakovlev, A. (eds.) ICATPN 2007. LNCS, vol. 4546, pp. 29–42. Springer, Heidelberg (2007)
5. Vanhatalo, J., Völzer, H., Leymann, F.: Faster and more focused control-flow analysis for business process models through SESE decomposition. In: Krämer, B.J., Lin, K.-J., Narasimhan, P. (eds.) ICSOC 2007. LNCS, vol. 4749, pp. 43–55. Springer, Heidelberg (2007)
6. van Dongen, B.F., Jansen-Vullers, M., Verbeek, H.M.W.E., van der Aalst, W.M.P.: Verification of the SAP reference models using EPC reduction, state-space analysis, and invariants. Comput. Ind. 58(6), 578–601 (2007)
7. Mendling, J., Neumann, G., van der Aalst, W.M.P.: Understanding the occurrence of errors in process models based on metrics. In: Meersman, R., Tari, Z. (eds.) OTM 2007, Part I. LNCS, vol. 4803, pp. 113–130. Springer, Heidelberg (2007)
8. Mendling, J.: Detection and Prediction of Errors in EPC Business Process Models. PhD thesis, Vienna University of Economics and Business Administration (May 2007)
9. Mendling, J., Verbeek, H.M.W.E., van Dongen, B.F., van der Aalst, W.M.P., Neumann, G.: Detection and prediction of errors in EPCs of the SAP reference model. Data Knowl. Eng. 64(1), 312–329 (2008)
10. Wynn, M., Verbeek, H.M.W.E., van der Aalst, W.M.P., ter Hofstede, A.H.M.: Business process verification: Finally a reality! Business Process Management Journal 15(1), 74–92 (2009)
11. Fahland, D.: Translating UML2 activity diagrams to Petri nets. Informatik-Berichte 226, Humboldt-Universität zu Berlin, Berlin, Germany (2008)
12. Desel, J., Esparza, J.: Free Choice Petri Nets. Cambridge University Press, New York (1995)
13. Kiepuszewski, B., ter Hofstede, A.H.M., van der Aalst, W.M.P.: Fundamentals of control flow in workflows. Acta Inf. 39(3), 143–209 (2003)
14. LoLA v1.11, http://service-technology.org/lola
15. Valmari, A.: Stubborn sets for reduced state space generation. In: Rozenberg, G. (ed.) APN 1990. LNCS, vol. 483, pp. 491–515. Springer, Heidelberg (1991)
16. UML2oWFN compiler, http://service-technology.org/uml2owfn
17. Murata, T.: Petri nets: Properties, analysis and applications. Proc. of the IEEE 77(4), 541–580 (1989)
18. Vanhatalo, J., Völzer, H., Koehler, J.: The refined process structure tree. In: Dumas, M., Reichert, M., Shan, M.-C. (eds.) BPM 2008. LNCS, vol. 5240, pp. 100–115. Springer, Heidelberg (2008)

Symbolic Abstraction and Deadlock-Freeness Verification of Inter-enterprise Processes

Kais Klai[1], Samir Tata[2], and Jörg Desel[3]

[1] LIPN, CNRS UMR 7030, Université Paris 13
99 avenue Jean-Baptiste Clément, F-93430 Villetaneuse, France
kais.klai@lipn.univ-paris13.fr
[2] Institut TELECOM, CNRS UMR Samovar
9 rue Charles Fourier 91011 Evry, France
Samir.Tata@int-edu.eu
[3] Department of Applied Computer Science
Catholic University of Eichstätt-Ingolstadt, 85071 Eichstätt, Germany
joerg.desel@ku-eichstaett.de

Abstract. The design of complex inter-enterprise business processes (IEBP) is generally performed in a modular way. Each process is designed separately from the others and then the whole IEBP is obtained by composition. Even if such a modular approach is intuitive and facilitates the design problem, it poses the problem that correct behavior of each business process of the IEBP taken alone does not guarantee a correct behavior of the composed IEBP (i.e. properties are not preserved by composition). Proving correctness of the (unknown) composed process is strongly related to the model checking problem of a system model. Among others, the *symbolic observation graph* based approach has proven to be very helpful for efficient model checking in general. Since it is heavily based on abstraction techniques and thus hides detailed information about system components that are not relevant for the correctness decision, it is promising to transfer this concept to the problem rised in this paper: How can the symbolic observation graph technique be adapted and employed for process composition? Answering this question is the aim of this paper.

1 Introduction

Business process composition and cooperation are two important research fields in the business process domain. The questions, what properties of a process has to be public so that potential partners can collaborate with the process without risking to have an ill-designed composed process, and what is the minimum necessary to be published, is a hot topic in the literature since many years (e.g. [2,14,13,9]). Also, one has to make sure that the composition of the processes has the desired behaviour. The importance of dealing with such inter-enterprise business processes (IEBP for short) on one hand and business process composition on the other hand is reflected in the literature by numerous publications [18,4,19,15].

U. Dayal et al. (Eds.): BPM 2009, LNCS 5701, pp. 294–309, 2009.

In general, an IEBP can be considered as the cooperation of several local processes designed separately. The activities of each process are formally of two kinds: internal activities and cooperative activities (interface activities). IEBP are often too large for formal analysis, and the details of the components are hidden to the public so that no party knows the entire process definition. Therefore, we defend the idea that the analysis should be on the local business process and, if necessary, on an abstraction of the composition partner or of the IEBP. In this paper, we propose a two steps abstraction technique: In the first step, an abstraction of each local process is built locally using a new variant of *symbolic observation graphs* (SOG for short) [6]. This abstraction has two advantages: the analysis of the corresponding process can be reduced to the analysis of its abstraction, and such an abstraction hides the internal structure and organization of the process, which is a desired requirement in the IEBP context. In the second step, the abstraction of the IEBP is obtained by composing the local abstractions (SOGs), leading to a global abstraction on which the analysis can be performed efficiently.

One of the most important properties an IEBP should enjoy is deadlock-freeness. In other words, assuming that the components itself are deadlock-free, it is undesirable that these components block each other. Taking the view of a single component, we want to identify situations where the other component is waiting for some message or action from this component while this component is waiting for some message or action from the other component. In such a situation, the other component does not do any visible action. However, since only the interface behaviour is visible, it is possible that internal actions of the other component do occur. So this behaviour, usually known as livelock, is as bad as deadlock behaviour. Hence, in this paper, we extend the deadlock notion by considering a deadlock state every state from which no cooperative action is possible in the future. One can check the deadlock-freeness on each SOG using efficient symbolic algorithms [6] (i.e. algorithms based on set operations). Since the deadlock freeness property is not preserved by composition, we supply a new algorithm for checking deadlock freeness of the synchronized product of the local SOGs. This algorithm is based on local information that can be made available once before the composition process. The deadlock freeness of the product guarantees correct cooperation between the underlying processes (i.e. a deadlock-free cooperation).

The composition of SOGs is immediately suitable for synchronous interorganizational processes. It can moreover be used for checking whether a cooperation between two processes, that communicate asynchronously, is deadlock-free. To this end, one can define an additional component that represents the asynchronous channel and has two observed actions: *receive* and *send*. Now the *send* action of the first component synchronizes with the *receive* action of the channel whereas the *receive* action of the second component synchronizes with the *send* action of the channel.

This paper is organized as follows. Section 2 adapts the structure of the *symbolic observation graph* in order to abstract business processes. Section 3 constitutes the core of the paper and shows how to build the symbolic observation graph of an IEBP and how to establish whether processes can be composed (or can collaborate) safely by checking the deadlock-freeness of the obtained composition of SOGs. A case study is used throughout these sections in order to illustrate our approach. Section 4 relates our work to other approaches. Finally, Section 5 summarizes the results and mentions some aspects of future work.

2   Process Abstraction

In this section, we show how the structure of the *symbolic observation graph* [6] (*SOG*) is used to abstract business processes. In [6], the authors have introduced the *SOG* as an abstraction of the *reachability graph* of concurrent systems and showed that the verification of an event-based formula of $LTL \setminus X$ (Linear-time Temporal Logic minus the next operator) on the *SOG* is equivalent to the verification on the original reachability graph. The construction of the *SOG* is guided by the set of actions occurring in the formula to be checked. Such actions are said to be observed while the other actions of the system are unobserved. The *SOG* is defined as a graph where each node is a set of states linked by unobserved actions and each arc is labeled with an observed action. Nodes of the *SOG* are called *meta-states* and may be represented and managed efficiently using decision diagram techniques (BDDs for instance [1]). In practice, due to the small number of actions in a typical formula, the SOG has a very moderate size and thus the time complexity of the verification process is negligible w.r.t. the building time of the SOG (see [6,8,7] for experimental results).

We propose to use a SOG to abstract a business process. The collaboration actions are observed while the internal ones are not. We will establish that such an abstraction is especially efficient for loosely coupled IEBPs.

2.1   Notations and Preliminary Results

The technique presented in this paper applies to different kinds of process models that can map to labeled transition systems, e.g. workflow Petri nets (WF-nets). For sake of simplicity and generality, we chose to present it for labeled transition systems, since this formalism is rather simple.

Definition 1 (Labeled Transition System)
A labeled transition system (LTS for short) is a 5-tuple $\langle \Gamma, Act, \rightarrow, I, F \rangle$ where:

- Γ *is a finite set of* states *;*
- *Act is a finite set of* actions *;*
- $\rightarrow \subseteq \Gamma \times Act \times \Gamma$ *is a* transition relation *;*
- $I \subseteq \Gamma$ *is a set of* initial states*;*
- $F \subseteq \Gamma$ *is a set of* final states.

In this paper, we distinguish LTS observed actions, denoted by a subset *Obs*, from unobserved actions, denoted by the subset *UnObs* (with $Obs \cup UnObs = Act$

and $Obs \cap UnObs = \emptyset$). Observed actions can represent cooperative (or interface) actions, while unobserved actions represent internal actions.

The following notations are used in this paper:

- For $s, s' \in \Gamma$ and $a \in Act$, we denote by $s \xrightarrow{a} s'$ that $(s, a, s') \in \to$.
- $s \xrightarrow{a}$ means that $\exists s' \in \Gamma$ s.t. $s \xrightarrow{a} s'$. If $\sigma = a_1 a_2 \cdots a_n$ is a sequence of actions, $\overline{\sigma}$ denotes the set of actions occurring in σ, while $|\sigma|$ denotes its length. Moreover, $s \xrightarrow{\sigma} s'$ denotes that $\exists s_1, s_2, \cdots s_{n-1} \in \Gamma \colon s \xrightarrow{a_1} s_1 \longrightarrow \cdots s_{n-1} \xrightarrow{a_n} s'$. $s \xrightarrow{*} s'$ denotes that s' is reachable from s (i.e. $\exists \sigma \in Act^*$ s.t. $s \xrightarrow{\sigma} s'$) and $s \xrightarrow{*}_T s'$ holds if $\overline{\sigma}$ is included in some subset of actions T.
- The set $Enable(s)$ denotes the set of actions a such that $s \xrightarrow{a}$. For a set of states S, $Enable(S)$ denotes $\bigcup_{s \in S} Enable(s)$.
- $\pi = s_0 \xrightarrow{a_1} s_1 \xrightarrow{a_2} \cdots$ is used to denote a path of an LTS. $\pi = s_0 \xrightarrow{a_1} \cdots \xrightarrow{a_n} s_n$ is said to be a *run* if $s_n \in F$ (i.e. s_n is a final state).
- A finite path $C = s_1 \xrightarrow{a_1} s_2 \xrightarrow{a_2} \cdots \xrightarrow{a_{n-1}} s_n$ is said to be a *cycle* if $s_n = s_1$. If $\{a_1, \ldots a_{n-1}\} \subseteq UnObs$ then C is said to be a *livelock*.
- $s \nrightarrow$, for $s \in (\Gamma \setminus F)$, denotes that s is a dead state i.e. $\nexists a \in Act \colon s \xrightarrow{a}$.
- $s \nRightarrow$, for $s \in (\Gamma \setminus F)$, denotes that no observed action can be enabled in the future starting from s, i.e., $\nexists o \in Obs, \tau \in UnObs^* \colon s \xrightarrow{\tau\, o}$.

If $s \nRightarrow$, for $s \in (\Gamma \setminus F)$, one can either reach a dead state using unobserved actions only, or a livelock. Such a livelock is said to be a *strong livelock*. In this paper we assume that a strong livelock behaviour is equivalent to a deadlock. In contrast, a cycle with states from which one can execute an observed action (possibly via an unobserved sequence) is said to be a *weak livelock*.

For checking LTL properties, livelock and deadlock behaviours have exactly the same interpretation. However, in the context of inter-organizational processes, we claim that only a strong livelock should be viewed as a deadlock, but not a weak livelock.

The set of states $Dead$ contains the states from which no action is enabled or from which no observed action is enabled in the future, i.e., $Dead := \{s \in (\Gamma \setminus F) \mid s \nRightarrow\}$ (we distinguish "dead" and "Dead"). The following definition characterizes deadlocks and strong livelocks in an homogenous way. We define a particular mapping applied to states of an LTS called *Observed behaviour*.

Definition 2 (Observed behaviour mapping)

Let $\mathcal{T} = \langle \Gamma, Obs \cup UnObs, \to, I, F \rangle$ be an LTS. The mapping $\lambda_{\mathcal{T}} : (\Gamma \setminus F) \to 2^{Obs}$ is defined by: $\lambda_{\mathcal{T}}(s) = \{o \in Obs \mid \exists s' \in \Gamma \text{ s.t. } s \xrightarrow{*}_{UnObs} s' \wedge s' \xrightarrow{o}\}$. $\mathcal{T}$ is Deadlock free iff $\lambda_{\mathcal{T}}(s) \neq \emptyset$ for each state s in $(\Gamma \setminus F)$

Informally, for each (non final) state s of an LTS $\mathcal{T}$, the observed behaviour of s, $\lambda_{\mathcal{T}}(s)$, stands for the set of observed actions which can be executed from s, possibly via a sequence of unobserved actions. This set is empty for a state s if and only if s is a $Dead$ state.

The observed behaviour mapping can be extended to sets of states: Given a set of states γ and a set of observed actions ψ, $\lambda(\gamma) = \psi$ iff $\forall s \in \gamma, \lambda(s) = \psi$.

The observed behaviour of a given state can be computed by the following two steps. First, compute $Sat(s)$, i.e., all the states reachable from s by executing unobserved actions only. Once such a set is saturated (no new state can be reached), the observed behaviour of s is $Enable(Sat(s)) \cap Obs$. One can improve the computation of $Sat(s)$ by storing the observed behaviours of already computed states.

2.2   Running Example

The example used in this paper is an adaptation of the one given in [18] which is inspired by electronic bookstores. In [18], local processes are modeled by workflow nets. Here, we use the "private" workflows of the involved models. Moreover, we modify these models, by removing some internal behaviours, in order to get manageable LTSs. There are four processes, modeling a customer, a bookstore, a publisher and a shipper. $c1$ (resp. $b1$, $p1$, $s1$) is the initial state of the customer's (resp. bookstore's, publisher's, shipper's) LTS.

The customer (Figure 1(a)) behaves as follows: First, he sends an order to a bookstore (c_order). Then the customer may receive a negative answer (c_reject) or be informed that his order is going to be handled (c_accept). After order handling, either the customer receives from a shipper the ordered book ($ship$) and from the bookstore a bill (c_bill), or he receives the bill first and then the book. After receiving the book and the bill the customer makes a payment (c_pay). Finally, the customer returns (c_init) to his initial state to order other books.

Figure 1(b) illustrates the bookstore's LTS that has no books in stock. Therefore, when the bookstore receives an order for a book, it transfers it to a publisher (b_order) and updates the customer profile ($update\_c\_profile$). Then

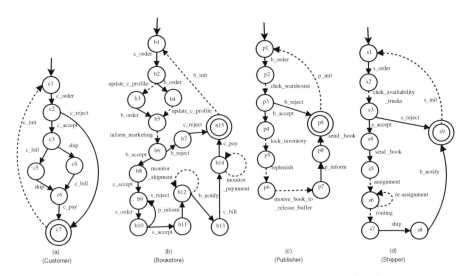

| (a) | (b) | (c) | (d) |
| (Customer) | (Bookstore) | (Publisher) | (Shipper) |

Fig. 1. LTS of a customer, a bookstore, a publisher and a shipper

it informs the marketing department. If the bookstore receives a negative answer (b_reject), i.e. its order was rejected, then it sends a negative response to the customer (c_reject). Otherwise, i.e. the bookstore receives a positive answer (b_accept), the customer is informed (c_accept) and the bookstore sends a request to a shipper (s_order). If the bookstore receives a negative answer (s_reject), it searches another shipper. This process is repeated until a shipper accepts (s_accept). When this happens, the bookstore informs the publisher (p_inform). After that, the bookstore waits for the shipper's notification (b_notify) and sends the bill to the customer (c_bill). Hence, the bookstore processes the payment (c_pay). Finally, after the payment or after an order reject the bookstore returns to its initial state (b_init).

Figure 1(c) presents the publisher's LTS. When receiving an order from a bookstore, the publisher evaluates the order and can either accept it (b_accept) or reject it (b_reject). After that, when the publisher is informed (p_inform) that a shipper was found, he sends the book to the shipper ($send\_book$). Finally, after shipment or a request reject, the publisher returns to its initial state (p_init).

Figure 1(d) presents the shipper's LTS. Notice that the original LTS contains 19 nodes and 31 arcs; here we present a reduced version of the graph. When receiving a request from a bookstore (s_order), the shipper evaluates the request and either accepts (s_accept) or rejects (s_reject) the shipping request. In case the shipper receives a book from the publisher ($send\_book$), he ships the book to the customer ($ship$) and then notifies the bookstore (b_notify). After shipment or request reject, the shipper returns to its initial state (s_init).

For each LTS of Figure 1, initial states are those having (no source) input arcs while final states are represented with double circles. The observed actions represent, for each component, the collaborative ones and are those labeling dotted arcs. None of these LTSs contains a *Deadlock* state.

2.3   The Symbolic Observation Graph

In this subsection, we first define formally what a *meta-state* is, before providing a formal definition of a SOG associated with an LTS and a set of observed actions. Our definitions are different from those given in [7] because, first, we do not distinguish deadlocks from strong livelocks (we do not pay attention to weak livelocks). Then, we distinguish *final* meta-states from others and, finally, the observed behaviour of the states belonging to a meta-state is stored in this meta-state (as a set of sets of observed actions). Meta-states have associated boolean attributes d and f which indicate whether a meta-state is *Dead* or not and whether it is final or not.

Definition 3 (Meta-state)
Let $\mathcal{T} = \langle \Gamma, Act, \rightarrow, I, F \rangle$ be a labeled transition system with $Act = Obs \cup UnObs$. A meta-state is a tuple $M = \langle S, d, f, \lambda \rangle$ defined as follows:

1. *S is a nonempty subset of Γ satisfying:*
   (a) $\forall s \in S \ \exists i \in I, \exists \sigma \in Act^* \ s.t. \ i \xrightarrow{\sigma} s$;
   (b) $\forall s \in S, \forall s' \in \Gamma, \forall \sigma \in UnObs^* : s \xrightarrow{\sigma} s' \Rightarrow s' \in S$;

2. $d \in \{true, false\}$. $d = true$ iff $\exists s \in S \setminus F$ s.t.$\lambda_{\mathcal{T}}(s) = \emptyset$;
3. $f \in \{true, false\}$. $f = true$ iff $S \cap F \neq \emptyset$;
4. $\lambda = \{\psi \subseteq Obs\}$ s.t. $\psi \in \lambda$ iff $\exists \gamma \subseteq S$ s.t. $\lambda(\gamma) = \psi$.

From now on, $M.S$, $M.d$, $M.f$ and $M.\lambda$ denote the corresponding attributes of a given meta-state M. Moreover, we introduce the following set of output states of M: $Out(M) = \{s \in M.S \mid \exists o \in Obs: s \xrightarrow{o} \}$. Notice that if the set $Out(M)$ is empty, then M necessarily contains a $Dead$ state.

Definition 4 (Symbolic Observation Graph)
The symbolic observation graph *(SOG($\mathcal{T}$)) associated with an LTS*
$\mathcal{T} = \langle \Gamma, Obs \cup UnObs, \rightarrow, I, F \rangle$ *is a 4-tuple* $\langle \Gamma', Act', \rightarrow', I' \rangle$ *such that:*

1. *Γ' is a finite set of meta-states;*
2. *$Act' = Obs$;*
3. *$\rightarrow' \subseteq \Gamma' \times Act' \times \Gamma'$ is a transition relation such that:*
   (a) For $M, M' \in \Gamma'$ and $a \in Act'$: $M \xrightarrow{a}' M'$ if and only if:
      i. $\forall s \in M.S, s' \in \Gamma: s \xrightarrow{a} s' \Rightarrow s' \in M'.S$,
      ii. $\forall o \in Out(M') \exists s \in M.S, \exists s' \in M'.S$ s.t. $s \xrightarrow{a} s' \wedge s' \xrightarrow{*}_{UnObs} o$,
      iii. $M'.d = true \Rightarrow$
         $(\exists l \in M'.S$ s.t. $\lambda_{\mathcal{T}}(l) = \emptyset) \wedge (\exists s \in M.S, \exists s' \in M'.S$ s.t. $s \xrightarrow{a} s' \wedge$
         $s' \xrightarrow{*}_{UnObs} l)$.
      iv. $M'.f = true \Rightarrow$
         $(\exists f \in M'.S \cap F$ s.t. $\forall s \in M.S, \forall s' \in M'.S : s \xrightarrow{a} s' \Rightarrow s' \xrightarrow{*}_{UnObs} f)$.
   (b) $\forall s, s' \in \Gamma \, \forall a \in Obs$
      $(s \xrightarrow{a} s' \Rightarrow \exists M, M' \in \Gamma': s \in M.S, s' \in M'.S \wedge M \xrightarrow{a} M')$,
4. *$I' = \{M_0\}$, where the meta-state M_0 satsifies $I \subseteq M_0.S$.*

Point 3a of the above definition requires explanation. An edge, labeled a, in the SOG is allowed between two meta-states M and M' iff: (3(a)i) each state $s' \in \Gamma$ reachable from some state $s \in M.S$, by action a, belongs to $M'.S$. If $S' = \{s' \in M'.S \mid \exists s \in M.S \wedge s \xrightarrow{a} s'\}$, then (3(a)ii) implies that each output state of M' is reachable from at least one state of S' (using unobserved actions only), while (3(a)iii) implies that when the Deadlock attribute of M' is true then one state l satisfying $\lambda_{\mathcal{T}}(l) = \emptyset$ in $M'.S$ is reachable from at least one state of S' using unobserved actions only. Finally, (3(a)iv) implies that if M' is a final meta-state, then some final *state* $s \in M'.S$ is reachable from each state of S' (defined below).

Figure 2 illustrates the SOGs associated with the LTSs of Figure 1. Final meta-state are represented by dotted circles. The SOG of the *customer* is isomorphic to its corresponding LTS (since all its actions are observed) while the SOG of the *bookstore* contains 12 nodes and 14 arcs (versus 15 nodes and 21 arcs in its corresponding LTS), the SOG of the *publisher* contains 5 nodes and 6 arcs (versus 9 nodes and 10 arcs in its corresponding LTS) and the SOG of the *shipper* contains 6 nodes and 7 arcs (versus 19 nodes and 31 arcs in its corresponding LTS). All of these SOGs are Deadlock-free. We give below the composition of some meta-states:

- $C1.S = \{c1\}$, $C2.S = \{c2\}$, $C3.S = \{c3\}$, $C4.S = \{c4\}$,
- $B2.S = \{b2, b3\}$, $B3.S = \{b4, b5, b6\}$, $B4.S = \{b8\}$, $B6.S = \{b9\}$,
- $P1.S = \{p1\}$, $P2.S = \{p1, p3\}$, $P3.S = \{p4, p5, p6, p7\}$, $P5.S = \{p9, p1\}$,
- $S1.S = \{s1\}$, $S2.S = \{s2, s3\}$, $S3.S = \{s4\}$, $S4.S = \{s5, s6, s7\}$.

Definition 5 (Deadlock-freeness property of a SOG)
An SOG $\langle \Gamma, Act, \rightarrow, I \rangle$ is said to be Deadlock-free iff $\nexists M \in \Gamma$ s.t. $M.d = true$.

The following result establishes that the Deadlock-freeness of a SOG is equivalent to the Deadlock-freeness of the corresponding LTS.

Proposition 1. *Let $\mathcal{T} = \langle \Gamma, Act = Obs \cup UnObs, \rightarrow, I, F \rangle$ be a labeled transition system and let $SOG(\mathcal{T})$ be the corresponding SOG. Then $\mathcal{T}$ is Deadlock-free if and only if $SOG(\mathcal{T})$ is Deadlock-free.*

Proof. The proof follows from Definition 3 and Definition 4: For each state s of $\mathcal{T}$ there exists a meta-state M of $SOG(\mathcal{T})$ containing s. Conversely, for each meta-state M, all states s in $M.S$ are reachable from some initial state of I in the LTS $\mathcal{T}$.

We claim that the SOG technique is suitable for abstracting processes for several reasons: First, the SOG allows to represent the language of the process projected on the cooperative transitions (i.e. the local behaviors are hidden) in addition to some particular internal behavior which can be relevant for the environment (Deadlock existence). It is a valid abstraction of a given process W because it preserves its privacy while supplying sufficient and necessary information to be known by a potential partner of W. The second reason is that this abstraction is suitable for checking whether two process represented by their $SOGs$ can be

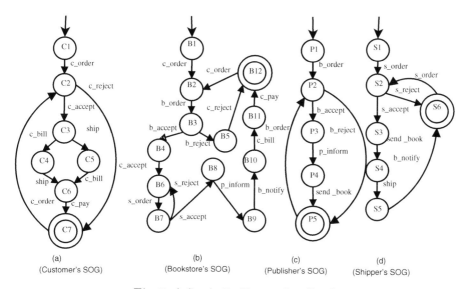

Fig. 2. A Symbolic Observation Graph

interconnected (see Section 3). Moreover, given a process, its *SOG* is built once and might be reused as long as local changes do not change its structure. Finally, the reduced size of the *SOG* (in most cases) makes the building and verification of the synchronized product of SOGs much cheaper than the building of the synchronized product of the original LTSs, especially when the involved models are loosely coupled.

3   Composition and Deadlock-Freeness Verification

This section constitutes the core of the paper. Starting from several LTSs which synchronize over a common set of actions, it shows how to synchronize the corresponding SOGs so that the obtained graph is Deadlock-free if and only if the synchronized product of the original LTSs is Deadlock-free.

We start with the standard method for synchronizing two LTSs, namely building their synchronized product. Each state of the resulting transition system is a pair of states, the first component indicating the respective state of the first LTS, the second component indicating the respective state of the second LTS. Each LTS can still do its private activities autonomously, i.e., only one component of the pair representing a state of the composed LTS is changed by such an action. For common activities, however, both components of the state are changed synchronously. Figure 3 shows a simple example of two LTSs (Module A and Module B) and their synchronization $A \times B$.

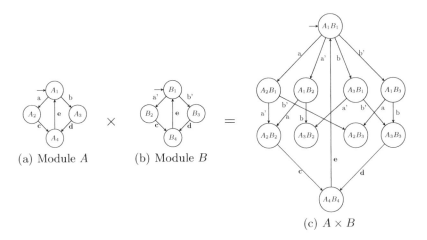

(a) Module A (b) Module B

(c) $A \times B$

Fig. 3. Synchronized product of two LTSs

3.1   Synchronization of LTSs

In the following, we define the synchronized product of two LTSs. The synchronized product of n LTSs (for $n > 2$) can be built by iterative multiplication.

Definition 6 (LTS synchronized product)

Let $\mathcal{T}_i = \langle \Gamma_i, Act_i, \rightarrow_i, I_i, F_i \rangle, i = 1, 2$ be two LTSs. The synchronized product of $\mathcal{T}_1$ and $\mathcal{T}_2$ is the LTS $\mathcal{T}_1 \times \mathcal{T}_2 = \langle \Gamma, Act, \rightarrow, I, F \rangle$ given by:

1. $\Gamma = \Gamma_1 \times \Gamma_2$;
2. $Act = Act_1 \cup Act_2$;
3. $\rightarrow$ is the transition relation, defined by:
$$\forall (s_1, s_2) \in \Gamma : (s_1, s_2) \xrightarrow{a} (s_1', s_2') \Leftrightarrow$$
$$\begin{cases} s_1 \xrightarrow{a}_1 s_1' \wedge s_2 \xrightarrow{a}_2 s_2' & \text{if } a \in Act_1 \cap Act_2 \\ s_1 \xrightarrow{a}_1 s_1' \wedge s_2 = s_2' & \text{if } a \in Act_1 \setminus Act_2 \\ s_1 = s_1' \wedge s_2 \xrightarrow{a}_2 s_2' & \text{if } a \in Act_2 \setminus Act_1 \end{cases}$$
4. $I = I_1 \times I_2$;
5. $F = F_1 \times F_2$.

The set of states is reduced to reachable states only, i.e. $\Gamma = \{(s_1, s_2) \in \Gamma_1 \times \Gamma_2 \mid \exists (i_1, i_2) \in I_1 \times I_2, \exists \sigma \in Act^* : (i_1, i_2) \xrightarrow{\sigma} (s_1, s_2)\}$. Similarly, the set of actions is reduced to those that can effectively take place in the synchronized product: $Act = \{a \in Act_1 \cup Act_2 \mid \exists s, s' \in \Gamma, (s, s') \xrightarrow{a}\}$.

It is well known that the Deadlock-freeness property is not preserved by composition. Given two Deadlock-free LTSs $\mathcal{T}_1$ and $\mathcal{T}_2$, their synchronized product is not guaranteed to be Deadlock-free. Figure 3 illustrates such a situation, where two modules (Figure 3(a) and Figure 3(b)) without Dead states lead, by synchronization over the set of observed actions {**c, d, e**}, to a synchronized product (Figure 3(c)) containing two Dead states ((A_2, B_3) and (A_3, B_2)).

We characterized Deadlock-freeness of an LTS by considering the *observed behaviour* of its states. In the following proposition, given a synchronized product $\mathcal{T}$ of two LTSs $\mathcal{T}_1$ and $\mathcal{T}_2$, we show how one can deduce the observed behaviour mapping $\lambda_{\mathcal{T}}$ from $\lambda_{\mathcal{T}_1}$ and $\lambda_{\mathcal{T}_2}$. This avoids the analysis of the paths of the synchronized product when such an analysis was already done locally in each involved LTS.

Proposition 2 (Observed behaviour mapping of an LTS synchronized product). Let $\mathcal{T} = \langle \Gamma, Act = Obs \cup UnObs, \rightarrow, I, F \rangle$ be the synchronized product of two LTSs, $\mathcal{T}_i = \langle \Gamma_i, Act_i = Obs_i \cup UnObs_i, \rightarrow_i, I_i, F_i \rangle, i = 1, 2$. Assume that $Act_1 \cap Act_2 = Obs_1 = Obs_2$. Then the observed behaviour mapping, named $\lambda_{\mathcal{T}}$, satisfies $\forall (s_1, s_2) \in \Gamma: \lambda_{\mathcal{T}}(s_1, s_2) = \lambda_{\mathcal{T}_1}(s_1) \cap \lambda_{\mathcal{T}_2}(s_2)$.

Proof. Let (s_1, s_2) be a state of $\mathcal{T}$.

Let us demonstrate that $\forall o \in Obs, o \in \lambda_{\mathcal{T}}(s_1, s_2) \Leftrightarrow o \in \lambda_{\mathcal{T}_1}(s_1) \cap \lambda_{\mathcal{T}_2}(s_2)$. Let $o \in \lambda_{\mathcal{T}}(s_1, s_2)$. Then there exists $\sigma \in UnObs^*$ s.t. $(s_1, s_2) \xrightarrow{\sigma o} (s_1', s_2')$. Let σ_1 and σ_2 be the projection of σ on $UnObs_1$ and $UnObs_2$, respectively. Knowing that $UnObs_1 \cap UnObs_2 = \emptyset$, we get that $(s_1, s_2) \xrightarrow{\sigma o} (s_1', s_2')$ means that in $\mathcal{T}_1$ and $\mathcal{T}_2$, $s_1 \xrightarrow{\sigma_1 o} s_1'$ and $s_2 \xrightarrow{\sigma_2 o} s_2'$ hold respectively. Thus, $o \in \lambda_{\mathcal{T}_1}(s_1) \cap \lambda_{\mathcal{T}_2}(s_2)$.

The synchronized product of the LTSs of Figure 1 contains 111 nodes and 264 edges and is too big to be presented here. It is Deadlock-free, like the different component LTSs.

3.2   Synchronization of SOGs

The above result allows to define the meta-state product $M = M_1 \times M_2$: a meta-state obtained by synchronizing two meta-states M_1 and M_2. Especially, the corresponding Deadlock attribute, $M.d$, can be computed by using the locally computed observed behaviours. Again, the meta-state product between n $(n > 2)$ meta-states can be easily deduced.

Definition 7 (Meta-state product)
Let $T_i = \langle \Gamma_i, Obs_i \cup UnObs_i, \rightarrow_i, I_i, F_i \rangle, i = 1, 2$ be two LTSs $T = T_1 \times T_2 = \langle \Gamma, Obs, \rightarrow, I, F \rangle$. Let $M_i = \langle S_i, d_i, f_i, \lambda_i \rangle$ be a meta-state of $SOG(T_i)$. The product meta-state $M = \langle S, d, f, \lambda \rangle = M_1 \times M_2$ is defined by:

- $S = S_1 \times S_2$,
- $d = true$ iff $\exists (s_1, s_2) \in \Gamma : (\lambda_T(s_1, s_2) = \emptyset)$,
- $f = true$ iff $f_1 = true$ and $f_2 = true$,
- $\lambda = \{\psi \subseteq Obs_i \cup UnObs_i\}$ s.t. $\psi \in \lambda$ iff $\exists \gamma \subseteq \Gamma : \lambda(\gamma) = \psi$.

Apart from dealing with meta-states instead of singular states, the definition of the synchronized product between two SOGs is identical to the synchronized product of two LTS (Definition 6).

Figure 4 shows the synchronized product of the SOGs of Figure 2. It contains 21 nodes and 24 edges (versus 111 nodes and 264 edges in the original synchronized LTS). The obtained SOG is not Deadlock-free (like each independent SOG).

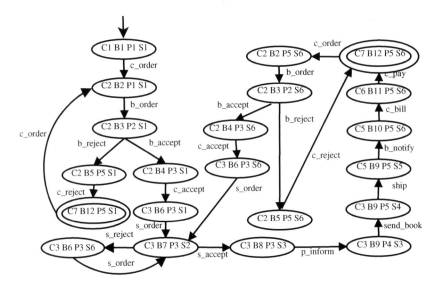

Fig. 4. Synchronized product between SOGs

Algorithm 1. Synchronized product of 2 SOGs

Require: $SOG(T_1, Obs_1)$ and $SOG(T_2, Obs_2)$
Ensure: $SOG(T_1, Obs_1) \times SOG(T_2, Obs_2)$
1: $Waiting \leftarrow \texttt{metastate}(I_1 \times I_2)$
2: **while** $Waiting \neq \emptyset$ **do**
3:    choose $M = M_1 \times M_2 \in Waiting$
4:    **for all** $a \in Act'_1 \cap Act'_2$ **do**
5:      **if** $M_1 \overset{a}{\longrightarrow}_1 M'_1 \wedge M_2 \overset{a}{\longrightarrow}_2 M'_2$ **then**
6:        $\texttt{metastate}(M'_1 \times M'_2)$
7:        $\texttt{arc}(M, a, M'_1 \times M'_2)$
8:      **end if**
9:    **end for**
10:   **for all** $a \in Act'_1 \setminus Act'_2$ **do**
11:     **if** $M_1 \overset{a}{\longrightarrow}_1 M'_1$ **then**
12:       $\texttt{metastate}(M'_1 \times M_2)$
13:       $\texttt{arc}(M, a, M'_1 \times M_2)$
14:     **end if**
15:   **end for**
16:   **for all** $a \in Act'_2 \setminus Act'_1$ **do**
17:     **if** $M_2 \overset{a}{\longrightarrow}_2 M'_2$ **then**
18:       $\texttt{metastate}(M_1 \times M'_2)$
19:       $\texttt{arc}(M, a, M_1 \times M'_2)$
20:     **end if**
21:   **end for**
22:   $Waiting \leftarrow Waiting \setminus \{M\}$
23: **end while**

The construction of the symbolic observation graph of a synchronized product of modules consists in first building the SOGs of the individual processes and then synchronizing them. Notice that the construction of the synchronized product of the SOGs aims mainly at establishing whether the underlying processes can collaborate safely (without being in a Deadlock). Checking the Deadlock-freeness of such a synchronized product is reduced to verifying that no (product) meta-state contains a Deadlock ($\forall M: M.d = false$) and that there exists a final (product) meta-state ($\exists M: M.f = true$).

Algorithm 1 implements the synchronized product of two symbolic observation graphs. This algorithm is very similar to the construction of the synchronized product of LTSs. Function $\texttt{metastate}$ ($M_1 \times M_2$) constructs the meta-state product $(M_1 \times M_2).S$. It assumes that the attributes of M_1 and M_2 are computed locally as well as the observed behaviours of their states. Then, it computes $M.d$ following Definition 7. In Subsection 3.3 an efficient way of computing the deadlock attribute of the product meta-state is discussed.

In the following, we establish the main result of this paper: given two LTSs, checking the Deadlock-freeness property on their synchronized product is equivalent to checking it on the synchronized product of the corresponding SOGs.

Proposition 3. *Let $\mathcal{T}_i = \langle \Gamma_i, Act_i = Obs_i \cup UnObs_i, \rightarrow_i, I_i \rangle$ ($i \in \{1, 2\}$) be two LTSs with $Act_1 \cap Act_2 \subseteq (Obs_1 \cap Obs_2)$. Then $SOG(\mathcal{T}_1 \times \mathcal{T}_2)$ and $SOG(\mathcal{T}_1) \times SOG(\mathcal{T}_2)$ are isomorphic.*

Proof. Follows from the construction.

Corollary 1. *Let $\mathcal{T}_i$ ($i \in \{1, 2\}$) be two LTSs, let $\mathcal{T}$ be their synchronized product, let $SOG(\mathcal{T}_i)$ be the SOGs of $\mathcal{T}_i$ and let $\mathcal{G}$ be their synchronized product. Then the following property holds: $\mathcal{T}$ is Deadlock-free $\Leftrightarrow$ $\mathcal{G}$ is Deadlock-free.*

Proof. Consequence of Proposition 1 and Proposition 3.

3.3   Checking Deadlock-Freeness on a SOGs Synchronized Product

According to the above results, the verification of Deadlock-freeness in a meta-state product is achieved by using the local observed behaviour mappings (i.e. $\lambda_{\mathcal{T}_1}$ and $\lambda_{\mathcal{T}_2}$). If we assume that Γ_1 and Γ_2 are the sets of states of the original LTSs $\mathcal{T}_1$ and $\mathcal{T}_2$ respectively, then the complexity of the Deadlock-freeness checking is polynomial with respect to the number of states in Γ_1 and Γ_2. However, in terms of efficiency, computing the value of these mappings for each state could reduce drastically the application of the SOG technique. In fact, the efficiency of this technique comes from the fact that it is suitable for symbolic implementation (based on set operations).

In the following, we propose two sufficient conditions for the existence of a Dead state within a meta-state product. Both conditions can be checked symbolically.

Proposition 4. *Let $SOG(\mathcal{T}_i)$, for $i = 1, 2$, be two SOGs corresponding to $\mathcal{T}_i = \langle \Gamma_i, Obs_i \cup UnObs_i, \rightarrow_i, I_i \rangle$. Let $M_i = \langle S_i, d_i \rangle$ be a meta-state of $SOG(\mathcal{T}_i)$ and let $M = \langle S, d \rangle = M_1 \times M_2$ be the product meta-state obtained by synchronizing M_1 and M_2. Then the following properties holds:*

1. *$M_1.d = true \vee M_2.d = true \Rightarrow M.d = true$*
2. *$Out(M_1) = \emptyset \vee Out(M_2) = \emptyset \Rightarrow M.d = true$*

Proof. Observe that both conditions imply $\exists (s_1, s_2) \in M_1.S \times M_2.S \colon \lambda_{\mathcal{T}_1}(s_1) \cap \lambda_{\mathcal{T}_2}(s_2) = \emptyset$.

In case the sufficient conditions of Proposition 4 are not satisfied, the following proposition establishes that the Deadlock-freeness of a meta-state product $M = M_1 \times M_2$ can be achieved by considering the projection of the λ mappings on the output states of M_1 and M_2 only instead of all states in $M_1.S$ and $M_2.S$, respectively. The number of output states (states enabling observed actions) is in general reduced with respect to the number of states of the system. This does not change the worst complexity of the observed behaviour mapping computation. However, in practice, it could significantly reduce the time and space consumption during the computation.

Proposition 5

Let $SOG(\mathcal{T}_i)$, be two SOGs corresponding to LTSs $= \langle \Gamma_i, Obs_i \cup UnObs_i, \rightarrow_i, I_i \rangle$, for $i = 1, 2$. Let $M_i = \langle S_i, d_i \rangle$ be a meta-state of $SOG(\mathcal{T}_i)$ and let $M = \langle S, d \rangle = M_1 \times M_2$ be the product meta-state obtained by synchronizing M_1 and M_2. When both conditions 1 and 2 of Proposition 4 are not satisfied, then the following property holds:

$$M.d = true \ \text{iff} \ \exists (s_1, s_2) \in Out(M_1.S) \times Out(M_2.S) \ s.t. \ \lambda_{\mathcal{T}_1}(s_1) \cap \lambda_{\mathcal{T}_2}(s_2) = \emptyset.$$

Proof. First, if $M.d = true$ then $\exists (s_1, s_2) \in M_1.S \times M_2.S \colon \lambda_{\mathcal{T}_1}(s_1) \cap \lambda_{\mathcal{T}_2}(s_2) = \emptyset$. Otherwise $M_i.d \neq true$ (for $i = 1, 2$) whence from s_1 (resp. s_2) one can reach an output state s'_1 (resp. s'_2) of M_1 (resp. M_2), with $\lambda_{\mathcal{T}_1}(s'_1) \subseteq \lambda_{\mathcal{T}_1}(s_1)$ (resp. $\lambda_{\mathcal{T}_2}(s'_2) \subseteq \lambda_{\mathcal{T}_2}(s_2)$). Then, $\lambda_{\mathcal{T}_1}(s'_1) \cap \lambda_{\mathcal{T}_2}(s'_2) = \emptyset$, which proves the proposition. ∎

To resume, the Deadlock attribute of a product meta-state can be deduced when one of the involved meta-states contains a Dead state. Otherwise, we only need to consider the observed behaviour of the output states.

4 Related Work

The importance of dealing with business processes on one hand and business process composition on the other hand is reflected in the literature by several publications.

In [17] the authors present various composition alternatives and their ability to preserve relaxed soundness [3]. The aim of this work was to analyze a list of significant composition techniques in terms of WF-nets and to prove that the composition of relaxed sound models is again relaxed sound. Hence, using these composition techniques does not preserve the deadlock-freeness property. In order to verify this property one has to explore the composed model, even though the component models are deadlock free. The approach we have presented in this paper allows verifying the deadlock-freeness property on the composition of abstract models (SOG).

In [5], the authors propose an approach for services retrieval based on behavioral specification. The idea consists in reducing the problem of service behavioral matching to a graph matching problem and then adapting existing algorithms for this purpose. The complexity of graph matchmaking algorithm used is $O(m^2 * n^2)$ in the best case and $O(m^n * n)$ in the worst case where m is the number of nodes of the request graph and n is the number of nodes of the advertised graph [5]. It is obvious that this approach is not suitable for workflow matching and composition when the number of advertised abstractions increases.

Another approach for workflow matchmaking was proposed in [10][11][12]. It assumes that two workflows match if they are equivalent. To reach this end, the author introduces the notion of communication graph *c-graph* and usability graph. If the *u-graph* of a workflow is isomorphic to the *c-graph* of another workflow, then the two workflows will be considered equivalent. However, the complexity of *c-graph* construction is exponential [10] in terms of the number of nodes. Moreover, it is well known that the subgraph isomorphism detection problem is NP-complete (see for example [16]). It is also obvious that this approach is not suitable for workflow matching when the number of advertised abstractions increases whereas the complexity of our matching algorithm is $O(m*n*l)$ where m and n are the number of meta-states of the corresponding abstractions to be matched and l is the number of the common cooperative transitions.

5   Conclusion

This paper addresses the problem of the abstraction and verification of inter-organizational processes. To preserve privacy of participating processes in an inter-organization process and to enhance verification, we have used the notion of symbolic observation graph to represent process abstractions. We have in addition shown how to build the symbolic observation graph of a composite (or inter-organizational) process and established whether processes can be composed (or can collaborate) safely by checking the deadlock freeness of the obtained symbolic observation graph. Our developed approach can be used for process advertisement, discovery and interconnection.

Several future works are envisaged. The first one would be to implement a tool for the abstraction and the Deadlock-freeness verification of inter-organizational processes. The extension of this work to checking $LTL \setminus X$ properties is direct since, by detecting divergent behaviours (Deadlocks) inside meta-state, the set of maximal paths is preserved. Moreover, we already started working on developping a graph-based registry for abstract process advertisement and discovery. We are going to extend process descriptions by ontology-based semantic descriptions. Our developed algorithms for service matching will be coupled with our presented algorithms to support semantic advertisement and discovery of processes for process composition at design time and for intra-enterprise use and for process cooperation to support inter-organizational processes.

References

1. Bryant, R.E.: Symbolic boolean manipulation with ordered binary-decision diagrams. ACM Computing Surveys 24(3), 293–318 (1992)
2. Bultan, T., Su, J., Fu, X.: Analyzing conversations of web services. IEEE Internet Computing 10(1), 18–25 (2006)
3. Dehnert, J., Rittgen, P.: Relaxed soundness of business processes. In: Dittrich, K.R., Geppert, A., Norrie, M.C. (eds.) CAiSE 2001. LNCS, vol. 2068, pp. 157–170. Springer, Heidelberg (2001)
4. Grefen, P., Aberer, K., Hoffner, Y., Ludwig, H.: Crossflow: Cross-organizational workflow management in dynamic virtual enterprises. International Journal of Computer Systems Science & Engineering 15(5), 277–290 (2000)
5. Grigori, D., Corrales, J.C., Bouzeghoub, M.: Behavioral matchmaking for service retrieval. In: ICWS 2006: Proceedings of the IEEE International Conference on Web Services, Washington, DC, USA, pp. 145–152. IEEE Computer Society Press, Los Alamitos (2006)

6. Haddad, S., Ilié, J.-M., Klai, K.: Design and evaluation of a symbolic and abstraction-based model checker. In: Wang, F. (ed.) ATVA 2004. LNCS, vol. 3299, pp. 196–210. Springer, Heidelberg (2004)
7. Klai, K., Petrucci, L.: Modular construction of the symbolic observation graph. In: Billington, J., Duan, Z., Koutny, M. (eds.) ACSD, pp. 88–97. IEEE, Los Alamitos (2008)
8. Klai, K., Poitrenaud, D.: MC-SOG: An LTL model checker based on symbolic observation graphs. In: van Hee, K.M., Valk, R. (eds.) PETRI NETS 2008. LNCS, vol. 5062, pp. 288–306. Springer, Heidelberg (2008)
9. Lohmann, N., Massuthe, P., Stahl, C., Weinberg, D.: Analyzing interacting ws-bpel processes using flexible model generation. Data Knowl. Eng. 64(1), 38–54 (2008)
10. Martens, A.: On Usability of Web Services. In: Calero, C., Daz, O., Piattini, M. (eds.) Proceedings of 1st Web Services Quality Workshop (WQW 2003), Rome, Italy (2003)
11. Martens, A.: Analyzing web service based business processes. In: Cerioli, M. (ed.) FASE 2005. LNCS, vol. 3442, pp. 19–33. Springer, Heidelberg (2005)
12. Martens, A.: Simulation and Equivalence between BPEL Process Models. In: Proceedings of the Design, Analysis, and Simulation of Distributed Systems Symposium (DASD 2005), Part of the 2005 Spring Simulation Multiconference (SpringSim 2005), San Diego, California (April 2005)
13. Martens, A., Simon, M., Achim, G., Karoline, F.: Analyzing compatibility of bpel processes. In: AICT-ICIW 2006: Proceedings of the Advanced Int'l Conference on Telecommunications and Int'l Conference on Internet and Web Applications and Services, Washington, DC, USA, p. 147. IEEE Computer Society Press, Los Alamitos (2006)
14. Massuthe, P., Wolf, K.: An Algorithm for Matching Nondeterministic Services with Operating Guidelines. Informatik-Berichte 202, Humboldt Universitat zu Berlin (2006)
15. Pankratius, V., Stucky, W.: A formal foundation for workflow composition, workflow view definition, and workflow normalization based on Petri nets. In: APCCM 2005: Proceedings of the 2nd Asia-Pacific conference on Conceptual modelling, Darlinghurst, Australia, Australia, pp. 79–88. Australian Computer Society, Inc. (2005)
16. Read, R., Corneil, D.: The Graph Isomorphism Disease. Graph Theory 1, 339–363 (1977)
17. Siegeris, J., Zimmermann, A.: Workflow model compositions preserving relaxed soundness. In: Dustdar, S., Fiadeiro, J.L., Sheth, A.P. (eds.) BPM 2006. LNCS, vol. 4102, pp. 177–192. Springer, Heidelberg (2006)
18. van der Aalst, W.M.P., Weske, M.: The P2P approach to interorganizational workflows. In: Dittrich, K.R., Geppert, A., Norrie, M.C. (eds.) CAiSE 2001. LNCS, vol. 2068, pp. 140–156. Springer, Heidelberg (2001)
19. van Dijk, A.: Contracting workflows and protocol patterns. In: van der Aalst, W.M.P., ter Hofstede, A.H.M., Weske, M. (eds.) BPM 2003. LNCS, vol. 2678, pp. 152–167. Springer, Heidelberg (2003)

Effect of Using Automated Auditing Tools on Detecting Compliance Failures in Unmanaged Processes

Yurdaer Doganata and Francisco Curbera

IBM T J Watson Research Center, 19 Skyline Drive, Hawthorne NY 10532
{yurdaer,curbera}@us.ibm.com

Abstract. The effect of using automated auditing tools to detect compliance failures in unmanaged business processes is investigated. In the absence of a process execution engine, compliance of an unmanaged business process is tracked by using an auditing tool developed based on business provenance technology or employing auditors. Since budget constraints limit employing auditors to evaluate all process instances, a methodology is devised to use both expert opinion on a limited set of process instances and the results produced by fallible automated audit machines on all process instances. An improvement factor is defined based on the average number of non-compliant process instances detected and it is shown that the improvement depends on the prevalence of non-compliance in the process as well as the sensitivity and the specificity of the audit machine.

Topics covered: BPM Governance and Compliance; Management Issues and Empirical Studies; Non-traditional BPM Scenarios.

1 Introduction

The operations of many businesses depend on business processes that rely heavily on human interactions, supported by collaboration software such as e-mail, calendar systems, and others. These processes are highly unstructured, often lack proper documentation and require human intervention as part of the process. The transitions between such unmanaged process activities are not always automated by software components; hence they cannot be fully controlled and monitored by utilizing process execution engines. In the absence of process automation software that can control and record resource and organizational access (who did what and when), compliance check is a costly and time consuming task performed manually by auditors.

Business provenance technology is proposed in [1] to increase the traceability of such unmanaged or partially managed processes. This technology provides for a generic data model, a middleware infrastructure to collect and correlate business events and a query interface to inspect which tasks are executed, when and what resources are involved. Information is selectively captured together with the context of in which it is used, and is then applied to detect compliance violations, in an interactive or automated way. Business provenance technology enables building automated

U. Dayal et al. (Eds.): BPM 2009, LNCS 5701, pp. 310–326, 2009.

auditing systems and tools to detect compliance failures continuously and reduce the cost of employing auditors significantly.

Automated continuous auditing systems provide for an almost cost-free auditing opportunity if the initial cost of building such a system is excluded. Such a system can run continuously and performs evaluation for all process instances without adding to the cost of auditing. While continuous audit systems eliminate or reduce the dependency on audit professionals, they are not infallible. The tools that are built to realize automated continuous auditing rely on information extraction from process events and information, including e-mail transactions between the people within the organizations. The extracted information about the processes may contain errors and due to these errors the decision on the compliance may be faulty. Moreover, the testing of a compliance condition may require a level of text analysis that is not yet available in automated systems. Hence, the automated systems can perform fast and extensive auditing of the internal control points at the cost of making mistakes. As a result, some compliance failures may be missed while some other cases that are compliant may be declared non-compliant.

There are many obvious reasons for organizations to worry about compliance in general. A business that has not taken adequate steps to achieve compliance may of course be subject to serious financial penalty as well as civil and penal consequences. Still, compliance has broader and practical impacts. On a more practical level, compliance ensures the quality of products and services and helps the organizations better control their operations and remain competitive. In short, the impact of non-compliance can be profound.

The cost of improving the status of compliance and reduce the risk of being non-compliant could run into millions of dollars for many organizations [2]. Auditing is a central component of compliance operations. Manual auditing involves the use of subject matter experts, but typically covers only a small set process instances because of time and cost constraints. Audits are performed in a quarterly or yearly basis, and cases are selected through statistical sampling. There is thus a trade of between the cost of sampling sufficient number of cases and the possibility of poor auditing which may cause missing opportunities for corrective action. While traditional audits are performed a few times a year, it is widely believed that compliance is an ongoing process that goes beyond testing and evaluating the internal controls of a sampled space. Thus many corporations focus on enhancing or implementing systems to ensure compliance on a continuous basis [3]. AMR Research survey reveals that the spending of companies on governance, risk management and compliance will increase 7.4% in 2008 and exceed $32B [4]. As a result, companies invest on implementing automated continuous audit systems [5] that would reduce the cost of compliance and would not be limited to selected instances of business processes due to budget constraints.

In this article, we introduce and measure the effectiveness of an automated continuous audit tool that is designed to detect compliance failures. We measure the effectiveness of the tool by its capacity to detect compliance failures during the execution of an unmanaged business process. This is accomplished by identifying a set internal control points and compare the number of non-compliance instances detected in the presence and in the absence of auditing tool. As a result of this comparison, we quantify how much the traditional auditing process performed by auditors under a budget constraint can be improved by employing auditing tool. Our approach is based on inferring the prevalence of non-compliance and the performance of the tool from a set of sample test results. We then use the inferred results to calculate the improvement as detailed in sections 5-7.

Next section briefly overviews the business provenance technology that we employed to implement the automated auditing tool. In Section 3, an e-mail based business process is described which is used to evaluate the effectiveness of the tool. This process is selected as a typical human centric process where an execution engine is not used to control and manage the process. Hence, traditionally auditing is done by employing subject matter experts. A set of internal key control points are defined to determine the status of compliance and as a basis for our comparative study. In Section 4, a mathematical model is presented for faulty auditing tools for which the statistical performance measures are inferred in section 5. A methodology is proposed to measure the effectiveness of the automated machines in Section 6 and the numerical results are presented in Section 7. We conclude in Section 8.

2 Related Work

Key control points within the business process help identifying risks throughout the organization before they cause integrity lapses. A risk classification associated with BPM cycle provided in [6] and mentions CobIT framework as a set of audit-oriented guidelines create control objectives aligned with the BPM life cycle concept. The key control points that we used to measure the effectiveness of automated audit tools are driven from the rules and regulations in business documents and specifications written in natural language.

A formal representation of these key control points are not within the scope of this paper, but we briefly mentioned the methodology we adopted below. Regardless, there has been a number of works in the literature focusing on formal representation of internal control points by using various rule languages. A compliance metamodel for formally capturing key control points and managing them in systematic lifecycle is presented in [7]. A formal system for business contract representation with reasoning about violations of obligations in the contacts is proposed in [10]. Various aspects of Business Contract Language (BCL) [9] are evaluated by using a logic-based formalism called Formal Contract Language (FCL) in [11] and the need to ensure compatibility between business processes and business contracts is addressed. In [13], a rule language, RuleML, is proposed to express business rules explicitly as a better alternative to other XML languages such as BCL and XrML [15]. In business provenance technology [1] that is employed in this paper, key control points are directly expressed in terms of the business provenance entities which are the nodes and edges of the provenance graph formed during the execution of business operations. While our approach does not require logical analysis of the business rules and is implemented by employing a simple SQL/XPATH based query interface to the provenance store, the rule developers are expected to aware of the runtime operational environment. Hence, it lacks the reusability and the flexibility of formal representation systems.

Similarly [12] proposes a framework to ensure semantic correctness of the process when ad-hoc process instance deviations occur or when modeling process templates. The domain knowledge is integrated into process management system as process constraints and each constraint is expressed in terms of source and target tasks, their orders and user defined parameters. The expressiveness of the presented constraints is, however, limited to task level, data is not included. We use the attributes of all business provenance data (data, task, resource, process) and their relation to express key control points.

A number of works [8], [14] advocate addressing control objectives early in design time and propose supporting mechanism for business process designers. A method is proposed in [8] to help the process designers to measure the compliance degree of a given process model against the set of objection. A language is introduced in [14] to express temporal rules about the obligations and permissions In order to help the designers at the process modeling time to validate and verify business contracts. In measuring the effectiveness of compliance tools, we do not assume that the control objectives were known at the time of process design. But the audit tools are designed based on the control objectives over existing business operations.

The method presented in this paper to measure the effectiveness of automated audit tools does not depend on a particular process tracking technology or control point representation. The methodology could be employed to cases that use other formal representations of control points and process tracking technologies.

3 Automated Auditing Tool

An automated auditing tool is a software system that captures information relevant to the internal control points of a business process, puts them into context and computes the compliance status for each control point. Auditing tools rely on correlating the data extracted from the underlying IT system to the relevant aspects of business control points effectively. Hence, relating the business goals to IT level data constitutes the core of this technology as described in [1]. Figure 1 outlines the step of building such a system which starts with converting business rules and regulations into compliance goals (Step ①). Compliance goals are identified by examining the business rules and deciding what action steps are needed. In other words, from the business rules expressed in the language of business people, compliance goals are identified (Step ②). This lays the ground work for setting up IT rules for compliance. Once the compliance goals are identified; tasks, activities, resources, artifacts and their relations that are relevant to the identified goal are determined and mapped onto a data model (Step ③). Recording probes collect business artifacts from the underlying information system and maps them onto provenance data (Step ④ and ⑥). A "provenance graph" is then formed with the data objects constituting the nodes and the relations among the data objects the edges. The data objects are correlated by using the compliance goals and the underlying data model (Step ⑤). Business control points are then expressed in terms of data entities extracted from the process execution trace as graph patterns (Step ⑦). Hence, control points provide a bridge between various components of the business operations and the actual data that could be consumed the IT system. A business control point that can be expressed in terms of the

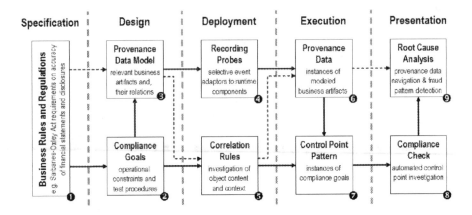

Fig. 1. Steps for compliance checking

data produced and consumed by the IT system can be computed to check compliance in step ⑧.   Root cause analysis of compliance failures can be done by querying the provenance graph in step ⑨.

4   Sample e-Mail Based Unmanaged Business Process

In order to measure the effectiveness an automated auditing tool in detecting compliance failures over an unmanaged business process, the following scenario is presented.   An e-hosting company manages the customer machines over the internet protected by a firewall and responsible for securing the information assets against unauthorized entry. The company security policy dictates that a firewall manager defines the firewall security controls and ensures on an ongoing basis that firewall policies are implemented using an auditable process. The process is called Firewall Rule Revalidation process and it involves in creating firewall rulesets and communicating these rulesets to the customer and receives approval. The objectives of the process are to ensure both the e-hosting account representatives and the customers understand what rules exist in the customer environment and ensure customer is aware of existing deviations from best practices defined by the e-hosting security policy. If such a process is not implemented, the customer may be at risk due to no longer needed protocols being available for transit traffic, or not being made aware of what protocols are in place and required for support of their environment. The e-hosting company may be held liable for insecure activities, if the customers is not informed of and signs off on the risk involved.

Figure 2 depicts the Firewall Rule Revalidation process where there are three actors of the process, information security advisor, account team and customer.

The responsibilities of these actors are defined as below:

Information Security Advisor (ISA): Prepares firewall rulesets according to the best security practices, modifies them as needed, sends them to the account team and copy to e-mail archive database.

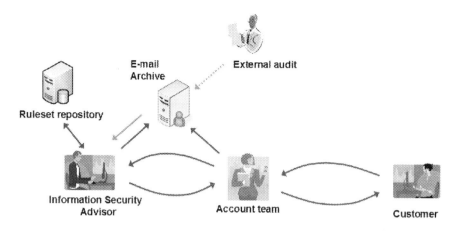

Fig. 2. Firewall Rule Revalidation Process flow

Account Team Member: Receives firewall rulesets from ISA and sends them to customer, records customer response into the e-mail archive.

Customer: Receives firewall rulesets from the account team, reviews them and replies with acceptance or change requests.

Firewall rule revalidation is done once a year. Before the revalidation cycle completes, ISA asks for the firewall rules from the network administrator and checks if the rules are consistent with customer requests and security policies and make modifications if necessary. Once the new rulesets is created, ISA attaches the ruleset to an e-mail and send the e-mail to the account team. Once the ruleset is ready for customer review the process starts. ISA attaches the e-mail to the ruleset and sends to the account team, reminding that it is time for yearly revalidation and ask account team.

4.1 Key-Control Points

In order to assure proper revalidation every year, several internal control points are defined. The compliance of key control points assures that firewall rule revalidation is completed successfully. The description of these internal key control points (KCPs) are described below:

KCP1: A process record exists with a copy of the email from ISA to account team with firewall ruleset is attached.
KCP2: The new ruleset is prepared before the revalidation period ends.
KCP3: A revalidation email must be sent by the account team to the customer within 5 days after email from ISA was received.
KCP4: An acceptance email response from customer must be within 10 days after the first email sent by the account team to the customer.
KCP5: The revalidation process completes within 30 days of being started
KCP6: The revalidation process was completed within the review interval after the prior revalidation completion.

Automated compliance checking process is based on analyzing all the emails in the e-mail archive and classifying them based who sends the e-mail and for what purpose. A data model is built that capture the relevant aspect of the process and according to the description given in [1]. Relationships among data items are extracted by using key control point definitions. As an example, in order to evaluate the status of KCPs, all the e-mail sent by ISAs, Account Team members and customers are examined. Text analysis is used to examine the unstructured parts of the e-mails such as body and subject; the e-mail addresses are extracted from "to" and "from" fields. Based on the extracted relations, each e-mail is scored and labeled as either "from ISA to Account Team" or "from Account Team to Customer" or "from Customer to Account Team". The relations between the e-mails, their context, receivers and senders are established. The dates of the labeled e-mail are extracted to check the compliance status of each control point.

5 Statistical Modeling Results

The problem of using automated audit machines to determine the compliance failures is equivalent to determining the prevalence of a medical condition through screening the population by using a medical diagnostic test which is not a gold standard [17]. The public health services in many cases use tests which are not 100 percent accurate to estimate the prevalence of the disease. Similarly, the prevalence of non-conformance in a business process can also be estimated by using automated auditing systems which are fallible in making classification for compliance. This is a binary classification problem where the instances of business processes are grouped into two on the basis that they satisfy certain key control points as compliant or not. A practical approach to this classification problem needs to consider two parameters, namely, quality of the classification decisions and cost. For the purpose of the work presented in this paper, we will assume that audits performed by experts always result in a correct classification decision. There is of course a human error factor that we are not factoring in the analysis presented here. On the other hand, there is a considerable cost associated with manual audits, which limits the number of process instances that can be audited this way. The cost of performing an evaluation of compliance by using an automated audit machine can be assumed negligible, allowing in many cases for full coverage of process instances. The results of automated classification, on the other hand are fallible.

In modeling this problem, we define the prevalence as the total number of non-compliant process instances in the population of all process instances. Mathematically, it is the probability that a case is marked as not-compliant by an audit expert, Pr $(I = 1)$. In medical field, this corresponds to the *prevalence, p,* of the disease. Prevalence cannot always be measured by using expert opinion because the cost of employing experts may be prohibitive. However, it is possible to draw inference about the prevalence, p, of non-conformance in a set of execution traces of process instances by using fallible automated auditing tool, if a measure of the auditing system's performance in identifying non-compliant instances is known. The performance of such an auditing tool is measured by its *sensitivity* and *specificity*. Sensitivity measures the proportions of actual positives (that is non-compliant cases) which are correctly identified, while specificity measures the proportions of negatives (compliant cases) which are correctly identified. We will refer the probability that a randomly selected instance is actually compliant as $Pr\ (I = 0)$, the probability that a fallible auditing tool labels an instance compliant as $Pr\ (F = 0)$ and non-compliant as $Pr\ (F = 1)$. Hence,

$$\eta,\ \text{Sensitivity: } TP/(TP + FN) = Pr\ (\ F=1/I=1) \tag{1}$$

$$\theta,\ \text{Specificity: } TN/(TN+FP) = Pr\ (\ F=0/I=0) \tag{2}$$

where TP is the number of non-compliant instances labeled as non-compliant, FN is the number of non-compliant instances labeled as compliant, TN is the number of compliant instances labeled as compliant, FP is the number of compliant instances labeled as non-compliant. The following joint probabilities of classification $p_{if} = P(I=i, F=f)$, where $i, f = 0, 1$ can be verified easily

$$p_{00} = P(I=0, F=0) = \theta.(1-p) \tag{3}$$
$$p_{10} = P(I=1, F=0) = (1-\eta).p \tag{4}$$

$$p_{01} = P(I=0, F=1) = (1-\theta).(1-p) \tag{5}$$
$$p_{11} = P(I=1, F=1) = \eta.p \tag{6}$$

Here p_{if} is the probability that a case is classified as f by an auditing tool when the infallible classifier, i. e audit expert, determines the case as i where $f, i \in \{0,1\}$.

In this section we will use Bayesian approach to estimate the distributions of η, θ and p based on the test results of the tool on a small set of process instances. Then, we will approximate p_{if} from equations **(3)** - **(6)** and compute the effectives of the tool by employing these estimations in sections 6 and 7. Following the Bayesian approach in the presence of misclassification, it is a common practice to assume that prior information is in the form of a beta density for prevalence, specificity and sensitivity [17]. The reason for selecting Beta distribution is that it is a flexible family of distribution and a wide verity of density shapes can be derived by changing the associated parameters of the beta distribution [21]. It is also a conjugate prior distribution for the binomial likelihood which simplifies the derivation of the posterior distribution significantly. The probability density function of a beta distribution with parameters (α, β) is given by

$$f(\varphi) = \begin{cases} \dfrac{1}{B(\alpha-1)}\varphi^{\alpha-1}(1-\varphi)^{\beta-1}, & 0 \leq \varphi \leq 1, \alpha, \beta \succ 0, \quad \text{and} \\ 0, & \text{otherwise} \end{cases} \tag{7}$$

We will hence assume that the prior information on p, η and θ is expressed through independent beta distributions as $Beta(\alpha, \beta)$, $Beta(\alpha_1, \beta_1)$ and $Beta(\alpha_2, \beta_2)$ respectively. Our purpose is to infer posterior distributions of p, η and θ after observing the compliance evaluation results of the auditing tool. Let the auditing tool with sensitivity η and specificity θ evaluates the compliance status of a key control point for N process instances, for which the truth values (actual compliance status) are known. As a result, let $n_{\bullet 1} = n_{11} + n_{01}$ instances are marked as non-compliant *(F=1)* and $n_{\bullet 0} = n_{10} + n_{00}$ instances are marked as compliant *(F=0)* where n_{11}, n_{10}, n_{01} and n_{00} are the number true positives, false negatives, false positives and true negatives respectively as shown in Table 1.

In the presence of the observed data, the posterior distributions of p, η and θ are still independent Beta distributions [21]. This can be shown directly by using the Bayesian theorem that the posterior joint distribution is the product of the likelihood function of the observed data *(binomial)* and the prior distribution *(beta)*. Hence, the joint posterior distributions of p, η and θ are found as

$$p \sim Beta(\alpha + N_1, \beta + N_0) \tag{8}$$

$$\eta \sim Beta(\alpha_1 + n_{11}, \beta_1 + n_{10}) \tag{9}$$

$$\theta \sim Beta(\alpha_2 + n_{00}, \beta_2 + n_{01}) \tag{10}$$

Equations (**9**) and (**10**) are the posterior distributions of the sensitivity and the specificity of the auditing tool that produced n_{11} true positives, n_{10} false negatives, n_{11} true positives and n_{01} false positives in a business environment where the prevalence of non-compliance is p. From these observed test results, inference about the marginal distributions for sensitivity and specificity is possible by using Gibbs sampler algorithm [23] as will be shown next.

6   Inference of Marginal Densities p, η, θ.

Inference about prevalence $p \sim Beta((\alpha, \beta)$, sensitivity $\eta \sim Beta(\alpha_1, \beta_1)$ and specificity $\theta \sim Beta(\alpha_2, \beta_2)$ can be drawn by running a test using the auditing tool and observing true positives and false negatives as depicted in Table 1. The technique is well known in the literature as Gibbs sampler algorithm [18]-[20], [23]. Gibbs sampler is an iterative Markov-chain Monte Carlo technique developed to approximate

Table 1. Test results of N sample process instances

		Truth (Audit expert)		
		+	−	
Test (Automated audit)	+	n_{11}	n_{01}	$n_{.1}$
	−	n_{10}	n_{00}	$n_{.0}$
		N_1	N_0	N

intractable posterior distributions. The algorithm uses the observed data to compute the posterior distributions of prevalence, specificity and sensitivity by applying Bayes' theorem and conversely computes the distributions of the observed data by using the prior distributions of prevalence, sensitivity and specificity as described in [23]. Gibbs sampler derives posterior probability distributions that best fit given prior distributions $Beta(\alpha,\beta)$, $Beta(\alpha_1,\beta_1)$ and $Beta(\alpha_2,\beta_2)$ and observed data, n_{11}, n_{10}, n_{11} and n_{01}. As described in [17], arbitrary starting values can be chosen for each parameter. If no prior knowledge or data is available for the initial distributions, α and β parameters are selected as 1 which corresponds to uniform distribution. Gibbs sampler converges to the true values of the posterior distributions after running tens of thousands of iterations.

Table 2. Input values for the prevalence calculator given in [22]

Input values for the prevalence calculator	
Test Results:	
Number of samples tested	N
Number of samples positive:	$n_{11} + n_{01}$
Alpha and Beta parameters for prior distributions:	
Prior prevalence, alpha	N_1
Prior prevalence, beta	$N - N_1$
Prior sensitivity, alpha	n_{11}
Prior sensitivity, beta	n_{10}
Prior specificity, alpha	n_{00}
Prior specificity, beta =	n_{01}
Simulation details:	
Number of iterations:	50K+
Number to discard: 5	5K+
Starting Values	
Number of true positives	n_{11}
Number of false negatives	n_{10}

Reference [22] provides for an on-line calculator to estimate the true prevalence based on testing of individual samples using a test with imperfect sensitivity and/ or specificity. The input values required for the calculator are listed in Table 2. These include the number of samples (process instances) tested, the number of samples labeled positive (non-compliant), α and β parameters for prior prevalence, sensitivity and specificity distributions, number of iterations to be simulated in the Gibbs sampler, number of iterations to be discarded to allow convergence of the model and initial number of true positives n_{11} and false negatives n_{10}. Table 2 is generated by using the notation given in Table 1. The initial α and β values for prior distributions are selected by using the fact that $\alpha /(\alpha + \beta)$ is the mean value of a beta distribution and by using the observed data given in Table 1. As an example, the mean value of the prior prevalence can be approximated as $N_1 /(N_1+N_0)$. Hence, the beta value for prior prevalence is approximated as N_0 and the alpha value is approximated as N_1.

7 Numerical Results

We run our auditing tool over 135 instances of the sample e-mail based firewall rule revalidation process to verify the compliance of the six key control points defined in section 3. 1. By using the prevalence calculator described above and the number of observed true positives, true negatives, false positives and false negatives for each key control point, we inferred the distributions for p, η and θ. The test results of the auditing tool and the associated mean values for p, η and θ are displayed for each key control point in Table 3.

Table 3. Average prevalence, sensitivity and specificity values for the auditing tool inferred for the six key control points

	KCP1	KCP2	KCP3	KCP4	KCP5	KCP6
n_{11} (TP)	7	64	55	41	42	40
n_{01} (FN)	1	5	5	12	5	9
n_{10} (FP)	8	20	31	41	18	6
n_{00} (TN)	119	46	44	41	70	80
$E(p)$	0.074	0.502	0.460	0.399	0.366	0.362
$E(\eta)$	0.826	0.924	0.847	0.759	0.790	0.793
$E(\theta)$	0.934	0.892	0.555	0.496	0.758	0.910

Table 3 implies that the performance of the auditing tool varies for each key control point. This is expected since the data used to compute the compliance of each control point is different. As a result, the effectiveness of the tool also varies for each control point. The inferred prevalence of non-compliance for different key control point shows that the rate of non-compliance is highest for KCP2 ($E(p)=0.502$) and lowest for KCP1 ($E(p) = 0.074$). This information can be used to identify the problematic points in the process.

Once we measure the performance of the auditing tool with its sensitivity and specificity values, we would like to understand how effectively we can use the tool to increase the rate of detecting non-compliant processes. In the next section, we will propose a method to measure the effectiveness of the automated auditing tool with given sensitivity and specificity in an environment where the prevalence of non-compliance is p.

8 Measuring the Effectiveness of Auditing Tool

Given a fallible auditing tool with sensitivity and specificity (η, θ), and given a fixed budget to fund the use of audit experts, we would like to find out how much we can improve the detection of non-compliant process instances. As discussed before, poor auditing may cause missing opportunities for corrective action. Hence, we would like to maximize the number of non-compliant cases detected as a result of auditing. On one hand we have a budget constraint which limits the number of cases we can audit by using an expert. On the other hand, we have a fallible automated audit machine which can be used to evaluate every process instance without incurring extra cost. The goal is to device a methodology for enabling to detect the largest number of non-compliant instances possible under these constraints. One possible methodology is to evaluate all process instances by using the automated audit machine and ask experts randomly re-evaluate M_1 cases among the ones marked as non-compliant (Region N) and M_2 among the ones marked as compliant (Region C) by the automated audit machine. This way the sample space that the experts operate is reduced. We assume that the budget permits the expert evaluation of only $M = M_1 + M_2$ cases. The effectiveness of the proposed methodology can be measured by comparing the expected number of non-compliant process instances detected. If the number is higher than what experts would have determined under budget constraint without using the methodology, then we can conclude that the methodology improves the auditing process in general.

The probability that a randomly selected process instance is labeled non-compliant, $P(F=1)$, by the auditing tool is ($p_{11} + p_{01}$). Given the condition that the auditors work only on instances labeled as non-compliant by the tool (Region N), probability that the auditors detect a non-compliant case is $Pr(I=1/F=1) = Pr(I=1, F=1)/P(F=1) = p_{11}/(p_{11} + p_{01})$. Similarly, the probability that an auditors detects a non-compliant cases among the ones labeled as compliant by the tool (Region C) is $Pr(I=1/F=0) = p_{10}/(p_{00} + p_{10})$. Hence, the average number of non-compliant cases detected by using this method can then be found as below where the function W is called the "worth" of this method.

$$W = M_1 \frac{p_{11}}{p_{11} + p_{01}} + M_2 . \frac{p_{10}}{p_{10} + p_{00}} \tag{11}$$

The worth function is maximized by making the experts work either in the region labeled as compliant (Region C) or as non-compliant (Region N) depending on the values of $p_{11}/(p_{11} + p_{01})$ and $p_{01}/(p_{00} + p_{01})$ provided that the budget constraint M is less than the size of both regions. This is a reasonable assumption since the size of process instances in both regions are usually much larger than M. Hence,

$$\max\{W\} = \begin{cases} M\dfrac{p_{11}}{p_{11}+p_{01}} & \dfrac{p_{01}}{p_{01}+p_{00}} \leq \dfrac{p_{11}}{p_{11}+p_{01}} \\[3mm] M\dfrac{p_{10}}{p_{10}+p_{00}} & \dfrac{p_{10}}{p_{10}+p_{00}} \succ \dfrac{p_{11}}{p_{11}+p_{01}} \end{cases} \tag{12}$$

In the absence of auditing tool, we would only rely on the efforts of the audit experts. The average worth of this practice would then be the product of M and the prevalence of non-compliance, p. Let W_0 be the worth of using only experts as auditors, the expected worth is then

$$W_0 = Mp \tag{13}$$

Potential improvement of using auditing tool can then be measured by the ratio of the worth functions $\max\{W\}$ and W_0. From **(12)** and **(13)** ,the improvement function I is found as:

$$I = \begin{cases} \dfrac{p_{11}}{(p_{11}+p_{01})p} & \dfrac{p_{01}}{p_{01}+p_{00}} \leq \dfrac{p_{11}}{p_{11}+p_{01}} \\[3mm] \dfrac{p_{10}}{(p_{10}+p_{00})p} & \dfrac{p_{10}}{p_{10}+p_{00}} \succ \dfrac{p_{11}}{p_{11}+p_{01}} \end{cases}, \tag{14}$$

9   Numerical Results for Improvement

In order to simplify the calculations, we will approximate the variables p_{11}, p_{01}, p_{10}, p_{00} with their mean values by using equations **(3)-(6)** and the fact that prevalence, sensitivity and specificity are independent beta distributions as Beta(α, β), Beta(α_1, β_1) and Beta(α_2, β_2) respectively as follows:

$$p_{00} \sim E(\theta.(1-p)) = \frac{\alpha_2\beta}{(\alpha_2+\beta_2)(\alpha+\beta)}, \tag{15}$$

$$p_{10} \sim E((1-\eta)p) = \frac{\alpha\beta_1}{(\alpha_1+\beta_1)(\alpha+\beta)}, \tag{16}$$

$$p_{01} \sim E((1-\theta).(1-p)) = \frac{\beta_2\beta}{(\alpha_2+\beta_2)(\alpha+\beta)}, \tag{17}$$

$$p_{11} \sim E(\eta.p) = \frac{\alpha_1\alpha}{(\alpha_1+\beta_1)(\alpha+\beta)}. \tag{18}$$

By using the expected sensitivity and specificity values displayed in Table 3 and employing equation **(15)-(18)**, the percentage improvement $(I-1)\times100$ of using auditing tool for each KCPs is calculated and the results are displayed in Table 4.

Table 4. Improvement obtained by using automated auditing tools for key control points

	KCP1	KCP2	KCP3	KCP4	KCP5	KCP6
$E(p)$	0.074	0.502	0.460	0.399	0.366	0.362
$E(\eta)$	0.826	0.924	0.847	0.759	0.790	0.793
$E(\theta)$	0.934	0.892	0.555	0.496	0.758	0.910
I	6.75	1.49	1.34	1.25	1.79	2.30
%	575	49	34	25	79	130

From equation (14) we can conclude that the improvement depends on the prevalence and the probability that a randomly selected process instance is found non-compliant by the expert within each particular region. Depending on the region (labeled compliant or non-compliant) experts work, this probability is either p_{11} / $(p_{11}+p_{01})$ or $p_{10}/(p_{10}+p_{00})$. It is clear from the equation that there is always going to be improvement as long as the prevalence is less than the probability of detecting a non-compliant process instance by using the tool. This is expected since the rate of detecting non-compliant instances by employing only auditors cannot be greater than p.

In order to understand the effect of sensitivity and specificity on the improvement as a function of prevalence, we approximate the improvement by using equation (15)-(18) as follows:

$$I = \frac{1}{p(1-\psi)+\psi},$$ (19)

where

$$\psi = \begin{cases} \dfrac{1-\theta}{\eta} & if \quad \dfrac{p_{01}}{p_{01}+p_{00}} \leq \dfrac{p_{11}}{p_{11}+p_{01}} \\ \dfrac{\theta}{1-\eta} & if \quad \dfrac{p_{01}}{p_{01}+p_{00}} > \dfrac{p_{11}}{p_{11}+p_{01}} \end{cases}.$$ (20)

In Figure 3, the percentage improvement is plotted as a function of ψ for various prevalence values changing between 0.1 and 0.9. The performance of the auditing tool for each key control point is also mapped on the same figure. As seen in the figure, the improvement is significantly greater in case of KCP1. On the other hand, using the tools does not give the same improvement for KCP4. In general, the improvement percentage is significantly higher when both the prevalence and ψ are small and it converges to zero as ψ converges to 1. In order to explain this behavior, without lack of generality, let's focus on the case where experts are asked to examine only process instances labeled as non-compliant (Region N) by the automated machine. Hence, equation (20) becomes $\psi = (1-\theta)/\eta$ and indicates that as the specificity and the sensitivity increases, the value of ψ decreases. In effect, the improvement percentage increases. This is expected since as the sensitivity of an audit machine improves, the likelihood of detecting non-compliant process instances by using the tool improves as

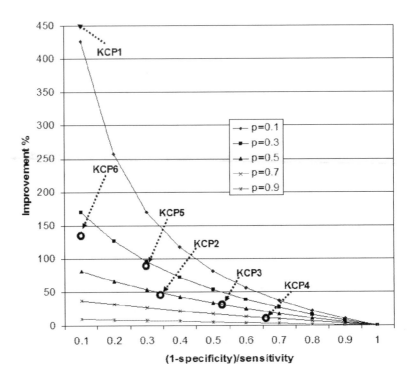

Fig. 3. Percentage of improvement as a function of $(1-\theta)/\eta$

well. This is why the improvement percentage is higher for smaller ψ values. On the other hand, as ψ approaches to 1, i. e., the sum of the specificity and the sensitivity approaches to one, improvement disappears. The reason for this is that in this case the likelihood of detecting non-compliant process instances approaches to, p, the prevalence. This can be explained as follows: When *sensitivity* is equal to (1 – *specificity*), the likelihood of having false positives becomes equal to likelihood of having true positives. In other words, the following holds

$$\eta = \frac{TP}{TP + FN} = 1 - \theta = 1 - \frac{TN}{TN + FP} = \frac{FP}{TN + FP}. \tag{21}$$

Equation (21) implies that

$$\frac{TP}{FP} = \frac{TP + FN}{TN + FP}, \tag{22}$$

where *TP, TN, FP* and *FN* are the number of true positive, true negative, false positive and false negative observations respectively. Further manipulation of equation **(22)** yields:

$$\frac{TP}{TP+FP} = \frac{TN+FP}{TN+FP+TP+FN} = p.$$ (24)

Here, TP, TN, FP and FN stand for the numbers of true positives, true negatives, false positives and false negatives. Note that this is also equal to the likelihood of detecting non-compliant instances in region N:

$$\frac{p_{11}}{(p_{11}+p_{01})} \approx \frac{TP}{TP+FP} = p \quad \rightarrow \quad I = 1$$ (25)

Equation (25) shows that working in region N does not give any advantage since the detecting a non-compliant process instances in this region is equivalent to the prevalence. The same argument holds for the other region without lack of generality. This means that labeling process instances with the auditing tool does not improve the rate of detecting non-compliant instances if the *sensitivity* of the automated machine is equal to 1 – *specificity*.

10 Conclusion

The level of compliance of a process, that is, the prevalence of non-compliant instances, can typically be reduced through automation by introducing a business processes management platform and other support middleware such as a content management system. A complementary approach is to increase the levels of compliance monitoring. Processes with low levels of automation, which are essentially unmanaged processes, must rely in an efficient auditing procedure as the only way to prevent systemic non-compliance. Automated auditing tools can be used to complement manual auditing by subject matter experts and expand the amount of process. In this article, we provide a methodology to estimate the effectiveness of these tools. We showed that the effectiveness depends on both the prevalence of non-compliant cases as well as the performance of the tool. The approach is expected to help businesses make smarter decision on employing subject matter experts and utilize automated audit tools.

Our future work will focus on optimizing the use of auditors by taking into account of all operational control points and their correlation.

References

1. Curbera, F., Doganata, Y., Martens, A., Mukhi, M., Slominski, A.: Business Provenance - A Technology to Increase Traceability of End-to-End Operations. In: OTM Conferences vol (1), pp. 100–119 (2008)
2. Greengard, S.: Compliance Software's Bonus Benefits. Business Finance Magazine (February 2004)
3. Gartner.: Simplifying Compliance: Best Practices and Technology, French Caldwell, (Business Process Management Summit (June 6, 2005)

4. Hagerty, J., Hackbush, J., Gaughan, D., Jacaobson, S.: The Governance, Risk Management, and Compliance Spending Report, 2008-2009, AMR Research Report, March 25 (2008)

5. Corfield, B.: Managing the cost of compliance, http://justin-taylor.net/webdocs/tip_of_the_iceberg.pdf

6. Zur Muehlen, M., Ho, D.T.: Risk Management in the BPM Lifecycle. In: Bussler, C.J., Haller, A. (eds.) BPM 2005. LNCS, vol. 3812, pp. 454–466. Springer, Heidelberg (2006)

7. Christopher, G., Müller, S., Pfitzmann, B.: From Regulatory Policies to Event Monitoring Rules: Towards Model-Driven Compliance Automation. IBM Research Report RZ 3662, IBM Zurich Research Laboratory (2006)

8. Lu, R., Sadiq, S., Governatori, G.: Compliance aware business process design. In: ter Hofstede, A.H.M., Benatallah, B., Paik, H.-Y. (eds.) BPM Workshops 2007. LNCS, vol. 4928, pp. 120–131. Springer, Heidelberg (2008)

9. Milosevic, Z., Gibson, S., Linington, J.C., Kulkarni, S.: On Design and implementation of a contract monitoring facility. In: Benatallah, B. (ed.) First IEEE International Workshop on Electronic Contracts, pp. 62–70. IEEE Press, Los Alamitos (2004)

10. Governatori, G., Milosevic, Z.: A Formal Analysis of a Business Contract Language. International Journal of Cooperative Information Systems 15(4), 659–685 (2006)

11. Governatori, G., Milosevic, Z., Sadiq, S.: Compliance checking between business processes and business contracts. In: Proceedings of the 10th IEEE Conference on Enterprise Distributed Object Computing (2006)

12. Ly, L.T., Rinderle, S., Dadam, P.: Integration and verification of semantic constraints in adaptive process management systems. Data and Knowledge Engineering 64(1), 3–23 (2008)

13. Governatori, G.: Representing Business Contracts in RuleML. International Journal of Cooperative Information Systems 14(2–3), 181–216 (2005)

14. Goedertier, S., Vanthienen, J.: Designing compliant business processes with obligations and permissions. In: Eder, J., Dustdar, S. (eds.) BPM Workshops 2006. LNCS, vol. 4103, pp. 5–14. Springer, Heidelberg (2006)

15. Lee, J.K., Sohn, M.M.: The eXtensible Rule Markup Language. Communications of ACM 46(5), 59–64 (2003)

16. Egizi, C.: High cost of compliance, http://www.cioupdate.com/career/article.php/3489431/The-High-Cost-of-Compliance.htm

17. Joseph, L., Gyorkos, T.W., Coupal, L.: Bayesian estimation of disease prevalence and the parameters of diagnostic tests in the absence of a gold standard. Am. J. Epidemiol (1995)

18. Gelfand, A.E., Smith, A.F.M.: Sampling-based approaches to calculating marginal densities. Journal American Statistics Assoc. 85, 348–409 (1990)

19. Gelfand, A.E., Hills, S.E., Racine-Poon, A., et al.: Illustration of Bayesian Inference in normal data using Gibbs sampling. Journal of American Statistics Assoc. 85, 972–985 (1990)

20. Tanner, M.A.: Tools for statistical inference. Springer, New York (1991)

21. Katsis, A.: Sample size determination of binomial data with the presence of misclassification. Metrika 63, 323–329 (2005)

22. Pooled Prevalence Calculator, http://www.ausvet.com.au/pprev/

23. Geman, S., Geman, D.: Stochastic Relaxation, Gibbs Distributions, and the Bayesian Restoration of Images. IEEE Transactions on Pattern Analysis and Machine Intelligence 6, 721–741 (1984)

Divide-and-Conquer Strategies
for Process Mining

Josep Carmona[1], Jordi Cortadella[1], and Michael Kishinevsky[2]

[1] Universitat Politècnica de Catalunya, Spain
[2] Intel Corporation, USA

Abstract. The goal of Process Mining is to extract process models from logs of a system. Among the possible models to represent a process, Petri nets is an ideal candidate due to its graphical representation, clear semantics and expressive power. The theory of regions can be used to transform a log into a Petri net, but unfortunately the transformation requires algorithms with high complexity. This paper provides techniques to overcome this limitation. Either by using decomposition techniques, or by clustering events in the log and working on projections, the proposed approach can be used to widen the applicability of classical region-based techniques.

1   Introduction

The goal of Process Mining [20] is to extract knowledge from event logs recorded in information systems. Several researchers have provided algorithms to mine formal models from logs, most of them included in the ProM framework [19].

The *synthesis problem* [10] is related to process mining: it consists in building a Petri net that has a behavior equivalent to a given transition system. The problem was first addressed by Ehrenfeucht and Rozenberg [11] introducing *regions* to model the sets of states that characterize marked places. In the area of synthesis, some techniques have been proposed to take the theory of regions into practice. In [2] polynomial algorithms for the synthesis of bounded nets were presented. These algorithms have been recently adapted for the problem of process mining in [3]. In [7], the theory of regions was applied for the synthesis of safe Petri nets with bisimilar behavior. Recently, the theory from [7] has been extended to bounded Petri nets [6].

Process mining differs from synthesis in the knowledge assumption: while in synthesis one assumes a complete description of the system, only a partial description of the system is assumed in process mining. However, synthesis can be adapted for process mining in two ways: either the log is encoded as a transition system (introducing state information, as described in [18]) and state-based methods for mining [5] are applied, or language-based methods are used directly on the log [3,21]. In this paper we follow the first approach.

Due to its complexity, it is clear that the theory of regions might become impractical when dealing with large logs. In this paper, we present methods to

U. Dayal et al. (Eds.): BPM 2009, LNCS 5701, pp. 327–343, 2009.

alleviate significantly the complexity of the region-based approach. Two techniques are presented to this end:

- A decomposition method to find a set of components (conservative Petri nets), each one describing a partial view of the log. This approach avoids the exhaustive computation of regions and instead applies local search of regions (inspired on the notion of *allocation* from Hack [13]) until a component is detected. The set of components can either be composed to form a unique Petri net or presented separately. It is described in Section 3.
- A *divide-and-conquer* method to split the log into pieces, by means of projection. The method selects groups of events tightly related in the log for which the decomposition technique will be applied, projecting the log on these events. When neither the classical region-based mining nor the decomposition approach are able to handle a large log, this aggressive technique has proven to be very successful. It is presented in Section 4.

In both approaches, the goal is to offer a set of partial views of the behavior observed in the log, by means of a set of Petri nets whose parallel composition can reproduce any trace observed in the log.

Let us illustrate the idea of the divide-and-conquer approach (see figure on the right): given a log L with set of events E, using some ordering relations of the events appearing in the log, derive a causal dependency graph of the set of the events. This graph is then cut into several pieces, each piece representing a set of events tightly related by causal dependencies (in the figure, the sets $E_1 \ldots E_n$ are found). Finding a good partitioning

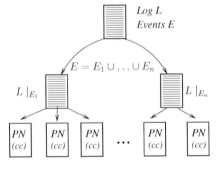

is a problem on its own, but several approaches can be used to this end, including *graph cut algorithms* [14, 12] or *spectral graph theory* [8]. Then the log is projected for each one of the sets of events. The decomposition method of this paper is then applied for each projection, obtaining a set of Petri nets (PN) that covers the traces in the log.

2   Basic Theory

2.1   Finite Transition Systems and Petri Nets

Definition 1 (Transition system). *A transition system* (TS) *is a tuple* (S, E, A, s_{in}), *where* S *is a set of* states, E *is an alphabet of* actions, $A \subseteq S \times E \times S$ *is a set of* (labelled) transitions, *and* $s_{in} \in S$ *is the* initial state.

We will use $s \xrightarrow{e} s'$ as a shortcut for $(s, e, s') \in A$, and the transitive closure of this relation will be denoted by $\xrightarrow{*}$. Let TS $= (S, E, A, s_{in})$ be a transition system. We consider connected TSs that satisfy the following axioms: i) S and E are finite sets, ii) every event has an occurrence and iii) every state is reachable from the initial state.

The *language* of a TS, $L(\mathsf{TS})$, is the set of traces feasible from the initial state. When $L(\mathsf{TS}_1) \subseteq L(\mathsf{TS}_2)$, we will denote TS_2 as an over-approximation of TS_1. Given a trace $\sigma \in L(\mathsf{TS})$ and a set $A \subseteq E$, $\sigma \mid_A$ is the trace resulting of removing from σ all events in $E - A$. Analogously, $\mathsf{TS} \mid_A$ is the TS that arises after contracting all transitions of events in $E - A$.

Definition 2 (Petri net [15]). *A Petri net (*PN*) is a tuple* (P, T, F, M_0) *where* P *and* T *represent finite and disjoint sets of places and transitions, respectively, and* $F \subseteq (P \times T) \cup (T \times P)$ *is the flow relation. The initial marking* $M_0 \subseteq P$ *defines the initial state of the system*[1].

The sets of input and output transitions of place p in PN N are denoted by ${}^{\bullet}_N p$ and $p^{\bullet}_N$, respectively (we omit the subscript indicating the net if the context is clear). The set of all markings reachable from the initial marking m_0 is called its Reachability Set. The *Reachability Graph* of PN (RG(PN)) is a transition system in which the set of states is the Reachability Set, the events are the transitions of the net and a transition (m_1, t, m_2) exists if and only if $m_1 \xrightarrow{t} m_2$. We use $L(\mathsf{PN})$ as a shortcut for $L(\mathsf{RG}(\mathsf{PN}))$.

2.2 Regions and Region-Based Synthesis

We now review the classical theory of regions for the synthesis of Petri nets [11, 10, 7]. Let S' be a subset of the states of a TS, $S' \subseteq S$. If $s \notin S'$ and $s' \in S'$, then we say that transition $s \xrightarrow{a} s'$ *enters* S'. If $s \in S'$ and $s' \notin S'$, then transition $s \xrightarrow{a} s'$ *exits* S'. Otherwise, transition $s \xrightarrow{a} s'$ *does not cross* S'.

The notion of a *region* is central for the synthesis of PNs. Intuitively, each region is a set of states that corresponds to a place in the synthesized PN, so that every state in the region models the marking of the place.

Definition 3 (Region). *A set of states* $r \subseteq S$ *in* $\mathsf{TS} = (S, E, A, s_{in})$ *is called a* region *if for each event* $e \in E$, *exactly one of the three predicates (*enters, exits *or* does not cross*) holds for all its transitions.*

Hence, a region is a subset of states in which *all* transitions labelled with the same event e have exactly the same "entry/exit" relation. This relation will become the predecessor/successor relation in the Petri net. Examples of regions are reported in Figure 1: from the TS of Figure 1(a), some regions are enumerated in Figure 1(b). For instance, for region r_2, event a is an exit event, event d is an entry event while the rest of events do not cross the region.

---

[1] For the sake of clarity, we restrict the region theory of this section to the class of *elementary net systems*: 1-bounded Petri nets without loops. The theory for the general case (k-bounded weighted Petri nets) is described in [6,5], and the theory of the rest of the paper is applicable for the general case, as demonstrated in [4].

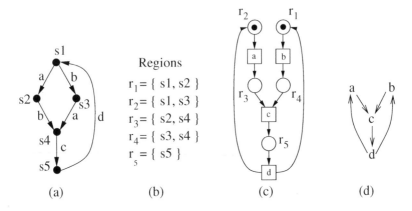

Fig. 1. (a) Transition system, (b) regions, (c) N_{TS}, (d) Causal dependency graph

---

Algorithm. PN synthesis on the set of regions R

- For each event $e \in E$ generate a transition labelled with e in the PN;
- For each region $r_i \in R$ generate a place r_i;
- Place r_i contains a token in the initial marking iff the corresponding region r_i contains the initial state of the TS s_{in};
- The flow relation is as follows: $e \in r_i \bullet$ iff r_i is a pre-region of e and $e \in \bullet r_i$ iff r_i is a post-region of e, i.e.,

$$F_R \stackrel{def}{=} \{(r,e)|r \in R_{TS} \wedge e \in E \wedge r \in {}^\circ e\}$$
$$\cup\{(e,r)|r \in R_{TS} \wedge e \in E \wedge r \in e^\circ\}$$

Fig. 2. Algorithm for Petri net synthesis from [11]

Each TS has two *trivial regions*: the set of all states, S, and the empty set. The set of non-trivial regions of TS will be denoted by R_{TS}. A region r is a *pre-region* of event e if there is a transition labelled with e which exits r. A region r is a *post-region* of event e if there is a transition labelled with e which enters r. The sets of all pre-regions and post-regions of e are denoted with ${}^\circ e$ and e°, respectively. By definition it follows that if $r \in {}^\circ e$, then all transitions labelled with e exit r. Similarly, if $r \in e^\circ$, then all transitions labelled with e enter r.

The algorithm given by [11] to synthesize a PN, $N_{TS} = (R, E, F_R, R_{s_{in}})$, from an *elementary transition system*[2] TS $= (S, E, A, s_{in})$ and a set of regions R, is illustrated in Figure 2. An example of the application of the algorithm is shown in Figure 1. The initial TS and a set of regions is reported in Figures 1(a) and (b), respectively. The synthesized PN is show in Figure 1(c). When the TS is

---

[2] Elementary transition systems are a proper subclass of the TS considered in this paper, where additional conditions to the ones presented in Section 2.1 are required.

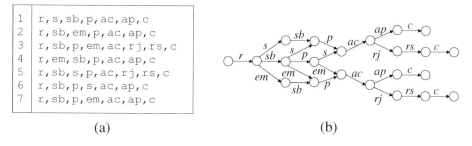

1	r,s,sb,p,ac,ap,c
2	r,sb,em,p,ac,ap,c
3	r,sb,p,em,ac,rj,rs,c
4	r,em,sb,p,ac,ap,c
5	r,sb,s,p,ac,rj,rs,c
6	r,sb,p,s,ac,ap,c
7	r,sb,p,em,ac,ap,c

(a)   (b)

Fig. 3. (a) event log, (b) corresponding transition system

elementary, running algorithm of Figure 2 on the set of non-trivial regions R_{TS} derives a PN such that $L(PN) = L(TS)$ [11].

Given an event e, $ER(e)$ denote the set of states where event e is enabled (Excitation Region), and $SR(e)$ the set of states reached when firing e in a state from $ER(e)$ (Switching Region)[3]. These sets will be used to compute the ordering relations between events (see below).

2.3 Deriving Transitions Systems from Logs

For a complete understanding of the approach presented in this paper, it is necessary to show how to transform a log into a TS, which is the starting point of our algorithms. The theory described in [18] presents many variants for solving this problem. The basic idea to incorporate state information is to look at the pre/post history of a subtrace in the log. Figure 3 shows an example, where states are decided by looking at the set of common prefixes.

2.4 Trigger Relations and Its Graph

In this section we present a relation on events, similar to the *log-based ordering relation* [20], but which is defined in the TS. It is based on the ER/SR sets.

Definition 4 (Causal Dependency Graph). *Given a* TS $= (S, E, A, s_{in})$, *and two events* $a, b \in E$:

1. *a triggers b $(a \rightarrow_{TS} b)$ if $SR(a) \cap ER(b) \neq \emptyset$ and $ER(a) \cap SR(b) = \emptyset$, and*
2. *a is concurrent to $(a \parallel_{TS} b)$ b if $SR(a) \cap ER(b) \neq \emptyset$ and $ER(a) \cap SR(b) \neq \emptyset$.*

The causal dependency graph *over* TS, *denoted* $CDG(TS)$, *is the directed graph* (E,M), *with* $M \subseteq E \times E$ *such that* $(a, b) \in M$ *iff* $a \rightarrow_{TS} b$ *or* $b \rightarrow_{TS} a$.

For instance, the causal dependency graph of the transition system of Figure 1(a) is depicted in Figure 1(d).

---

[3] Excitation and switching regions are not regions in the terms of Definition 3. The terms are used due to historical reasons.

3   Computation of Conservative Components

The goal of this section is, given a TS, derive a set of conservative components whose parallel composition contains all the traces possible in the TS. For the sake of simplicity, we will restrict the definitions for the case of conservative 1-bounded nets, known as *state machines* [15]. Formal proofs of the main results of this section can be found in [4].

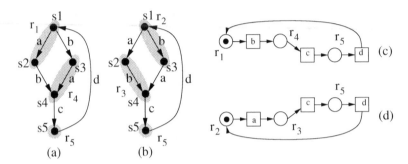

Fig. 4. Example of conservative components decomposition for the example of Figure 1: (a) Partition of the transition system on regions r_1, r_4 and r_5, and (b) for regions r_2, r_3 and r_5. The corresponding state machines are drawn in (c) and (d), respectively.

Let us illustrate the theory of this section revisiting the example of Figure 1. From the set of regions reported ($r_1 \ldots r_5$), there are two subsets that correspond to *partitions* of the set of states in the transition system of Figure 1(a) (depicted in Figures 4(a) and (b)). For instance, the subset r_1, r_4 and r_5 forms a partition. The main idea is: when a subset R of regions is a partition, then the synthesis algorithm from Figure 2 applied on R will derive a conservative Petri net, i.e. a Petri net where the number of tokens is preserved. Figure 4(c) and (d) show the two Petri nets corresponding to each partition, respectively.

3.1   State Machines and Its State-Based Representation

First we define formally the concepts of subnet and state machine component:

Definition 5 (Subnet). *A triple $N' = (P', T', F')$ is a subnet of a net $N = (P, T, F)$ if $P' \subseteq P$, $T' \subseteq T$ and $F' = F \cap ((P' \times T') \cup (T' \times P'))$.*

Definition 6 (State Machine Component). *A state machine component (SMC) $N' = (P', T', F')$ of a net N is a subnet of N such that*

1. *for every $t \in T' : |{}^{\bullet}_{N'}t| = |t^{\bullet}_{N'}| = 1$, and*
2. *for every $p \in P'$, $({}^{\bullet}_N p \cup p^{\bullet}_N) \subseteq T'$*

An SMC of a PN (N, M_0) is a pair (N', M'_0) such that N' is a SMC of N, for every $p \in P' : M'_0(p) = M_0(p)$ and $\sum_{p \in P'} M'_0(p) = 1$.

Algorithm 1. SMCComputation

Input: Transition system $\mathsf{TS} = (S, E, A, s_{in})$, event $ev \in E$
Output: Set of regions R forming a partition of S

```
1  begin
2  │   R ⟵ ∅
3  │   Evs ⟵ {ev}
4  │   r_i ⟵ PickOneRegion({r|r ∈ °ev})
5  │   r_j ⟵PickOneRegion({r|r ∈ ev° ∧ r ∩ r_i = ∅}
6  │   PendingRegs ⟵ {r_i, r_j}
7  │   Part ⟵ {r_i, r_j}
8  │   repeat
9  │   │   r ⟵ RemoveOneRegion(PendingRegs)
10 │   │   forall e ∈ E − Evs : e ∈ °r ∪ r° do
11 │   │   │   r_i ⟵ PickOneRegion({r|r ∈ °e ∧ r ∩ Part = ∅})
12 │   │   │   r_j ⟵ PickOneRegion({r|r ∈ e° ∧ r ∩ Part = ∅})
13 │   │   │   if r_i ≠ ∅ ∨ r_j ≠ ∅ then
14 │   │   │   │   Evs ⟵ Evs ∪ {e}
15 │   │   │   │   R ⟵ R ∪ {r_i, r_j}
16 │   │   │   │   PendingRegs ⟵ PendingRegs ∪ {r_i, r_j}
17 │   │   │   │   Part ⟵ Part ∪ {r_i, r_j}
18 │   │   │   end
19 │   │   end
20 │   until PendingRegs = ∅ ∨ Part = S
21 │   if Part ⊂ S then R ⟵ {S}
22 end
```

The following theorem states the main result of this section:

Theorem 1. *Let* $\mathsf{TS} = (S, E, A, s_{in})$, *and consider the net* $N_{\mathsf{TS}} = (R_{\mathsf{TS}}, E, F_{R_{\mathsf{TS}}}, R_{\mathsf{TS}_{s_{in}}})$ *obtained by the algorithm of Figure 2 on* R_{TS}. *Given a set of regions* $R \subseteq R_{\mathsf{TS}}$, *if* R *forms a partition of* S, *then algorithm of Figure 2 on* R *defines an SMC of* N_{TS}.

3.2   Allocation-Based SMC Computation

In Hack's thesis [13], the idea of *allocation* was introduced to decompose a Free-choice Petri net into a set of safe and conservative components (S-components). The idea is to select a-priori, among the places in the pre-set of a transition, the one that will be in the pre-set of the transition in the constructed S-component.

Following the idea of allocation from Hack's thesis, we present a method to derive an SMC from a given TS. Algorithm 1 describes the iterative process of finding regions until a partition of the states in TS is computed. Due to Theorem 1, the set of regions R forms an SMC. The idea of the algorithm is: starting from an initial event ev and two arbitrary regions in $°ev$ and $ev°$ (lines 4-5 of the algorithm), keep growing a partition by iteratively including pre-post regions of new events until the partition equals the set of states in TS or no more

---

Algorithm 2. SMCDecomposition

Input: Transition system $\mathsf{TS} = (S, E, A, s_{in})$
Output: $SMC_1 = (R_1, E_1, F_1, M_{0,1}) \ldots SMC_n = (R_n, E_n, F_n, M_{0,n})$
1 **begin**
2 | $X \longleftarrow E$
3 | $i \longleftarrow 1$
4 | **repeat**
5 | | $ev \longleftarrow \text{RemoveOneEvent}(X)$
6 | | $R_i \longleftarrow \text{SMCComputation}(\mathsf{TS}, ev)$
7 | | $E_i \longleftarrow \{e | e \in (°r \cup r°) \ \wedge \ r \in SMC_i \ \wedge \ r \subset S\} \cup \{ev\}$
8 | | $F_i \longleftarrow \{(r, e) | e \in r° \wedge r \in R_i \wedge e \in E_i\} \cup \{(e, r) | e \in °r \wedge r \in R_i \wedge e \in E_i\}$
9 | | $M_{0,i} \longleftarrow \forall r \in R_i : M_{0,i}(r) = r(s_{in})$
10 | | $X \longleftarrow X - E_i$
11 | | $i \longleftarrow i + 1$
12 | **until** $X = \emptyset$
13 **end**

---

regions can be found (lines 8-20). If the set of regions found are not enough as to form a partition of S, the trivial region S is returned and therefore the SMC will simply be the initial event with a self-loop place (line 21).

The general method to find a set of SMCs that cover every event of the TS is described in Algorithm 2. At each iteration i, it tries to find a new SMC SMC_i that covers one of the events still not covered by any SMC_j, for $j < i$. When an event ev can only be covered by the trivial region S, then $E_i = \{ev\}$, and therefore the SMC_i derived will be a self-loop place on event ev.

Property 1. Algorithm 1 derives an SMC, and $L(\mathsf{TS}) \subseteq L(SMC)$.

Property 2. Given the set of regions $R_1 \ldots R_n$ found by Algorithm 2, $\bigcup_{i=1 \ldots n} R_i \subseteq R_{\mathsf{TS}}$.

Property 2 ensures that the set of regions needed to cover the events is at most the set of non-trivial regions. A complexity alleviation (with respect to classical synthesis methods) can be obtained when the set of regions computed by Algorithm 2 is a proper subset. Section 5 shows examples of this.

Informally, the parallel composition of n PNs is a PN where every transition with the same label in two or more components represents a synchronization point for the components [22]. The following theorem can be proven:

Theorem 2. *Let* $SMC_1 = (R_1, E_1, F_1, M_{0,1}) \ldots SMC_n = (R_n, E_n, F_n, M_{0,n})$ *be the set of components found by Algorithm 2 on* $\mathsf{TS} = (S, E, A, s_{in})$. *Then* $L(\mathsf{TS}) \subseteq L(SMC_1 \| \ldots \| SMC_n)$.

Algorithm 2 is nondeterministic: depending on the order of events selected, a different set of state machines can arise. This has an impact both in the quality of the overapproximation obtained and in the complexity of the method, measured in the number of regions needed. In the future, more elaborated strategies can be build on top of the approach presented to address these concerns[4].

3.3   Covering the Causal Dependency Graph

The causal dependency graph can be used to improve the quality of the generated parallel composition: if some causal dependency between a pair of events is not transferred to an SMC with a shared place of the corresponding transitions, one can try to derive a new SMC that contains this relation.

Let $a \longrightarrow_{TS} b$ be an ordering relation found in the TS. If the set $\{r \mid SR(a) \cup ER(b) \subseteq r \wedge r \in R_{TS}\}$ is not empty, then any region of this set may be used to try to find an SMC covering the ordering relation $a \longrightarrow_{TS} b$. Algorithm 1 can be adapted to search for some region in this set that derives a non-trivial SMC (i.e. different from the self-loop place SMC) containing the causality relation between a and b. This is done by adapting the PickOneRegion predicates to search for regions r in the set above.

The theory presented in this section has been generalized to arbitrary k-bounded PNs, as it is shown in [4] (Section 3.4). Due to the lack of space, we only show the experiments on this extension in Section 5.

4   A Divide-and-Conquer Approach for Petri Net Mining

To face the complexity required for dealing with large TSs, an approach is presented to project the TS into tightly related events, obtaining smaller TSs. These smaller TSs can then be handled by computationally expensive Petri net mining methods. In this section we show how the decomposition approach of Section 3 can be applied on the TSs obtained by the projection technique to derive a set of Petri nets. These nets can be combined to form a unique Petri net that covers the traces of the initial log. Formal proofs of the main results can be found in [4].

4.1   Introductory Example

Let us illustrate the idea with the example from Figure 5, representing the behavior $(A; ((B; E) \parallel (C; F) \parallel (D; G)); H)$, in a TS having 28 states. In Figure 6(a), we depict the causal dependency graph. Our goal is to find balanced partitions of the causal dependency graph by means of cuts. Figure 6(b,top) reports a minimal cut from the graph of Figure 6(a), namely $\{C, F\}$. Notice that, provided that we are interested in conservative components that are synchronized with common events, when projecting the behavior of the initial TS into the set of events found in the cut we include the events outside of the cut which are adjacent to vertices in the cut, e.g. events A and H in the figure (these events are called *border* events). From each one of the sets of events found, the TS from Figure 5 is projected onto them and a conservative component covering the events in the projection is found (this is shown in the bottom part of each cut).

---

[4] Notice that the parallel composition might derive a general PN , i.e. no restriction on the class of PNs after composition is assumed in this paper.

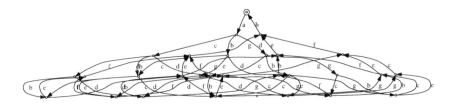

Fig. 5. Transition system

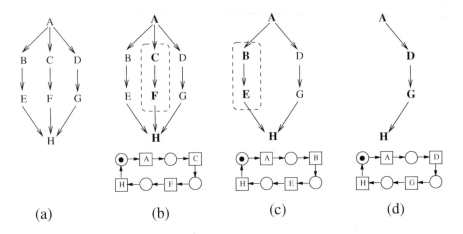

(a)   (b)   (c)   (d)

Fig. 6. (a) Causal dependency graph, (b)-(d) (Top) Consecutive cuts of the causal dependency graph, (Bottom) State machines covering each cut

4.2   Causal Dependency Graph Partitioning

There exist several techniques for the partitioning of a graph into a set of clusters [14,12]. In this section we show one of these techniques, that does not have to be the most efficient nor the optimal, but it was easy to implement. It consists on iteratively finding bi-partitions until some halting criterion is reached.

In order to find a balanced partition of $CDG(\mathsf{TS}) = (E, M)$ into two sets, let us use the well known RatioCut metric. Given a partition $E_1 \ldots E_n$ of the set E, the metric is defined as:

$$RatioCut(E_1 \ldots E_n) = \sum_{i=1}^{n} \frac{cut(E_i, \overline{E_i})}{|E_i|}$$

where $\overline{E_i}$ denotes the complement of set E_i, and $cut(A, B) = |\{(i,j)|(i,j) \in M \wedge i \in A, j \in B\}|$. If only two sets (i.e. a bi-partition) are used in the previous formula, the following optimization problem can be considered:

$$\min_{A \subset E} RatioCut(A, \overline{A})$$

Algorithm 3. DivideAndConquerMining

Input: TS $= (S, E, A, s_{in})$, MaxSize
Output: Set of SMCs
$$SMC_1 = (R_1, E_1, F_1, M_{0,1}) \ldots SMC_n = (R_n, E_n, F_n, M_{0,n})$$

1 **begin**
2   Compute $\longrightarrow_{TS}$, $\|_{TS}$ event relations
3     $(E_1 \ldots E_n) \longleftarrow$ GraphPartition(CDG(TS),MaxSize)
4   **forall** E_i **do**
5        $E_i \longleftarrow$ AddBorderEvents(TS,E_i,E)
6   SMCDecomposition(TS $|_{E_i}$)
7   **end**
8 **end**

---

i.e. finding the best bi-partition for the given graph. A way to approximate the optimal solution to this optimization problem is by using the *Fielder* vector, which is the eigenvector corresponding to the second smallest eigenvalue of the (unnormalized) Laplacian matrix $L = D - A$, where D is the *degree matrix* of the nodes in the causal dependency graph, and A its adjacency matrix [8].

More concretely, if $f \in \mathbb{R}^{|E|}$ is the Fielder vector, then a bipartition (E_1, E_2) can be obtained as follows: $e \in E_1$ if $f_e \geq 0$, and $e \in E_2$ otherwise.

By iteratively finding bi-partitions, one can derive n partitions of the set of events of the causal dependency graph, as it has been done for the causal dependency graph from the example of Section 4.1. This iteration can be terminated using a halting criterion: number of events in the projection, size of the log, CPU time-limit for the region computation, among others. In our approach, provided that we are interested in the mining of conservative components, the degree of concurrency between the events and the maximal size allowed has been used to decide if further partitioning is required.

4.3 Divide-and-Conquer Approach

Algorithm 3 presents the approach. First it computes the causal and concurrent relations (see Definition 4) present in the TS (line 2). Then the causal dependency graph is partitioned into n sets (n is an output of the method, and is dependant on the MaxSize parameter). Finally, the computation of SMCs covering each projection is applied (lines 4-7). Notice that in order to avoid the derivation of independent SMCs, i.e. SMCs without common events, each set E_i is augmented with border events, i.e. events in $E - E_i$ that are adjacent in the causal dependency graph to some event in E_i (line 5). The following theorem provides the main result of this section (see [4] for the formal proof):

Theorem 3. *Let* $SMC_1 = (R_1, E_1, F_1, M_{0,1}) \ldots SMC_n = (R_n, E_n, F_n, M_{0,n})$ *be the set of components found by Algorithm 3 on* TS $= (S, E, A, s_{in})$. *Then* $L(TS) \subseteq L(SMC_1 \| \ldots \| SMC_n)$.

5   Experiments

The theory described in Sections 3 and 4 has been incorporated into the tool Genet [6,5]. The first experiments were conducted to test the ability to rediscover conservative components from well-structured descriptions, i.e. to apply Algorithms 1 and 2, and its corresponding generalizations (as described in [4]). To this end, the TS of some k-bounded Petri nets was used (see Figures 7(a)-(c)). Table 1 reports the first experiment: comparing mining (-pm) versus conservative components derivation (-cc). For each benchmark, the size of the transition system considered (states and arcs), together with the number of places and transitions derived by the k-bounded mining method described in [5] is given. Finally, the number of conservative components found by Algorithm 2 and the sum of all the places found in the components is reported (the number of transitions in the conservative components derivation is equal to the number of transitions in the mining approach and is not reported). The CPU time is provided for each one of the approaches. For each example, Figure 7 provides gray boxes with the conservative components found (some of the boxes share transitions, i.e. they will synchronize on the firing of the transition in the parallel composition of the components). In conclusion, the derivation of conservative components might overcome the complexity problems of the region-based method, sometimes without the inclusion of extra behavior in the mined Petri net.

The second experiment was to have some confidence on the quality of the approach presented in Section 3. For that end, we used the *fitness* factor, described in [17]. Fitness evaluates whether the mined net complies with the log, and it is one of the main measures provided by the *Conformance checker* within ProM. Numerically, fitness ranges from 1 (good) to 0 (bad). The table on

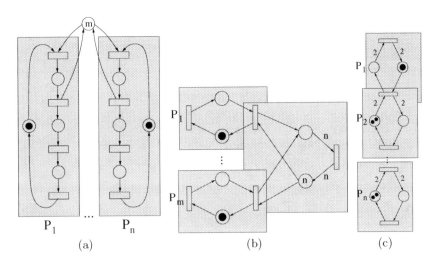

Fig. 7. Parameterized benchmarks: (a) n processes competing for m shared resources, (b) m producers and n consumers, (c) a 2-bounded pipeline of n processes. Each box represents a conservative component found.

the right reports the fitness of some miners within ProM, and the fitness of the net corresponding the parallel composition of the SMCs computed by Algorithm 2. The three logs used are the illustrative logs described in [17]. In summary, numbers in the table

Log	α	$\alpha++$	DWS	Heuristic	Genet-cc
L1	0.83	0.80	0.84	0.84	0.85
L2	0.84	0.81	0.84	0.85	0.86
L3	0.63	0.55	0.62	0.62	0.58

are promising for our approach, and we believe they can improve if techniques like the ones presented in Section 3.3 are additionally applied.

Table 1. Synthesis versus derivation of conservative components

benchmark	$|S|$	$|E|$	Genet-pm			Genet-cc		
			$|P|$	$|T|$	CPU	$|CC|$	$|P|$	CPU
SHAREDRESOURCE(5,2)	918	4320	21	20	0s	5	20	0s
SHAREDRESOURCE(4,3)	255	1016	17	16	0s	4	16	0s
SHAREDRESOURCE(6,4)	4077	24372	25	24	18s	5	24	5s
SHAREDRESOURCE(7,5)	16362	114408	29	28	25m	7	28	47s
PRODUCERCONSUMER(3,3)	32	92	8	7	0s	4	8	0s
PRODUCERCONSUMER(4,3)	64	240	10	9	0s	5	10	0s
PRODUCERCONSUMER(6,3)	256	1408	14	13	0s	7	14	0s
PRODUCERCONSUMER(8,3)	1024	7424	18	17	2s	9	18	0s
PRODUCERCONSUMER(8,5)	1536	11520	18	17	1h10m	9	18	25m
BOUNDEDPIPELINE(6)	729	1539	12	7	6s	6	12	4s
BOUNDEDPIPELINE(7)	2187	5103	14	8	48s	7	14	40s
BOUNDEDPIPELINE(8)	6561	16767	16	9	12m	8	16	11m
BOUNDEDPIPELINE(9)	19683	54765	18	10	1h50m	9	18	1h30m

The third experiment was to test the divide-and-conquer mining approach described in Section 4 (-rec). We have used two types of examples: logs from [1], and a real-life system modelling a complex module that controls the operation of optical lithography process for mass chip production [16]. Both types of benchmarks are difficult to mine using the region-based mining approach described in [5]. Table 2 compares the classical region-based mining and divide-and-conquer mining for these benchmarks. We report the size of the transition system, and columns $|P|$, $||S||$ report the number of places and size of the corresponding reachability graph of the mined Petri net. For the divide-and-conquer mining, columns $|Bis|$, k, $|CC|$, $|P|$ and $|T_U|$ report the number of bisections performed on the causal dependency graph (see Section 4.3), the bound used in the conservative component generation, the total number of conservative components found, the total number of places found and the number of events not covered by any place (the less events uncovered, the better), respectively. We use mem to report that the approach aborted due to memory problems. The conclusion from Table 2 is the superiority to handle large systems for the divide-and-conquer approach when compared to the classical region-based mining.

Table 2. Mining versus divide-and-conquer mining

				Genet-pm safe			Genet-pm 2-bounded			Genet-rec k-bounded					
benchmark	$|S|$	$|A|$	$|E|$	$|P|$	$|[S]|$	CPU	$|P|$	$|[S]|$	CPU	$|Bis|$	k	$|CC|$	$|P|$	$|T_U|$	CPU
pn_ex_10	233	479	11	13	281	0s	16	145	4s	3	2	3	9	0	0s
a12f0n50_1	78	77	11	17	80	0s	39	63	52s	3	2	4	23	0	0s
a12f0n50_2	151	150	11	21	92	0.5s	119	96	15m	3	2	8	19	0	5s
a12f0n50_3	188	187	11	21	92	0.5s	178	102	21m	1	2	4	13	0	5s
a22f0n00_1	1209	1208	20	16	78	9m	–	–	mem	0	1	4	30	0	5s
a22f0n00_2	3380	3379	20	16	78	15m	–	–	mem	3	1	6	24	1	4s
a22f0n00_3	5334	5333	20	16	78	32m	–	–	mem	3	1	7	32	1	7s
WaferStepper	55043	289443	27	–	–	mem	–	–	mem	3	6	9	28	5	5m

Table 3. Comparison for large logs from [21]

						Parikh		DWS		Genet-rec	
benchmark	# cases	# events	$|T|$	$|S|$	$|A|$	a'_B	CPU	a'_B	CPU	a'_B	CPU
a22f0n00_5	900	16952	22	676	1469	0.949	37s	0.935	4s	0.979	1s
t32f0n00_1	200	16358	32	1590	2339	0.992	7m 47s	0.863	10s	0.858	6s
a32f0n00_5	900	23195	32	2517	5907	0.933	3m	0.935	6s	1.000	9s
a42f0n00_5	900	26169	42	11170	21528	0.715	59m 29s	0.889	10s	0.962	1m 33s

Finally, we compared the divide-and-conquer technique presented in this paper with the *DWS miner*, also a clustering method presented in [9] (see Section 6 for a qualitative comparison). To this end, we used some of the largest logs that were used in [21] for a numerical analysis of the Parikh miner. We also report some other conformance measure, the *advanced behavioral appropriateness*, that gives an estimation of the degree of accuracy in which the model describes the log [17]. This measure ranges accuracy from 0 (low) to 1 (high). The results are provided in Table 3, were columns report the benchmark, number of cases, number of events, number of different tasks in the log, size of the corresponding TS[5], and for each approach we report the conformance estimation (a'_B) and the cpu time . For the benchmarks considered, one can see that the approach of this paper has similar complexity and appropriateness than the *DWS miner*.

6   Related Work

Together with the approach presented in [9], to the best of our knowledge there is no other approach for Petri net mining like the one presented in this paper. The differences are:

---

[5] Although benchmarks in Table 3 and 2 are produced from the same family of logs (being the benchmarks in Table 3 considerably larger), the settings of the FSM miner [18] used to create in each table the TS were different.

1. In this approach we give a special emphasis into the mining of conservative components, i.e. Petri nets that describe sequential and conflict dependencies between events. This sequential views can be good for visualization.
2. In [9] the partition is on the set of instances (traces) of the log, i.e. the log is horizontally partitioned, whereas in our approach the separation is done on the set of events hence the log is vertically partitioned.
3. The partition approach presented in this paper is related to the Petri net derivation applied afterwards, in the sense that events tightly related by causal dependencies are likely to become in the same conservative component. In contrast, the partition approach presented in [9] uses a different principle: each trace is projected into the most relevant features (computed previously) and associated with a vector of values. Then the *k-means* algorithm is used to partition the vectorial space defined by the traces.

The divide-and-conquer technique presented in this paper can be used in combination with the region-based approaches for Petri net mining [5,21,3] to improve their applicability in two dimensions: firstly, to allow their application for large logs, and second, to avoid the problem of *overfitting*: in our experiments the resulting model (after the parallel composition) is often more general than the one obtained from a single application of the mining approach. The technique presented in this paper is suitable when the log contains a significant amount of different tasks, thus allowing the partition phase to be applied extensively.

7   Conclusions

High-level and decomposition approaches are usually required to solve large problems. This paper shows that the region-based technique for process mining can also be solved using these type of approaches. First, the decomposition approach enables the search for sequential views of the process that might be more useful than the complete process itself. Second, when the size of the log forbids the application of classical or decomposition mining, the divide-and-conquer method presented in this paper alleviates the complexity of computing regions by projecting the TS into the events that are likely to be related, thus decreasing considerably its size. Both approaches have been presented in combination with theoretical results that guarantee a covering of the initial log with respect to the parallel composition of the obtained nets.

Acknowledgements

We would like to thank A. Rozinat and E. Verbeek for the continuous help and guidance on using ProM. This work has been supported by the CICYT project FORMALISM (TIN2007-66523), and a grant by Intel Corporation.

References

1. Process mining, http://www.processmining.org
2. Badouel, E., Bernardinello, L., Darondeau, P.: Polynomial algorithms for the synthesis of bounded nets. In: Mosses, P.D., Schwartzbach, M.I., Nielsen, M. (eds.) CAAP 1995, FASE 1995, and TAPSOFT 1995. LNCS, vol. 915, pp. 364–383. Springer, Heidelberg (1995)
3. Bergenthum, R., Desel, J., Lorenz, R., Mauser, S.: Process mining based on regions of languages. In: Proc. 5th Int. Conf. on Business Process Management, September 2007, pp. 375–383 (2007)
4. Carmona, J., Cortadella, J., Kishinevsky, M.: Divide-and-conquer strategies for process mining. Technical Report LSI-08-35-R, Software Department, Universitat Politécnica de Catalunya (2008)
5. Carmona, J., Cortadella, J., Kishinevsky, M.: A region-based algorithm for discovering Petri nets from event logs. In: Dumas, M., Reichert, M., Shan, M.-C. (eds.) BPM 2008. LNCS, vol. 5240, pp. 358–373. Springer, Heidelberg (2008)
6. Carmona, J., Cortadella, J., Kishinevsky, M., Kondratyev, A., Lavagno, L., Yakovlev, A.: A symbolic algorithm for the synthesis of bounded Petri nets. In: 29th International Conference on Application and Theory of Petri Nets and Other Models of Concurrency (June 2008)
7. Cortadella, J., Kishinevsky, M., Lavagno, L., Yakovlev, A.: Deriving Petri nets from finite transition systems. IEEE Transactions on Computers 47(8), 859–882 (1998)
8. Cvetković, D., Rowlinson, P., Simić, S.: Eigenspaces of Graphs. Cambridge University Press, Cambridge (1997)
9. de Medeiros, A.K.A., Guzzo, A., Greco, G., van der Aalst, W.M.P., Weijters, A.J.M.M.T., van Dongen, B.F., Saccà, D.: Process mining based on clustering: A quest for precision. In: ter Hofstede, A.H.M., Benatallah, B., Paik, H.-Y. (eds.) BPM Workshops 2007. LNCS, vol. 4928, pp. 17–29. Springer, Heidelberg (2008)
10. Desel, J., Reisig, W.: The synthesis problem of Petri nets. Acta Inf. 33(4), 297–315 (1996)
11. Ehrenfeucht, A., Rozenberg, G.: Partial (Set) 2-Structures. Part I, II. Acta Informatica 27, 315–368 (1990)
12. Fiduccia, C.M., Mattheyses, R.M.: A linear-time heuristic for improving network partitions. In: DAC 1982: Proceedings of the 19th conference on Design automation, Piscataway, NJ, USA, 1982, pp. 175–181. IEEE Computer Society Press, Los Alamitos (1982)
13. Hack, M.: Analysis of production schemata by Petri nets. M.s. thesis, MIT (Feburary 1972)
14. Kernighan, B.W., Lin, S.: An efficient heuristic procedure for partitioning graphs. The Bell system technical journal 49(1), 291–307 (1970)
15. Murata, T.: Petri Nets: Properties, analysis and applications. Proceedings of the IEEE, 541–580 (April 1989)
16. Pretorius, A.J.: Visualization of State Transition Graphs. PhD thesis, Technical University of Eindhoven (2008)
17. Rozinat, A., van der Aalst, W.M.P.: Conformance checking of processes based on monitoring real behavior. Inf. Syst. 33(1), 64–95 (2008)
18. van der Aalst, W.M.P., Rubin, V., Verbeek, H., van Dongen, B., Kindler, E., Günther, C.: Process mining: A two-step approach to balance between underfitting and overfitting. Technical Report BPM-08-01, BPM Center (2008)

19. van der Aalst, W.M.P., van Dongen, B.F., Günther, C.W., Mans, R.S., de Medeiros, A.K.A., Rozinat, A., Rubin, V., Song, M., Verbeek, H.M.W(E.), Weijters, A.J.M.M.T.: ProM 4.0: Comprehensive support for *real* process analysis. In: Kleijn, J., Yakovlev, A. (eds.) ICATPN 2007. LNCS, vol. 4546, pp. 484–494. Springer, Heidelberg (2007)
20. van der Aalst, W.M.P., Weijters, T., Maruster, L.: Workflow mining: Discovering process models from event logs. IEEE Trans. Knowl. Data Eng. 16(9), 1128–1142 (2004)
21. van der Werf, J.M.E.M., van Dongen, B.F., Hurkens, C.A.J., Serebrenik, A.: Process discovery using integer linear programming. In: van Hee, K.M., Valk, R. (eds.) PETRI NETS 2008. LNCS, vol. 5062, pp. 368–387. Springer, Heidelberg (2008)
22. Vogler, W.: Modular Construction and Partial Order Semantics of Petri Nets. LNCS, vol. 625. Springer, Heidelberg (1992)

Discovering Reference Models by Mining Process Variants Using a Heuristic Approach

Chen Li[1,*], Manfred Reichert[2], and Andreas Wombacher[1]

[1] Computer Science Department, University of Twente, The Netherlands
lic@cs.utwente.nl, a.wombacher@utwente.nl
[2] Institute of Databases and Information Systems, Ulm University, Germany
manfred.reichert@uni-ulm.de

Abstract. Recently, a new generation of adaptive Process-Aware Information Systems (PAISs) has emerged, which enables structural process changes during runtime. Such flexibility, in turn, leads to a large number of process variants derived from the same model, but differing in structure. Generally, such variants are expensive to configure and maintain. This paper provides a heuristic search algorithm which fosters learning from past process changes by mining process variants. The algorithm discovers a reference model based on which the need for future process configuration and adaptation can be reduced. It additionally provides the flexibility to control the process evolution procedure, i.e., we can control to what degree the discovered reference model differs from the original one. As benefit, we cannot only control the effort for updating the reference model, but also gain the flexibility to perform only the most important adaptations of the current reference model. Our mining algorithm is implemented and evaluated by a simulation using more than 7000 process models. Simulation results indicate strong performance and scalability of our algorithm even when facing large-sized process models.

1 Introduction

In today's dynamic business world, success of an enterprise increasingly depends on its ability to react to changes in its environment in a quick, flexible and cost-effective way. Generally, process adaptations are not only needed for configuration purpose at build time, but also become necessary for single process instances during runtime to deal with exceptional situations and changing needs [11,17]. In response to these needs adaptive process management technology has emerged [17]. It allows to configure and adapt process models at different levels. This, in turn, results in large collections of *process variants* created from the same process model, but slightly differing from each other in their structure. So far, only few approaches exist, which utilize the information about these variants and the corresponding process adaptations [4].

---

* This work was done in the MinAdept project, which has been supported by the Netherlands Organization for Scientific Research under contract number 612.066.512.

U. Dayal et al. (Eds.): BPM 2009, LNCS 5701, pp. 344–362, 2009.

Fig. 1 describes the goal of this paper. We aim at learning from past process changes by "merging" process variants into one generic process model, which covers these variants best. By adopting this generic model as new *reference process model* within the Process-aware Information System (PAIS), need for future process adaptations and thus cost for change will decrease. Based on the two assumptions that (1) process models are well-formed (i.e., block-structured like in WS-BPEL) and (2) all activities in a process model have unique labels, this paper deals with the following fundamental research question: *Given a reference model and a collection of process variants configured from it, how to derive a new reference process model by performing a sequence of change operations on the original one, such that the average distance between the new reference model and the process variants becomes minimal?*

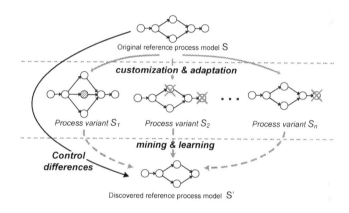

Fig. 1. Discovering a new reference model by learning from past process configurations

The distance between the reference process model and a process variant is measured by the number of high-level change operations (e.g., to insert, delete or move activities [11]) needed to transform the reference model into the variant. Clearly, the shorter the distance is, the less the efforts needed for process adaptation are. Basically, we obtain a new reference model by performing a sequence of change operations on the original one. In this context, we provide users the flexibility to control the distance between old reference model and newly discovered one, i.e., to choose how many change operations shall be applied. Clearly, the most relevant changes (which significantly reduce the average distance) should be considered first and the less important ones last. If users decide to ignore less relevant changes in order to reduce the efforts for updating the reference model, overall performance of our algorithm with respect to the described research goal is not influenced too much. Such flexibility to control the difference between the original and the discovered model is a significant improvement when compared to our previous work [5,9].

Section 2 gives background information for understanding this paper. Section 3 introduces our heuristic algorithm and provides an overview on how it can be used for mining process variants. We describe two important aspects of our heuristics algorithm (i.e., fitness function and search tree) in Sections 4 and 5. To evaluate its performance, we conduct a simulation in Section 6. Section 7 discusses related work and Section 8 concludes with a summary.

2   Backgrounds

Process Model. Let $\mathcal{P}$ denote the set of all sound process models. A particular *process model* $S = (N, E, \ldots) \in \mathcal{P}$ is defined as Well-structured Activity Net[1][11]. N constitutes the set of process activities and E the set of control edges (i.e., precedence relations) linking them. To limit the scope, we assume Activity Nets to be block-structured (like BPEL). Examples are depicted in Fig. 2.

Process change. A *process change* is accomplished by applying a sequence of *change operations* to the process model S over time [11]. Such change operations modify the initial process model by altering the set of activities and their order relations. Thus, each application of a change operation results in a new process model. We define *process change* and *process variants* as follows:

Definition 1 (Process Change and Process Variant). *Let $\mathcal{P}$ denote the set of possible process models and $\mathcal{C}$ be the set of possible process changes. Let $S, S' \in \mathcal{P}$ be two process models, let $\Delta \in \mathcal{C}$ be a process change expressed in terms of a high-level change operation, and let $\sigma = \langle \Delta_1, \Delta_2, \ldots \Delta_n \rangle \in \mathcal{C}^*$ be a sequence of process changes performed on initial model S. Then:*

- *$S[\Delta\rangle S'$ iff Δ is applicable to S and S' is the (sound) process model resulting from application of Δ to S.*
- *$S[\sigma\rangle S'$ iff $\exists\, S_1, S_2, \ldots S_{n+1} \in \mathcal{P}$ with $S = S_1$, $S' = S_{n+1}$, and $S_i[\Delta_i\rangle S_{i+1}$ for $i \in \{1, \ldots n\}$. We denote S' as variant of S.*

Examples of high-level change operations include *insert activity, delete activity,* and *move activity* as implemented in the ADEPT change framework [11]. While *insert* and *delete* modify the set of activities in the process model, *move* changes activity positions and thus the order relations in a process model. For example, operation *move(S,A,B,C)* shifts activity A from its current position within process model S to the position after activity B and before activity C. Operation *delete(S,A)*, in turn, deletes activity A from process model S. Issues concerning the correct use of these operations, their generalizations, and formal pre-/post-conditions are described in [11]. Though the depicted change operations are discussed in relation to our ADEPT approach, they are generic in the sense that they can be easily applied in connection with other process meta models as well [17]; e.g., life-cycle inheritance known from Petri Nets [15]. We refer to ADEPT since it covers by far most high-level change patterns and change support features [17], and offers a fully implemented process engine.

---

[1] A formal definition of a Well-structured Activity Net contains more than only node set N and edge set E. We omit other components since they are not relevant in the given context [18].

Definition 2 (Bias and Distance). *Let* S, $S' \in \mathcal{P}$ *be two process models.* ***Distance*** $d_{(S,S')}$ *between* S *and* S' *corresponds to the minimal number of high-level change operations needed to transform* S *into* S'; *i.e., we define* $d_{(S,S')} := min\{|\sigma| \mid \sigma \in \mathcal{C}^* \wedge S[\sigma\rangle S'\}$. *Furthermore, a sequence of change operations* σ *with* $S[\sigma\rangle S'$ *and* $|\sigma| = d_{(S,S')}$ *is denoted as* ***bias*** $B_{(S,S')}$ *between* S *and* S'.

The *distance* between S and S' is the minimal number of high-level change operations needed for transforming S into S'. The corresponding sequence of change operations is denoted as *bias* $B_{(S,S')}$ between S and S'.[2] Usually, such distance measures the complexity for model transformation (i.e., configuration). As example take Fig. 2. Here, distance between model S and variant S_1 is 4, i.e., we minimally need to perform 4 changes to transform S into S' [7]. In general, determining bias and distance between two process models has complexity at $\mathcal{NP} - hard$ level [7]. We consider high-level change operations instead of change primitives (i.e., elementary changes like adding or removing nodes / edges) to measure distance between process models. This allows us to guarantee soundness of process models and provides a more meaningful measure for distance [7,17].

Trace. A *trace* t on process model $S = (N, E, \ldots) \in \mathcal{P}$ denotes a valid and complete execution sequence $t \equiv < a_1, a_2, \ldots, a_k >$ of activity $a_i \in N$ according to the control flow set out by S. All traces S can produce are summarized in trace set $\mathcal{T}_S$. $t(a \prec b)$ is denoted as precedence relation between activities a and b in trace $t \equiv < a_1, a_2, \ldots, a_k >$ iff $\exists i < j : a_i = a \wedge a_j = b$.

Order Matrix. One key feature of any change framework is to maintain the structure of the unchanged parts of a process model [11]. To incorporate this in our approach, rather than only looking at direct predecessor-successor relation between activities (i.e., control edges), we consider the transitive control dependencies for each activity pair; i.e., for given process model $S = (N, E, \ldots) \in \mathcal{P}$, we examine for every pair of activities $a_i, a_j \in N$, $a_i \neq a_j$ their transitive order relation. Logically, we determine order relations by considering all traces the process model can produce. Results are aggregated in an order matrix $A_{|N| \times |N|}$, which considers four types of control relations (cf. Def. 3):

Definition 3 (Order matrix). *Let* $S = (N, E, \ldots) \in \mathcal{P}$ *be a process model with* $N = \{a_1, a_2, \ldots, a_n\}$. *Let further* $\mathcal{T}_S$ *denote the set of all traces producible on* S. *Then: Matrix* $A_{|N| \times |N|}$ *is called* ***order matrix*** *of* S *with* A_{ij} *representing the order relation between activities* $a_i, a_j \in N$, $i \neq j$ *iff:*

- $A_{ij} = '1'$ *iff* $[\forall t \in \mathcal{T}_S$ *with* $a_i, a_j \in t \Rightarrow t(a_i \prec a_j)]$. *If for all traces containing activities* a_i *and* a_j, a_i *always appears* BEFORE a_j, *we denote* A_{ij} *as* '**1**', *i.e.,* a_i *always precedes* a_j *in the flow of control.*
- $A_{ij} = '0'$ *iff* $[\forall t \in \mathcal{T}_S$ *with* $a_i, a_j \in t \Rightarrow t(a_j \prec a_i)]$. *If for all traces containing activities* a_i *and* a_j, a_i *always appears* AFTER a_j, *we denote* A_{ij} *as a* '**0**', *i.e.* a_i *always succeeds* a_j *in the flow of control.*

---

[2] Generally, it is possible to have more than one minimal set of change operations to transform S into S', i.e., given process models S and S' their bias does not need to be unique. A detailed discussion of this issue can be found in [15,7].

- $A_{ij} = $ '*' iff $[\exists t_1 \in \mathcal{T}_S,$ with $a_i, a_j \in t_1 \wedge t_1(a_i \prec a_j)] \wedge [\exists t_2 \in \mathcal{T}_S,$ with $a_i, a_j \in t_2 \wedge t_2(a_j \prec a_i)].$ If there exists at least one trace in which a_i appears before a_j and another trace in which a_i appears after a_j, we denote A_{ij} as '*', i.e. a_i and a_j are contained in different parallel branches.
- $A_{ij} = $ '-' iff $[\neg \exists t \in \mathcal{T}_S : a_i \in t \wedge a_j \in t].$ If there is no trace containing both activity a_i and a_j, we denote A_{ij} as '-', i.e. a_i and a_j are contained in different branches of a conditional branching.

Given a process model $S = (N, E, \ldots) \in \mathcal{P}$, the complexity to compute its order matrix $A_{|N| \times |N|}$ is $\mathcal{O}(2|N|^2)$ [7]. Regarding our example from Fig. 2, the order matrix of each process variant S_i is presented on the top of Fig. 4.[3] Variants S_i contain four kinds of control connectors: AND-Split, AND-Join, XOR-Split, and XOR-join. The depicted order matrices represent all possible order relations. As example consider S_4. Activities H and I never appear in same trace since they are contained in different branches of an XOR block. Therefore, we assign '-' to matrix element A_{HI} for S_4. If certain conditions are met, the order matrix can uniquely represent the process model. Analyzing its order matrix (cf. Def. 3) is then sufficient in order to analyze the process model [7].

It is also possible to handle loop structures based on an extension of order matrices, i.e., we need to introduce two additional order relations to cope with loop structures in process models [7]. However, since activities within a loop structure can run an arbitrarily number of times, this complicates the definition of order matrix in comparison to Def. 3. In this paper, we use process models without loop structures to illustrate our algorithm. It will become clear in Section 4 that our algorithm can easily be extended to also handle process models with loop structures by extending Def. 3.

3   Overview of Our Heuristic Search Algorithm

Running Example. An example is given in Fig. 2. Out of original reference model S, six different process variants $S_i \in \mathcal{P}$ $(i = 1, 2, \ldots 6)$ are configured. These variants do not only differ in structure, but also in their activity sets. For example, activity X appears in 5 of the 6 variants (except S_2), while Z only appears in S_5. The 6 variants are further weighted based on the number of process instances created from them; e.g., 25% of all instances were executed according to variant S_1, while 20% ran on S_2. We can also compute the distance (cf. Def. 2) between S and each variant S_i. For example, when comparing S with S_1 we obtain distance 4 (cf. Fig. 2); i.e., we need to apply 4 high-level change operations $[move(S, \text{H}, \text{I}, \text{D}), move(S, \text{I}, \text{J}, endFlow), move(S, \text{J}, \text{B}, endFlow)$ and $insert(S, \text{X}, \text{E}, \text{B})]$ to transform S into S_1. Based on weight w_i of each variant S_i, we can compute average weighted distance between reference model S and its variants. As distances between S and S_i we obtain 4 for $i = 1, \ldots, 6$ (cf. Fig. 2). When considering variant weights, as average weighted distance, we obtain $4 \times 0.25 + 4 \times 0.2 + 4 \times 0.15 + 4 \times 0.1 + 4 \times 0.2 + 4 \times 0.1 = 4.0$. This means we need

---

[3] Due to lack of space, we only depict order matrices for activities H,I,J,X,Y and Z.

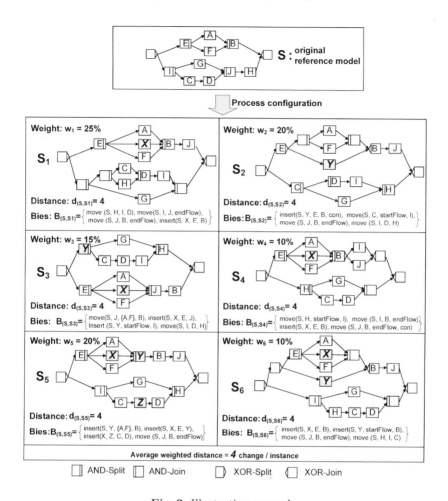

Fig. 2. Illustrating example

to perform on average 4.0 change operations to configure a process variant (and related instance respectively) out of S. Generally, *average weighted distance* between a reference model and its process variants represents how "*close*" they are. The goal of our mining algorithm is to discover a reference model for a collection of (weighted) process variants with minimal average weighted distance.

Heuristic Search for Mining Process Variants. As measuring distance between two models has $\mathcal{NP} - hard$ complexity (cf. Def. 2), our research question (i.e., finding a reference model with minimal average weighted distance to the variants), is a $\mathcal{NP} - hard$ problem as well. When encountering real-life cases, "finding "the optimum" would be either too time-consuming or not feasible. In this paper, we therefore present a *heuristic search algorithm* for mining variants.

Heuristic algorithms are widely used in various fields, e.g., artificial intelligence [10] and data mining [14]. Although they do not aim at finding the "real

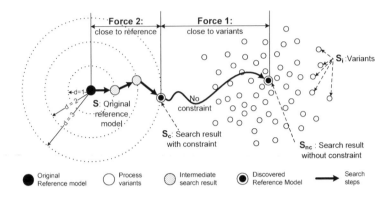

Fig. 3. Our heuristic search approach

optimum" (i.e., it is neither possible to theoretically prove that discovered results are optimal nor can we say how close they are to the optimum), they are widely used in practice. Particularly, they can nicely balance goodness of the discovered solution and time needed for finding it [10].

Fig. 3 illustrates how heuristic algorithms can be applied for the mining of process variants. Here we represent each process variant S_i as single node (white node). The goal for variant mining is then to find the "center" of these nodes (bull's eye S_{nc}), which has minimal average distance to them. In addition, we want to take original reference model S (solid node) into account, such that we can control the difference between the newly discovered reference model and the original one. Basically, this requires us to balance two forces: one is to bring the newly discovered reference model closer to the variants; the other one is to "move" the discovered model not too far away from S. Process designers should obtain the flexibility to discover a model (e.g., S_c in Fig. 3), which is closer to the variants on the one hand, but still within a limited distance to the original model on the other hand.

Our heuristic algorithm works as follows: *First*, we use original reference model S as starting point. As *Step 2*, we search for all neighboring models with distance 1 to the currently considered reference process model. If we are able to find a model S_c with lower average weighted distance to the variants, we replace S by S_c. We repeat Step 2 until we either cannot find a better model or the maximally allowed distance between original and new reference model is reached.

For any heuristic search algorithm, two aspects are important: the *heuristic measure* (cf. Section 4) and the *algorithm* (Section 5) that uses heuristics to search the state space.

4   Fitness Function of Our Heuristic Search Algorithm

Any fitness function of a heuristic search algorithm should be quickly computable. Average weighted distance itself cannot be used as fitness function, since complexity for computing it is $\mathcal{NP} - hard$. In the following, we introduce a fitness function computable in polynomial time, to approximately measure "closeness" between a candidate model and the collection of variants.

4.1 Activity Coverage

For a candidate process model $S_c = (N_c, E_c, \ldots) \in \mathcal{P}$, we first measure to what degree its activity set N_c covers the activities that occur in the considered collection of variants. We denote this measure as *activity coverage* $AC(S_c)$ of S_c. Before we can compute it, we first need to determine *activity frequency* $g(a_j)$ with which each activity a_j appears within the collection of variants. Let $S_i \in \mathcal{P}$ $i = 1, \ldots, n$ be a collection of variants with weights w_i and activity sets N_i. For each $a_j \in \bigcup N_i$, we obtain $g(a_j) = \sum_{S_i : a_j \in N_i} w_i$. Table 1 shows the frequency of each activity contained in any of the variants in our running example; e.g., X is present in 80% of the variants (i.e., in S_1, S_3, S_4, S_5, and S_6), while Z only occurs in 20% of the cases (i.e., in S_5).

Table 1. Activity frequency of each activity within the given variant collection

Activity	A	B	C	D	E	F	G	H	I	J	X	Y	Z
$g(a_j)$	1	1	1	1	1	1	1	1	1	1	0.8	0.65	0.2

Let $M = \bigcup_{i=1}^{n} N_i$ be the set of activities which are present in at least one variant. Given activity frequency $g(a_j)$ of each $a_j \in M$, we can compute *activity coverage* $AC(S_c)$ of candidate model S_c as follows:

$$AC(S_c) = \frac{\sum_{a_j \in N_c} g(a_j)}{\sum_{a_j \in M} g(a_j)} \tag{1}$$

The value range of $AC(S_c)$ is $[0, 1]$. Let us take original reference model S as candidate model. It contains activities A, B, C, D, E, F, G, H, I, and J, but does not contain X, Y and Z. Therefore, its activity coverage $AC(S)$ is 0.858.

4.2 Structure Fitting

Though $AC(S_c)$ measures how representative the activity set N_c of a candidate model S_c is with respect to a given variant collection, it does not say anything about the structure of the candidate model. We therefore introduce *structure fitting* $SF(S_c)$, which measures, to what degree a candidate model S_c structurally fits to the variant collection. We first introduce *aggregated order matrix* and *coexistence matrix* to adequately represent the variants.

Aggregated Order Matrix. For a given collection of process variants, first, we compute the order matrix of each process variant (cf. Def. 3). Regarding our running example from Fig. 2, we need to compute six order matrices (see to

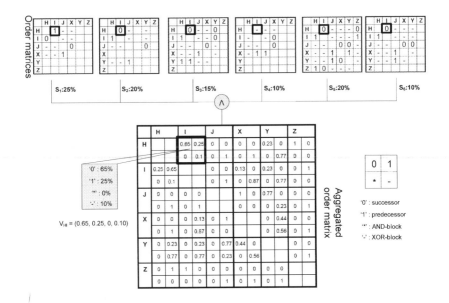

Fig. 4. Aggregated order matrix based on process variants

of Fig. 4). Note that we only show a partial view on the order matrices here
(activities H, I, J, X, Y and Z) due to space limitations. As the order relation
between two activities might be not the same in all order matrices, this analysis
does not result in a fixed relation, but provides a distribution for the four types
of order relations (cf. Def. 3). Regarding our example, in 65% of all cases H
succeeds I (as in S_2, S_3, S_4 and S_6), in 25% of all cases H precedes I (as in S_1),
and in 10% of all cases H and I are contained in different branches of an XOR
block (as in S_4) (cf. Fig. 4). Generally, for a collection of process variants we
can define the order relation between activities a and b as 4-dimensional vector
$V_{ab} = (v_{ab}^0, v_{ab}^1, v_{ab}^*, v_{ab}^-)$. Each field then corresponds to the frequency of the
respective relation type ('0', '1', '*' or '-') as specified in Def. 3. For our example
from Fig. 2, for instance, we obtain $V_{HI} = (0.65, 0.25, 0, 0.1)$ (cf. Fig. 4). Fig. 4
shows aggregated order matrix V for the process variants from Fig. 2.

Coexistence Matrix. Generally, the order relations computed by an aggregated order matrix may not be equally important; e.g., relationship V_{HI} between
H and I (cf. Fig. 4) would be more important than relation V_{HZ}, since H and
I appear together in all six process variants while H and Z only show up together in variant S_5 (cf. Fig. 2). We therefore define *Coexistence Matrix CE*
in order to represent the importance of the different order relations occurring
within an aggregated order matrix V. Let S_i ($i = 1 \ldots n$) be a collection of
process variants with activity sets N_i and weight w_i. The *Coexistence Matrix*
of these process variants is then defined as 2-dimensional matrix $CE_{m \times m}$ with
$m = |\bigcup N_i|$. Each matrix element CE_{jk} corresponds to the relative frequency
with which activities a_j and a_k appear together within the given collection of

	H	I	J	X	Y	Z
H		1	1	0.8	0.65	0.2
I	1		1	0.8	0.65	0.2
J	1	1		0.8	0.65	0.2
X	0.8	0.8	0.8		0.45	0.2
Y	0.65	0.65	0.65	0.45		0.2
Z	0.2	0.2	0.2	0.2	0.2	

Fig. 5. Coexistence Matrix

variants. Formally: $\forall a_j, a_k \in \bigcup N_i, a_j \neq a_k : CE_{jk} = \sum_{S_i : a_j \in N_i \wedge a_k \in N_i} w_i$. Table 5 shows the coexistence matrix for our running example (partial view).

Structure Fitting $SF(S_c)$ of Candidate Model S_c. Since we can represent candidate process model S_c by its corresponding order matrix A_c (cf. Def. 3), we determine *structure fitting* $SF(S_c)$ between S_c and the variants by measuring how similar order matrix A_c and aggregated order matrix V (representing the variants) are. We can evaluate S_c by measuring the order relations between every pair of activities in A_c and V. When considering reference model S as candidate process model S_c (i.e., $S_c = S$), for example, we can build an aggregated order matrix V^c purely based on S_c, and obtain $V_{HI}^c = (1,0,0,0)$; i.e., H always succeeds I. Now, we can compare $V_{HI} = (0.65, 0.25, 0, 0.1)$ (representing the variants) and V_{HI}^c (representing the candidate model).

We use Euclidean metrics $f(\alpha, \beta)$ to measure closeness between vectors $\alpha = (x_1, x_2, ..., x_n)$ and $\beta = (y_1, y_2, ..., y_n)$: $f(\alpha, \beta) = \frac{\alpha \cdot \beta}{|\alpha| \times |\beta|} = \frac{\sum_{i=1}^{n} x_i y_i}{\sqrt{\sum_{i=1}^{n} x_i^2} \times \sqrt{\sum_{i=1}^{n} y_i^2}}$. $f(\alpha, \beta) \in [0, 1]$ computes the cosine value of the angle θ between vectors α and β in Euclidean space. The higher $f(\alpha, \beta)$ is, the more α and β match in their directions. Regarding our example we obtain $f(V_{HI}, V_{HI}^c) = 0.848$. Based on $f(\alpha, \beta)$, which measures *similarity* between the order relations in V (representing the variants) and in V^c (representing candidate model), and Coexistence matrix CE, which measures *importance* of the order relations, we can compute *structure fitting* $SF(S_c)$ of candidate model S_c as follows:

$$SF(S_c) = \frac{\sum_{j=1}^{n} \sum_{k=1, k \neq j}^{n} [f(V_{a_j a_k}, V_{a_j a_k}^c) \cdot CE_{a_j a_k}]}{n(n-1)} \quad (2)$$

$n = |N_c|$ corresponds to the number of activities in candidate model S_c. For every pair of activities $a_j, a_k \in N_c, j \neq k$, we compute similarity of corresponding order relations (as captured by V and V_c) by means of $f(V_{a_j a_k}, V_{a_j a_k}^c)$, and the importance of these order relations by $CE_{a_j a_k}$. Structure fitting $SF(S_c) \in [0, 1]$ of candidate model S_c then equals the average of the similarities multiplied by the importance of every order relation. For our example from Fig. 2, structure fitting $SF(S)$ of original reference model S is 0.749.

4.3   Fitness Function

So far, we have described the two measurements *activity coverage* $AC(S_c)$ and *structure fitting* $SF(S_c)$ to evaluate a candidate model S_c. Based on them, we can compute *fitness* $Fit(S_c)$ of S_c : $Fit(S_c) = AC(S_c) \times SF(S_c)$.

As $AC(S_c) \in [0, 1]$ and $SF(S_c) \in [0, 1]$, value range of $Fit(S_c)$ is [0,1] as well. Fitness value $Fit(S_c)$ indicates how "close" candidate model S_c is to the given collection of variants. The higher $Fit(S_c)$ is, the closer S_c is to the variants and the less configuration efforts are needed. Regarding our example from Fig. 2, fitness value $Fit(S)$ of the original reference process model S corresponds to $Fit(S) = AC(S) \times SF(S) = 0.858 \times 0.749 = 0.643$.

5   Constructing the Search Tree

5.1   The Search Tree

Let us revisit Fig. 3, which gives a general overview of our heuristic search approach. Starting with the current candidate model S_c, in each iteration we search for its direct "neighbors" (i.e., process models which have distance 1 to S_c) trying to find a better candidate model S_c' with higher fitness value. Generally for a given process model S_c, we can construct a neighbor model by applying *ONE* insert, delete, or move operation to S_c. All activities $a_j \in \bigcup N_i$, which appear at least in one variant, are candidate activities for change. Obviously, an insert operation adds a new activity $a_j \notin N_c$ to S_c, while the other two operations delete or move an activity a_j already present in S_c (i.e., $a_j \in N_c$).

Generally, numerous process models may result by changing one particular activity a_j on S_c. Note that the positions where we can insert ($a_j \notin N_c$) or move ($a_j \in N_c$) activity a_j can be numerous. Section 5.2 provides details on how to find all process models resulting from the change of one particular activity a_j on S_c. First of all, we assume that we have already identified these neighbor models, including the one with highest fitness value (denoted as the *best kid* S_{kid}^j of S_c when changing a_j). Fig. 6 illustrates our search tree (see [8] for more details). Our search algorithm starts with setting the original reference model S as the initial state, i.e., $S_c = S$ (cf. Fig. 6). We further define AS as *active activity set*, which contains all activities that might be subject to change. At the beginning, $AS = \{a_j | a_j \in \bigcup_{i=1}^{n} N_i\}$ contains all activities that appear in at least one variant S_i. For each activity $a_j \in AS$, we determine the corresponding best kid S_{kid}^j of S_c. If S_{kid}^j has higher fitness value than S_c, we mark it; otherwise, we remove a_j from AS (cf. Fig. 6). Afterwards, we choose the model with highest fitness value S_{kid}^{j*} among all best kids S_{kid}^j, and denote this model as *best sibling* S_{sib}. We then set S_{sib} as the first intermediate search result and replace S_c by S_{sib} for further search. We also remove a_{j*} from AS since it has been already considered.

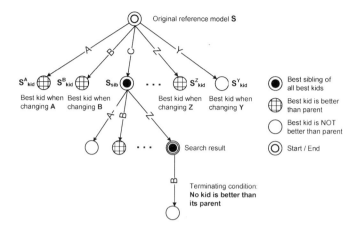

Fig. 6. Constructing the search tree

The described search method goes on iteratively, until termination condition is met, i.e., we either cannot find a better model or the allowed search distance is reached. The final search result S_{sib} corresponds to our discovered reference model S' (the node marked by a bull's eye and circle in Fig. 6).

5.2 Options for Changing One Particular Activity

Section 5.1 has shown how to construct a search tree by comparing the best kids S_{kid}^j. This section discusses how to find such best kid S_{kid}^j, i.e., how to find all "neighbors" of candidate model S_c by performing *one* change operation on a particular activity a_j. Consequently, S_{kid}^j is the one with highest fitness value among all considered models. Regarding a particular activity a_j, we consider three types of basic change operations: *delete, move* and *insert* activity. The neighbor model resulting through deletion of activity $a_j \in N_c$ can be easily determined by removing a_j from the process model and the corresponding order matrix [7]; furthermore, movement of an activity can be simulated by its deletion and subsequent re-insertion at the desired position. Thus, the basic challenge in finding neighbors of a candidate model S_c is to apply one activity *insertion* such that *block structuring* and *soundness* of the resulting model can be guaranteed. Obviously, the positions where we can (correctly) insert a_j into S_c are our subjects of interest. Fig. 7 provides an example. Given process model S_c, we would like to find all process models that may result when inserting activity X into S_c. We apply the following two steps to "simulate" the insertion of an activity.

Step 1 (Block-enumeration): First, we enumerate all possible blocks the candidate model S_c contains. A block can be an atomic activity, a self-contained part of the process model, or the process model itself. Let S_c be a process model with activity set $N_c = \{a_1, \ldots, a_n\}$ and let further A_c be the order matrix of S_c. Two activities a_i and a_j can form a block if and only if $[\forall a_k \in N_c \setminus \{a_i, a_j\}:$

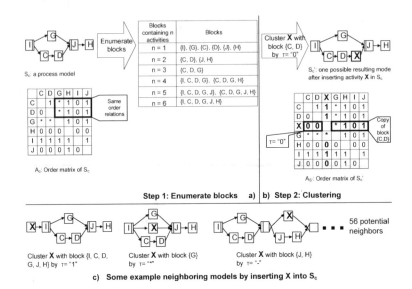

Fig. 7. Finding the neighboring models by inserting X into process model S

$A_{ik} = A_{jk}]$ holds (i.e., iff they have exactly same order relations to the remaining activities). Consider our example from Fig. 7a. Here C and D can form a block {C, D} since they have same order relations to remaining activities G, H, I and J. In our context, we consider each block as set rather than as process model, since its structure is evident in S_c. As extension, two blocks B_j and B_k can merge into a bigger one iff $[(a_\alpha, a_\beta, a_\gamma) \in B_j \times B_k \times (N \setminus B_j \bigcup B_k) : A_{\alpha\gamma} = A_{\beta\gamma}]$ holds; i.e., two blocks can merge into a bigger block iff all activities $a_\alpha \in B_j$, $a_\beta \in B_k$ show same order relations to the remaining activities outside the two blocks. For example, block {C, D} and block {G} show same order relations in respect to remaining activities H, I and J; therefore they can form a bigger block {C, D, J}. Fig. 7a shows all blocks contained in S_c (see [8] for a detailed algorithm).

Step 2 (Cluster Inserted Activity with One Block): Based on the enumerated blocks, we describe where we can (correctly) insert a particular activity a_j in S_c. Assume that we want to insert X in S_c (cf. Fig. 7b). To ensure the block structure of the resulting model, we "cluster" X with an enumerated block, i.e., we replace one of the previously determined blocks B by a bigger block B' that contains B as well as X. In the context of this clustering, we set order relations between X and all activities in block B as $\tau \in \{0, 1, *, -\}$ (cf Def. 3). One example is given in Fig. 7b. Here inserted activity X is clustered with block {C, D} by order relation $\tau =$ "0", i.e., we set X as successor of the block containing C and D. To realize this clustering, we have to set order relations between X on the one hand and activities C and D from the selected block on the other hand to "0". Furthermore, order relations between X and the remaining activities are same as for {C, D}. Afterwards these three activities form a new block {C, D, X} replacing the old one $\{C, D\}$. This way, we obtain a sound and block-structured process model S'_c by inserting X into S_c.

We can guarantee that the resulting process model is sound and block-structured. Every time we cluster an activity with a block, we actually add this activity to the position where it can form a bigger block together with the selected one, i.e., we replace a self-contained block of a process model by a bigger one. Obviously, S'_c is not the only neighboring model of S_c. For each block B enumerated in Step 1, we can cluster it with X by any one of the four order relations $\tau \in \{0, 1, *, -\}$. Regarding our example from Fig. 7, S contains 14 blocks. Consequently, the number of models that may result when adding X to S_c equals $14 \times 4 = 56$; i.e., we can obtain 56 potential models by inserting X into S_c. Fig. 7c shows some neighboring models of S_c.

5.3   Search Result for Our Running Example

Regarding our example from Fig. 2, we now present the search result and all intermediate models we obtain when applying our algorithm (see Fig. 8).

The first operation $\Delta_1 = move(S, J, B, endFlow)$ changes original reference model S into intermediate result model R_1 which is the one with highest fitness value among all neighboring models of S. Based on R_1, we discover R_2 by change $\Delta_2 = insert(R_1, X, E, B)$, and finally we obtain R_3 by performing change $\Delta_3 = move(R_2, I, D, H)$ on R_2. Since we cannot find a "better" process model by changing R_3 anymore, we obtain R_3 as final result. Note that if we only allow to change original reference model by maximal d change operations, the final search result would be: R_d if $d \leq 3$ or R_3 if $d \geq 4$.

We further compare original reference model S and all (intermediate) search results in Fig. 9. As our heuristic search algorithm is based on finding process models with higher fitness values, we observe improvements of the fitness values for each search step. Since such fitness value is only a "reasonable guessing", we also compute average weighted distance between the discovered model and the variants, which is a precise measurement in our context. From Fig. 9, average

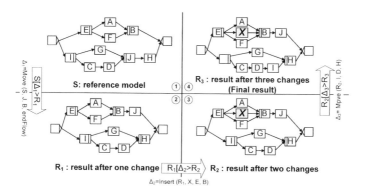

Fig. 8. Search result after every change

	S	R_1	R_2	R_3
Fitness	0.643	0.814	0.854	0.872
Average weighted distance	4	3.2	2.6	2.35
Delta-fitness		0.171	0.04	0.017
Delta-distance		0.8	0.6	0.25

Fig. 9. Evaluation of the search results

weighted distance also drops monotonically from 4 (when considering S) to 2.35 (when considering R_3).

Additionally, we evaluate *delta-fitness* and *delta-distance*, which indicate relative change of fitness and average weighted distance for every iteration of the algorithm. For example, Δ_1 changes S into R_1. Consequently, it improves fitness value (delta-fitness) by 0.0171 and reduces average weighted distance (delta-distance) by 0.8. Similarly, Δ_2 reduces average weighted distance by 0.6 and Δ_3 by 0.25. The monotonic decrease of delta-distance indicates that important changes (reducing average weighted distance between reference model and variants most) are indeed discovered at beginning of the search.

Another important feature of our heuristic search is its ability to automatically decide on which activities shall be included in the reference model. A predefined threshold or filtering of the less relevant activities in the activity set is not needed; e.g., X is automatically inserted, but Y and Z are not added. The three change operations (insert, move, delete) are automatically balanced based on their influence on the fitness value.

5.4   Proof-of-Concept Prototype

The described approach has been implemented and tested using Java. We use our ADEPT2 Process Template Editor [12] as tool for creating process variants. For each process model, the editor can generate an XML representation with all relevant information (like nodes, edges, blocks). We store created variants in a variants repository which can be accessed by our mining procedure. The mining algorithm has been developed as stand-alone service which can read the original reference model and all process variants, and generate the result models according to the XML schema of the process editor. All (intermediate) search results are stored and can be visualized using the editor.

6   Simulation

Of course, using one example to measure the performance of our heuristic mining algorithm is far from being enough. Since computing average weighted distance is at $\mathcal{NP} - hard$ level, fitness function is only an approximation of it. Therefore, the first question is *to what degree delta-fitness is correlated with delta-distance?* In addition, we are interested in knowing to what degree important change operations are performed at the beginning. If biggest distance reduction is obtained with the first changes, setting search limitations or filtering out the change operations performed at the end, does not constitute a problem. Therefore, the second research question is: *To what degree are important change operations positioned at the beginning of our heuristic search?*

We try to answer these questions using *simulation*; i.e., by generating thousands of data samples, we can provide a statistical answer for these questions. In our simulation, we identify several parameters (e.g., size of the model, similarity of the variants) for which we investigate whether they influence the performance of our heuristic mining algorithm (see [8] for details). By adjusting these parameters, we generate 72 groups of datasets (7272 models in total) covering different scenarios. Each group contains a randomly generated *reference process model* and a collection of *100 different process variants*. We generate each variant by configuring the reference model according to a particular scenario.

We perform our heuristic mining to discover new reference models. We do not set constraints on search steps, i.e., the algorithm only terminates if no better model can be discovered. *All (intermediate) process models* are documented (see Fig. 8 as example). We compute the *fitness* and *average weighted distance* of each intermediate process models as obtained from our heuristic mining. We additionally compute *delta-fitness* and *delta-distance* in order to examine the influence of every change operation (see Fig. 9 for an example).

Improvement on average weighted distances. In 60 (out of 72) groups we are able to discover a new reference model. The average weighted distance of the discovered model is *0.765* lower than the one of the original reference model; i.e., we obtain a reduction of *17.92%* on average.

Execution time. The number of activities contained in the variants can significantly influence execution time of our algorithm. Search space becomes larger for bigger models since the number of candidate activities for change and the number of blocks contained in the reference model become higher. The average run time for models of different size is summarized in Fig. 10.

Correlation of delta-fitness and delta-distance. One important issue we want to investigate is how delta-fitness is correlated to delta-distance. Every change operation leads to a particular change of the process model, and consequently creates a delta-fitness x_i and delta-distance y_i. In total, we have performed *284* changes in our simulation when discovering reference models. We use Pearson correlation to measure correlation between delta-fitness and delta-distance [13]. Let X be delta-fitness and Y be delta-distance. We obtain n data samples (x_i, y_i), $i = 1, \ldots, n$. Let $\bar{x}$ and $\bar{y}$ be the mean of X and Y, and let s_x and s_y be the standard deviation of X and Y. The Pearson correlation r_{xy} then equals $r_{xy} = \frac{\sum x_i y_i - n\bar{x}\bar{y}}{(n-1)s_x s_y}$ [13]. Results are summarized in Fig. 10. All correlation coefficients are *significant* and *high* (> 0.5). The high positive correlation between delta-fitness and delta-distance indicates that when finding a model with higher fitness value, we have very high chance to also reduce average weighted distance. We additionally compare these three correlations. Results indicate that

	Execution time information			Correlation analysis		
	# of activity per variant	Average search steps	Average execu-tion time (s)	# of data	Correlation	Significant?
Small-sized	10 ~ 13	1.83	0.148	33	0.762	Yes
Medium-sized	20 ~ 26	3.52	4.568	74	0.589	Yes
Large-sized	50 ~ 65	8.43	805.539	177	0.623	Yes

Fig. 10. Execution time and correlation analysis of groups with different sizes

they do not show significant difference from each other, i.e., they are statistically same (see [8]). This implies that our algorithm provides search results of similar goodness *independent* of the number of activities contained in the process variants.

Importance of top changes. Finally, we measure to what degree our algorithm applies more important changes at the beginning. In this context, we measure to what degree the top $n\%$ changes have reduced the average weighted distance. For example, consider search results from Fig. 9. We have performed in total 3 change operations and reduced the average weighted distance by 1.65 from 4 (based on S) to 2.35 (based on R_3). Among the three change operations, the first one reduces average weighted distance by 0.8. When compared to the overall distance reduction of 1.65, the top 33.33% changes accomplished $0.8/1.65 = 48.48\%$ of our overall distance reduction. This number indicates how important the changes at beginning are. We therefore evaluate the distance reduction by analyzing the top 33.3% and 50.0% change operations. On average, the top 33.3% change operations have achieved *63.80%* distance reduction while the top 50.0% have achieved *78.93%*. Through this analysis, it becomes clear that the changes at beginning are *a lot more important* than the ones performed at last.

7   Related Work

Though heuristic search algorithms are widely used in areas like data mining [14] or artificial intelligence [10], only few approaches use heuristics for process variant management. In process mining, a variety of techniques have been suggested including heuristic or genetic approaches [19,2,16]. As illustrated in [6], traditional process mining is different from process variant mining due to its different goals and inputs. There are few techniques which allow to learn from process variants by mining recorded change primitives (e.g., to add or delete control edges). For example, [1] measures process model similarity based on change primitives and suggests mining techniques using this measure. Similar techniques for mining change primitives exist in the field of association rule mining and maximal sub-graph mining [14] as known from graph theory; here common edges between different nodes are discovered to construct a common sub-graph from a set of graphs. However, these approaches are unable to deal with silent activities and also do not differentiate between AND- and XOR-branchings. To mine high level change operations, [3] presents an approach using process mining algorithms to discover the execution sequences of changes. This approach simply considers each change as individual operation so the result is more like a visualization of changes rather than mining them. None of the discussed approaches is sufficient in supporting the evolution of reference process model towards an easy and cost-effective model by learning from process variants in a controlled way.

8   Summary and Outlook

The main contribution of this paper is to provide a heuristic search algorithm supporting the discovery of a reference process model by learning from a collection of (block-structured) process variants. Adopting the discovered model as new reference process model will make process configuration easier. Our heuristic algorithm can also take the original reference model into account such that the user can control how much the discovered model is different from the original one. This way, we cannot only avoid spaghetti-like process models but also control how many changes we want to perform. We have evaluated our algorithm by performing a comprehensive simulation. Based on its results, the fitness function of our heuristic algorithm is highly correlated with average weighted distance. This indicates good performance of our algorithm since the approximation value we use to guide our algorithm is nicely correlated to the real one. In addition, simulation results also indicate that the more important changes are performed at the beginning - the first $1/3$ changes result in about $2/3$ of overall distance reduction. Though results look promising, more work needs to be done. As our algorithm takes relatively long time when encountering large process models, it would be useful to further optimize it to make search faster.

References

1. Bae, J., Liu, L., Caverlee, J., Rouse, W.B.: Process mining, discovery, and integration using distance measures. In: ICWS 2006, pp. 479–488 (2006)
2. Alves de Medeiros, A.K.: Genetic Process Mining. PhD thesis, Eindhoven University of Technology, NL (2006)
3. Günther, C.W., Rinderle-Ma, S., Reichert, M., van der Aalst, W.M.P., Recker, J.: Using process mining to learn from process changes in evolutionary systems. Int'l Journal of Business Process Integration and Management 3(1), 61–78 (2008)
4. Hallerbach, A., Bauer, T., Reichert, M.: Managing process variants in the process lifecycle. In: Proc. 10th Int'l Conf. on Enterprise Information Systems (ICEIS 2008), pp. 154–161 (2008)
5. Li, C., Reichert, M., Wombacher, A.: Discovering reference process models by mining process variants. In: ICWS 2008, pp. 45–53. IEEE Computer Society Press, Los Alamitos (2008)
6. Li, C., Reichert, M., Wombacher, A.: Mining process variants: Goals and issues. In: IEEE SCC (2), pp. 573–576. IEEE Computer Society Press, Los Alamitos (2008)

7. Li, C., Reichert, M., Wombacher, A.: On measuring process model similarity based on high-level change operations. In: Li, Q., Spaccapietra, S., Yu, E., Olivé, A. (eds.) ER 2008. LNCS, vol. 5231, pp. 248–264. Springer, Heidelberg (2008)
8. Li, C., Reichert, M., Wombacher, A.: A heuristic approach for discovering reference models by mining process model variants. Technical Report TR-CTIT-09-08, University of Twente, NL (2009)
9. Li, C., Reichert, M., Wombacher, A.: What are the problem makers: Ranking activities according to their relevance for process changes. In: ICWS 2009. IEEE Computer Society Press, Los Alamitos (to appear, 2009)
10. Luger, G.F.: Artificial Intelligence: Structures and Strategies for Complex Problem Solving. Pearson Education, London (2005)
11. Reichert, M., Dadam, P.: ADEPTflex -supporting dynamic changes of workflows without losing control. J. of Intelligent Information Sys. 10(2), 93–129 (1998)
12. Reichert, M., Rinderle, S., Kreher, U., Dadam, P.: Adaptive process management with ADEPT2. In: ICDE 2005, pp. 1113–1114. IEEE Computer Society Press, Los Alamitos (2005)
13. Sheskin, D.J.: Handbook of Parametric and Nonparametric Statistical Procedures. CRC Press, Boca Raton (2004)
14. Tan, P.N., Steinbach, M., Kumar, V.: Introduction to Data Mining. Addison-Wesley, Reading (2005)
15. van der Aalst, W.M.P., Basten, T.: Inheritance of workflows: an approach to tackling problems related to change. Theor. Comput. Sci. 270(1-2), 125–203 (2002)
16. van der Aalst, W.M.P., Weijters, T., Maruster, L.: Workflow mining: Discovering process models from event logs. IEEE TKDE 16(9), 1128–1142 (2004)
17. Weber, B., Reichert, M., Rinderle-Ma, S.: Change patterns and change support features - enhancing flexibility in process-aware information systems. Data and Knowledge Engineering 66(3), 438–466 (2008)
18. Weber, B., Reichert, M., Wild, W., Rinderle-Ma, S.: Providing integrated life cycle support in process-aware information systems. Int'l Journal of Cooperative Information Systems (IJCIS), 19(1) (2009)
19. Weijters, A.J.M.M., van der Aalst, W.M.P.: Rediscovering workflow models from event-based data using little thumb. Integr. Com.-Aided Eng. 10(2), 151–162 (2003)

Author Index